Coastal Research Library

Volume 15

Series Editor
Charles W. Finkl
Department of Geosciences
Florida Atlantic University
Boca Raton, FL
USA

The aim of this book series is to disseminate information to the coastal research community. The Series covers all aspects of coastal research including but not limited to relevant aspects of geological sciences, biology (incl. ecology and coastal marine ecosystems), geomorphology (physical geography), climate, littoral oceanography, coastal hydraulics, environmental (resource) management, engineering, and remote sensing. Policy, coastal law, and relevant issues such as conflict resolution and risk management would also be covered by the Series. The scope of the Series is broad and with a unique cross-disciplinary nature. The Series would tend to focus on topics that are of current interest and which carry some import as opposed to traditional titles that are esoteric and non-controversial. Monographs as well as contributed volumes are welcomed.

More information about this series at http://www.springer.com/series/8795

Rafael Riosmena-Rodríguez • Wendy Nelson
Julio Aguirre
Editors

Rhodolith/Maërl Beds: A Global Perspective

Springer

Editors
Rafael Riosmena-Rodríguez (deceased)
Departamento de Biología Marina
Universidad de Baja California Sur
La Paz, BCS, Mexico

Julio Aguirre
Departamento de Estratigrafía y Paleontología
Universidad de Granada
Granada, Spain

Wendy Nelson
National Institute of Water & Atmospheric Research
Wellington, New Zealand

School of Biological Sciences
University of Auckland
Auckland, New Zealand

ISSN 2211-0577 ISSN 2211-0585 (electronic)
Coastal Research Library
ISBN 978-3-319-29313-4 ISBN 978-3-319-29315-8 (eBook)
DOI 10.1007/978-3-319-29315-8

Library of Congress Control Number: 2016950722

Printed on acid-free paper

This Springer imprint is published by Springer Nature
The registered company is Springer International Publishing AG Switzerland

Tribute to Rafael Riosmena-Rodríguez

In Memoriam

On March 15th 2016 we heard the sad news that our colleague and friend Rafael Riosmena-Rodríguez passed away suddenly as a result of heart failure. Rafael Riosmena, Rafa, was an internationally renowned scientist who dedicated 25 years working on many different aspects of marine biology, but particularly on the taxonomy and biology of red coralline algae.

Rafael was born in 1966 in León, Guanajuato. He began his degree in marine biology at the Universidad Autónoma de Baja California Sur (UABCS), in La Paz (México), then a Master's degree from San Jose State University (SJSU) in the United States, and a Ph.D. degree in marine botany at La Trobe University in Australia. He was currently a professor and researcher at UABCS, level III member of the National Research System of México (SNI).

He focused his work on the taxonomy of rhodolith-forming coralline red algae and the ecological significance of rhodolith beds. Rafa has left behind a valuable scientific legacy of more than 100 scientific papers published in peer-reviewed journals. He also edited many books on different aspects of the marine biology of the Gulf of California. Rafael was the driving force behind this book, his latest contribution to the knowledge of rhodoliths and rhodolith beds.

One of his most important contributions to science was to build a consolidated group of many young people, that now are spread in respected institutions all over the world, continuing his studies. His enthusiasm, generosity, and friendly behaviour facilitated an excellent atmosphere in which to work, and great collegiality with scientists internationally. Thanks to this tireless work training good researchers, Rafa's scientific school is now continuing his legacy.

Rafa wasn't only a highly regarded phycologist. He was also deeply committed to the defence and protection of the environment, also denouncing corrupt activities in México. His wide knowledge of the marine environments of the Gulf of California and his untiring work trying to unravel the mechanisms behind ecological interactions led him into contact with institutions trying to find the way to manage and protect these environments. We all depend upon marine ecosystems, so we have to learn how to interact with wildlife maintaining the equilibrium and seeking sustainable approaches. Rafa provided scientific data to guide non-scientists at different levels in the protection of these precious environments.

In addition to all his scientific accomplishments and contributions, Rafa was a good friend, with kind words and always willing to help anyone. Our memories will be of Rafa smiling. We are indebted to him because, thanks to his constancy and firm conviction, he established a group of active scientists researching rhodoliths that share more than scientific results. We will miss him, but we know that his legacy will live in his lovely family and in all his students around the world. Nada nos detiene! Thanks Rafa!

Contents

Part I
The Role of Rhodolith/Maërl Beds in Modern Oceans

Chapter 1
Natural History of Rhodolith/Maërl Beds: Their Role in Near-Shore Biodiversity and Management

Rafael Riosmena-Rodríguez

Abstract Rhodolith/maërl beds are living and dead aggregations of free-living non-geniculate coralline algae that cover extensive benthic areas in recent oceans and are common in fossil deposits. They are slow growing organisms and can be long-lived (>100 years), distributed over a wide depth range from intertidal sites to 270 m. Rhodolith/maërl beds are a common feature of modern and ancient carbonate shelves worldwide that represent a sedimentary transition from sandy/muddy areas to the rocky substrate. They are bioengineers and provide a three-dimensional habitat for associated species. It has been demonstrated that rhodolith/maërl grounds are a suitable habitat for multispecies recruitment and provide refuge for juvenile life stages of commercially important shellfish species. Rhodoliths are resilient to a variety of environmental disturbances, but can be severely impacted by harvesting these commercial species, ocean acidification or global warming. The value of rhodoliths as a unique biotope around the world is under threat from different kinds of human activities. Despite the importance of rhodolith/maërl beds in the marine environment, a major limitation for protection is the lack of a clear definition of an ecosystem. A thorough review of the literature revealed a total of 12 vernacular/scientific terms that have been applied to free-living coralline red algae and these should be treated as synonyms. The Challenger Expedition (1872–1876) was one of the first voyages that promoted the understanding of the rich flora and fauna associated with coralline deposits. During the nineteenth century additional surveys in other areas of the world have confirmed the value of this ecosystem. During twentieth and twenty-first centuries many researchers have produced a vast scientific literature, documenting the importance of rhodolith/maërl, to understand their relevance regarding biodiversity in nearshore habitats. The relevance includes the description of new species or where the distribution of poorly known species has been extended, but more importantly the high number of associated species which

R. Riosmena-Rodríguez (deceased) (✉)
Departamento de Biología Marina, Universidad de Baja California Sur,
La Paz, BCS 23080, Mexico
e-mail: Riosmena@uabcs.mx

R. Riosmena-Rodríguez et al. (eds.), *Rhodolith/Maërl Beds: A Global Perspective*, Coastal Research Library 15, DOI 10.1007/978-3-319-29315-8_1

includes species under protection, species ecologically relevant or species which are part of a formal fisheries. As a consequence of the concern about the state of the ecosystems in Europe at the end of the twentieth century, the EU developed a network of protected areas known as Natura 2000 sites. A series of publications on the conservation status of the maërl/rhodoliths in Atlantic and Mediterranean waters, Brittany, Gulf of California, and their relationship with fisheries, stated clearly that the health of rhodolith habitats in some areas of the world is decreasing, and there is an urgent need for management strategies. The combination of the interest in developing rhodolith/maërl conservation in other countries, the decline of the French Atlantic maërl deposits, and the correlation of rhodolith/maërl presence in or near oil deposits has motivated the exploration of rhodoliths in other areas such as Brazil, México, Australia and New Zealand. Understanding is increasing about the ecological role of rhodoliths in nearshore environments worldwide, the biodiversity associated with rhodoliths, and how human activities are having an increasing impact. The recognition of the importance of rhodolith beds as biodiversity centers has increased with the number of published papers and the growth in knowledge about the taxonomic status of the associated species.

1.1 Introduction

Rhodolith/maërl beds are living and dead aggregations of free-living non-geniculate coralline algae that cover extensive benthic areas in recent oceans and are common in fossil deposits. They are distributed worldwide, with a wide depth range from intertidal sites to 270 m, the deepest known record for a marine plant (Littler and Littler 1984). Individual rhodolith/maërl can be monospecific or multispecific (Riosmena-Rodríguez et al. 2010; Fig. 1.1a, see Villas-Bôas et al. 2014). Rhodoliths develop around skeletal and non-skeletal nuclei (Fig. 1.1b) although they can also grow without any material in the center, usually producing a concentric arrangement of the algal thalli (Rivera et al. 2004, Fig. 1.1c). Rhodoliths have a wide variety of growth forms ranging from fruticose to lumpy, and they can be discoid or irregularly shaped (Fig. 1.1d) averaging approximately 6 cm in diameter (Foster 2001).

Rhodoliths are slow growing and can be long-lived (>100 years), being resilient to varying environmental disturbances (Bosence 1983a, b, c; Steller et al. 2003). The growth, distribution and survivorship of rhodoliths are regulated by many factors, of which light, temperature, and sedimentation are considered the main ones (Wilson et al. 2004). Occasional rotation is necessary but does not need to be frequent, to prevent sedimentation and to allow light to reach all surfaces of the thalli. This rotation is the result of natural disturbances, stochastic phenomena recognized as important determinants of the presence and abundance of associated species. However, Marrack (1999) has shown that bioturbation is more relevant for rhodolith movement than currents. Steller (2003) documented the disturbance caused by fisheries to rhodolith beds and the associated community in Bahía Concepción

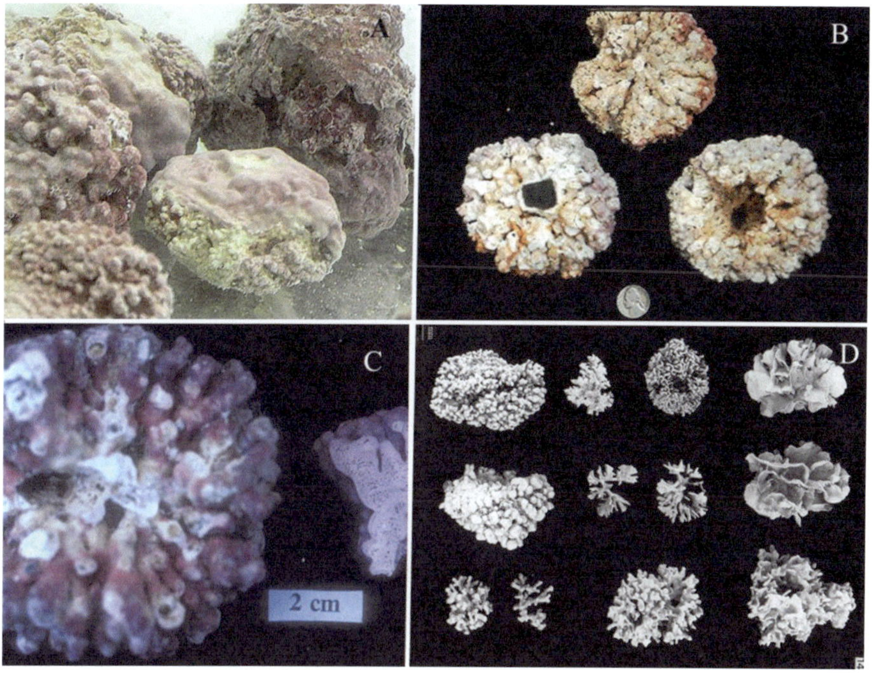

Fig. 1.1 (**a**) Several species growing on a single rhodolith. (**b**) Central core made from different materials (Photo courtesy MS Foster). (**c**) Rhodolith plant in which growth bands can be observed (*right*). (**d**) Range of growth forms present in rhodolith beds

(México). She forecast the recovery process that might eventually return it to its original state, and this process can take from days up to years.

Rhodolith/maërl beds are common features of modern and ancient carbonate shelves worldwide (Nelson 2009) that represent a sedimentary transition from sandy/muddy areas to the rocky substrate (Foster 2001). They can be very extensive, with typically 30–100 % cover, mostly in coarse gravel and clean sands, or in muddy mixed sediments, which occur either on the open coast or in tide-swept channels of marine inlets (Foster et al. 1997). Bosence (1983a, b) outlined the relevance of rhodolith/maërl deposits around the world, initially describing a detailed morphological classification system (Bosence 1976, 1983a), probably the first attempt before the excellent system proposed by Woelkerling et al. (1993). Bosence (1983b) described the ecological settings where rhodolith/maërl are distributed based on a review of the literature (Bosence 1983b – Figs. 1.2 and 1.3), and this set the standard for the map in Foster (2001). Bosence (1983b) stated that rhodolith beds can be present in (1) tropical reefs in tropical areas, (2) seagrass meadows in both tropical and temperate areas, (3) kelp forests in temperate areas, and, (4) forming large beds themselves in both temperate and tropical areas. Freiwald et al. (1991) supported this classification, and Kendrick and Brearley (1997) and Foster et al. (2007) described the same relationship with *Sargassum* forests in subtropical and tropical

Fig. 1.2 (**a**) Rhodolith bed associated with seagrasses. (**b**) Rhodolith bed in a coral reef (part known as rubble). (**c**) Holdfast of *Undaria* attached to rhodoliths in Korea (Photo courtesy SM Boo)

zones. Bosence (1983b) also noted the strong influence of water depth and geographic distribution on bed composition.

Rhodolith beds are bioengineers and provide a three-dimensional habitat for associated species (review by Nelson 2009). Rhodoliths harbor distinctive high-diversity faunal and floral assemblages by virtue of their branching and interlocking nature (Steller et al. 2003; Teichert 2013). Rhodoliths also provide stable habitats onto which a variety of species can attach, including other algae, commercial species such as clams, pen shells and corals. Rhodoliths are resilient to a variety of environmental disturbances, but can be severely impacted by harvesting these commercial species (Steller et al. 2003; Hall-Spencer et al. 2010), ocean acidification or global warming (Martin and Gatusso 2009).

Rhodoliths have been used to obtain recent paleoecological and paleoclimatic information (Kamenos et al. 2008) but also deep-time paleoecological-paleoclimatological inferences (Aguirre et al. this volume). Rhodolith communities contribute significantly to the global calcium carbonate budget (Nelson 2009; Amado-Filho et al. 2012a, b). Rhodoliths can be the main carbonate sediment producers, often forming extensive deposits such as beaches or sand dunes (Sewell et al. 2007; Johnson et al. 2012). They have been exploited as a source of calcium carbonate and used for a wide variety of applications including as fertilizer and soil

Fig. 1.3 (**a**) Subtidal bed with high diversity of red algae (Photo taken by AN Suárez-Castillo). (**b**) Rich sponge-rhodolith assemblage. (**c**) High biomass producers such as *Caulerpa* and *Codium* along with a seastar (Photo courtesy MS Foster). (**d**) Rich cyanobacteria gas producer associated with rhodolith beds

ameliorator, biological denitrifier, drinking water potabilizer, toxin eliminator, as an animal fodder additive, in pharmaceutics, cosmetics, bone surgery, and even in the nuclear industry (López-Benito 1963; Blunden et al. 1980, 1997; Gray et al. 2000; Barbera et al. 2003; Grall and Hall-Spencer 2003; Riul et al. 2008).

The value of rhodoliths as a unique biotope around the world is under threat from different kinds of human activities. The lack of regulation of small and large-scale trawling, reduction of water quality, modification of water circulation, and any other activities that affect rhodolith stability and disturbance are of particular concern (Hall-Spencer et al. 2010; Figueiredo et al. 2012, 2015).

In several parts of the world rhodolith beds and rhodolith-forming species have been recognised as important to conserve, for example as Special Areas of Conservation (UK and Europe), or are considered in the development of Marine Parks (BIOMAËRL 1999, 2003) under the European Community Habitats Directive (Annex V) (Hall-Spencer et al. 2007). Maërl beds have also been included in the UK Biodiversity Action Plan (Anon 1995; Birkett et al. 1998), the Natura 2000 sites, in the OSPAR list, and in the Mediterranean Red Book of threatened habitats (Boudouresque et al. 1990; Hall-Spencer et al. 2010). These management strategies led to the prohibition of their commercial exploitation across Europe, and ultimately rhodolith beds have been proposed as critical marine habitats for conservation in several areas of the world (IBAMA 1996; Birkett et al. 1998; De Grave 1999; De Grave and Whitaker 1999; Hall-Spencer 2005; Metri 2006; Metri and Rocha 2008; Rocha et al. 2006; Hall-Spencer et al. 2010; Sanchez-Ibarra et al. 2013; Peña et al. 2014; as well as the links http://www.petronoticias.com.br/archives/62197; http://www.europapress.es/illes-balears/noticia-oceana-celebra-proteccion-contra-pesca-arrastre-dos-montanas-sumergidas-canal-mallorca-20140811143359.html). It is necessary to take into consideration the response of these assemblages to both natural and artificial disturbances in the development and implementation of coastal management plans around the world and develop monitoring activities in marine protected areas.

In New Zealand and Australia, rhodolith beds have also been recognized as important biogenic habitats for fishery productivity and different monitoring programs are currently being developed to integrate these habitats into management systems, which will help resource users and managers strike a better balance between resource extraction and ecosystem integrity and resilience. It is also known that rhodolith forming-species have been included in the New Zealand list of "sensitive marine habitats" by the Ministry for the Environment (MacDiarmid et al. 2013).

Despite the importance of rhodolith/maërl beds in the marine environment, a major limitation to protection is the lack of a clear definition of an ecosystem. There have been proposed some vernacular and scientific names used for the same or similar concepts, and there is confusion about the meaning of these terms. Some names have been used for living material only, based on their form, shape or origin, whereas other names have been applied to the sediment produced by the organisms, and additional names have been applied to the environment in general. Without a detailed analysis of the all possible vernacular names, the understanding of their role in coastal ecosystems is limited: reliable nomenclature is necessary for their inclusion in legislation or to enable them to be protected.

1.2 Vernacular and Scientific Names

Understanding vernacular/scientific names applied to rhodoliths is relevant to appreciating the natural history of rhodolith/maërl beds, their historical context and how people have recognized this habitat, and also how to correctly define and interpret what is a rhodolith, and how they are distributed in the world oceans. A thorough review of the literature revealed a total of 12 vernacular/scientific terms that have been applied to free-living coralline red algae (Table 1.1). Rhodolith and maërl are the most common terms used for free-living coralline red algae that live on sediments. These names are synonyms (Steller et al. 2003) because they represent the same ecological function. There are other ten names that have been used for the same concept and should be treated as orthographic synonyms.

The term "maërl" was documented by Crouan and Crouan (1867) as a Breton word for unattached, branched corallines, living or dead, loose-lying, often occurring

Table 1.1 An exhaustive list of terms and definitions that have been used for free-living coralline red algae in scientific literature

Scientific term	References
1. Boxwork rhodolith	Basso (1998)
2. Coating	Bressan and Babbini (2003)
3. Coralline algal nodules	Siesser (1972)
4. Maërl	Crouan and Crouan (1867), Lemoine (1910), Cabioch (1969), Bosence (1983a, b), Steneck (1986), Irvine and Chamberlain (1994), Freiwald (1995), Cabioch in Dauvin (1997), Atabey (1998), Birkett et al. (1998), and Hall-Spencer et al. (2010)
5. Nucleated rhodoliths	Lemoine (1910)
6. Nodules	Steneck (1986)
7. Prâlines	Basso (1998)
8. Oncolith/onkoid	Steneck (1986)
9. Rhodoid	Steneck (1986)
10. Rhodolith	Bosellini and Ginsburg (1971), Siesser (1972) Bates and Jackson (1983), Bosence (1983a, b), Steneck (1986), Irvine and Chamberlain (1994), Freiwald (1995), Foster et al. (1997), Atabey (1998), Birkett et al. (1998), Foster (2001), Bressan and Babbini (2003), and Hall-Spencer et al. (2010)
11. Rubble	Steneck (1986)
12. Unattached branches	Steneck (1986)

in extensive deposits or algal gravels found off the northwest coast of France (maërl beds or banks). Cabioch (in Dauvin 1997), described that the name came from "marga" (calcareous sedimentary outcrops), "marle" or "marne", and that it is used generally for non-algal calcareous sedimentary lacustrine deposits. This name is probably based on the latin word "margella" (= coral). Modern definitions for Maërl include: (1) Cabioch (1969) – as Corallinacées arbusculaires; (2) Steneck (1986) – as free-living corallines composed of loose-lying branches of a single species, and, (3) Hall-Spencer et al. (2010) – a collective term for various species of non-jointed coralline red algae (Corallinaceae) that live unattached.

The etymological meaning of rhodolith is "red stone" (Bosence 1983a), derived from the Greek words rose-like (ῥόδειος) and stone (λίθος). Rhodoliths, described as nodules and unattached branched growths with a nodular form composed principally of coralline algae (Bosence 1983a), are often restricted to structures with a nonalgal core (Irvine and Chamberlain 1994). Those coralline algal nodules with a nucleus have been referred to as nucleated rhodoliths by Lemoine (1910). In addition, the term rhodolith is used for nodules composed of a single species of coralline alga or multiple species or boxwork (Basso 1998, Fig. 1.1). Due to fragmentation of branched rhodoliths, their occurrences are generally linked with maërl-type (algal gravel) carbonate deposits (Freiwald 1995).

Rubble is a term commonly used in coral reef ecology to describe broken pieces of coral reefs (Shannon et al. 2013) that are covered with coralline and noncoralline algal species. However, after some time these pieces become covered with coralline algae and thus might be considered equivalent to rhodoliths, as they form beds from these broken pieces (Guzman et al. 2004).

The remaining nine terms that have been used in the literature are defined and discussed in great detail by Aguirre et al. (this volume). Evaluating these terms about their ecological significance is necessary.

1.3 Relevance to Nearshore Biodiversity

The Challenger Expedition (1872–1876) was one of the first voyages that promoted the understanding of the rich flora and fauna associated with coralline deposits. Kjellman (1883: 96) was the first researcher to discover the maërl beds in the Arctic Ocean and Gran (1893) described the associated flora and fauna. Later, Pruvot (1897) described the diverse fauna in La Mancha Occidentale. In the Indian Ocean, three major expeditions (German Tiefsee- Expedition -1898–1899, Siboga Expedition -1899–1900, and Percy Sladen Trust Expedition -1905) investigated not only the coralline algal flora, but also the associated biota. As part of these cruises the flora and fauna were studied by scientists documenting and describing new species including Weber van Bosse and Foslie (1904), Weber van Bosse (1913), and Gardiner (1920). Miranda (1934) provided the first account about the maërl and associated biodiversity in the Galician Rias, and Feldmann (1943) described the flora associated with deep maërl banks in Algeria, one of the few studies about this specific habitat worldwide in war time. In the Pacific Ocean, the studies of Dawson (1944, 1960) in Baja California Peninsula and Segawa (1959) for Japan reported the presence of beds and described the rhodolith species composition.

The discovery and description of the kelp species *Laminaria rodriguezii* in the Mediterranean associated with maërl beds triggered interest in improving understanding of the flora associated with this system. Pérès and Piccard (1955) sketched the spatial organization of marine communities based on the diversity of benthic habitats in the Mediterranean. They used the term "binominia" for each benthic habitat (Pérès and Piccard 1955) and maërl was considered a type of ecosystem. Jacoutte (1962) provided the first description of the flora and invertebrate fauna associated with maërl beds in the Mediterranean, showing the biodiversity of the beds. In this study, 64 seaweed species were determined, and the Rhodophyta reported as the most diverse group.

The best represented invertebrate phyla in rhodolith beds are crustaceans, echinoderms, molluscs and polychaetes. In maërl beds in the Atlantic Ocean, Donze (1968) reported a few subtidal species, but this work was a starting point to understand the rich underwater flora: the habitat produced by rhodolith beds is not only the area supporting biodiversity but also the sediment produced from the rhodoliths. This realization led researchers to describe and analyze the fauna and flora in the sediments derived from maërl/rhodoliths in different regions, such as Algeria (Falconeti 1970), Ireland (Bosence 1979; Keegan 1974), Britain (Hardiman et al. 1976), and France (Augier and Boudouresque 1978). Blunden et al. (1977) proposed the use of maërl commercially not only to promote the use of maërl but also to highlight the need to understand the associated species.

As part of the characterization of the benthic communities along the Brazilian coastal shelf, Kempf et al. (1968), Kempf (1970) stated the "remarkable predominance of coralline algae" in areas not dominated by riverine sediments. The relevance of Kempf's (1970) observations was in relation to the spatial description of the beds, that reached 100 m in depth but also 120 nautical miles off shore, as the dominant habitat along the Brazilian continental shelf. In addition to the physical characterization of the ecosystem, considerable efforts were made to understand the rich flora and fauna (Lavradoi and Ignacio 2006).

Rhodolith/maërl deposits were clearly recognized as important ecosystems in Europe in the late 1970s, and the inventory efforts showed their importance as highly biodiverse habitats. Mora-Bemúdez (1980) documented the distribution of carbonate deposits with high diversity of infauna in Galician Rias. Later, Myers and McGrath (1980, 1983) described new amphipod species associated with maërl beds in several areas of the world. At the same time, studies on molluscs from Mulroy Bay (Nunn 1993) documented the presence of species that were also part of fisheries. Maggs (1983) continued with the description of the flora associated with maërl beds in Irish waters. In this study, two different types of beds were determined based on wave exposure. This was the first example of a quantitative sampling study using SCUBA to determine rhodolith abundance. Strong seasonal changes were found rather than spatial changes. Bosence (1985) carried out the first comparison of the fauna in dead and living maërl, finding strong differences in species composition and density.

A similar situation was found in the Central Pacific nearby Penguin Bank, where Agegian and Abbott (1985) described three slightly different mesophotic algal zones based on submersible surveys. The shallow zone (45–70 m) contained a diverse assemblage of algae (*Lobophora variegata*, *Dictyota friabilis*, *Halimeda* spp., coralline algal rhodoliths, *Mesophyllum mesomorphum*, and *Peyssonnelia rubra*). The mid-depth zone (90–110 m) was characterized by *Codium mamillosum* and crustose coralline algae (CCA), while the deep zone (110–182 m) was composed primarily of CCA. Ballesteros (1989) also studied this kind of banding in the Mediterranean seaweed assemblages in maërl, concluding that spatial differences are more important than temporal changes.

Rowe et al. (1990) documented the infauna associated with maërl beds in southwest England. They found higher richness in the maërl beds than in surrounding areas. Basso (1992) recognized 93 species in the phytobenthic communities in circalitoral soft bottoms. Using Bray Curtis analyses (to evaluate compositional dissimilarity between two different sites), Rowe et al. (1990) found that all sites dominated by rhodolith beds were closely related, and were very different from the other substrates. Therefore, the presence of rhodoliths was a factor controlling the distribution of associated species. Ramos-Esplá (1992) developed an extensive taxonomic analysis of the ascidians in Spain and found a high diversity of species associated with maërl/rhodolith beds.

Freiwald (1993) described the close relationship of small gastropods, foraminifera, diatoms, ostracods and bacteria associated with rhodolith beds and their trophic relationships with sea urchins and their impacts on kelp populations. Freiwald (1995) found that the surface of rhodoliths are dominated by vagile herbivores and stimulate metamorphosis of echinoderm species. Later, Friewald (1998) illustrated that fleshy macroalgae are involved in the inshore and offshore sediment transportation by kelp drafting, and Freiwald and Mostafawi (1998) described a close relationship of an ostracod community to sediment type, being the most diverse community that associated with rhodolith beds.

Grall and Glémarec (1997a, b) have evaluated the overall biodiversity value of maërl/rhodolith beds in Bretagne, and have used these communities to calculate indices of perturbation (Grall and Glémarec 1997a). Their contribution was to use these indices to evaluate impacted (where exploitation of maërl was happening) versus unimpacted (control) sites (Grall and Glémarec 1997b). Their functional approach provided a strong baseline for future impact analysis. Hall-Spencer (1998) further demonstrated the relevance of maërl as critical habitats for associated molluscs in Scotland, and Basso and Brusoni (2004) in the Mediterranean.

Hall-Spencer et al. (1998) have shown the impact of Rapido trawling on scallops in benthic habitats and on the abundance of slow moving benthos. Hall-Spencer and Atkinson (1999) have shown the relevance of the maërl for one species (*Upgebia deltaura*) in most of its geographical distribution. They also found that this species functions ecologically as a bioturbator in the sediment removal for the development of their deep holes that may survive storms. The architecture of maërl habitats also provides good shelter for invertebrate predators from larger predators. De Grave (1999) found that 75 % of all macrobenthic species within specific maërl beds are crustaceans, the majority of which are amphipoda. Hall-Spencer (2001) has documented the maërl in the Firth of Clyde, suggesting that this site is well preserved for investigations into process and function of the habitat demonstrating the value of maërl. De Grave and Whitaker (1999) have documented high species richness associated with high functional diversity that shifts from omnivorous crustaceans to filter feeders after dredging. Veale et al. (1999) have documented a direct relationship between the number of individuals in the community in relation to the increase of fishing effort on selective removal of species. Hall-Spencer and Moore (2000a, b) have shown that scallop fishing impacted 70 % of maërl habitats and no signs of recovery were evident in the following years, providing further evidence of the vulnerability of this habitat to specific demersal fishing practices.

During the 1990s research about rhodolith beds started in the Gulf of California as part of an international collaboration between USA and México, and was funded by a series of agencies (InterAmerican Global Change Agency, National Geography, Packard Foundation, Consejo Nacional de Ciencia y Tecnología México, Comisión Nacional para el uso y estudio de la Biodiversidad, National Science Foundation). Studies in Western Australia yielded a series of critical papers about this habitat: Kendrick and Brearley (1997) demonstrated the relevance of rhodolith beds for

Sargassum species demography and dispersal; Collins (1988) and James et al. (1999) documented their distribution between 35 and 60 m depth in the Rottnest Shelf; while Sim and Townsend (1999) documented their presence around Rottnest Island. Steller and Foster (1995) have shown the value of measuring branch density in relation to wave action and how this is independent of species. Foster et al. (1997) reinforced the idea that rhodolith beds develop in two type of environments, wave and current beds, and this information might be able to be used to interpret paleoenvironments.

A regional workshop in La Paz in 1998 led to increasing collaboration between México and USA. It was also the starting point for establishing a new collaborative link with Brazil. Derived from these collaborations, several papers were produced showing the relevance of rhodolith beds in the nearshores of the Gulf of California. Reyes-Bonilla et al. (1998) showed that rhodolith beds also might be a habitat for the rich coral fauna in the Gulf of California as described for free living corals in the Western Pacific by Scoffin et al. (1985). In addition, these rhodolith beds are a feeding ground for sea urchins (James 2000). Studies of rhodolith beds in deeper waters have yielded new species of chitons (Clark 2000), as well as new records and new species of seaweeds (Cho and Riosmena-Rodríguez 2008; Cho et al. 2003; Leon-Cisneros et al. 2009; Hernández-Kantun et al. 2010). The value of rhodoliths as fish habitat was clearly presented by Aburto-Oropeza and Balart-Paez (2001) and was used to delimit Marine Protected Areas by Sala et al. (2002).

The first International Workshop on 'The Conservation and Management of Maërl', 23–27 February, 2001 at University Marine Biological Station, Millport (Isle of Cumbrae, Scotland), was dedicated to (1) exchange ideas produced as part of the BIOMAËRL Project among the European countries (Barbera et al. 2003), (2) evaluate the conservation status in relation to fisheries (Grall and Hall-Spencer 2003; Hall-Spencer et al. 2003), (3) evaluate carbonate production and accumulation (Bosence and Wilson 2003), and, (4) evaluate for the first time the impact of trawling over maërl beds in the Mediterranean (Bordehore et al. 2003).

In the second rhodolith workshop in La Paz BCS, in 2006, seven papers were produced as special section of the journal Ciencias Marinas (Riosmena-Rodríguez et al. 2007). Later, the meeting was held in Buzios, Brazil, in 2009 and 12 papers were published in a special section of the Journal of Coastal Research. Finally, the meeting in Granada, Spain, in 2012 promoted a special issue of Cryptogamie Algologie (Aguirre and Peña 2014) with six papers. In addition to the monographs derived from these international meetings, papers on the systematics of coralline algae were published in a special issue of Phytotaxa about the Corallinophyceae of the Atlantic Ocean (Riosmena-Rodríguez et al. 2015). All of the above efforts have promoted and encouraged more work on recognizing the biodiversity associated with maërl beds and their role in ecological processes.

Steller et al. (2003) compared the species richness and density between rhodolith beds and the surrounding sandflats and demonstrated strong differences between the two habitats but also the value of rhodolith beds in terms of biodiversity. A complementary analysis was presented by Hinojosa-Arango and Riosmena-Rodríguez (2004) who found that faunal abundance differed significantly in relation to rhodolith-forming species, but no significant differences were observed between different growth-forms in the Gulf of California. Crustaceans, polychaetes and molluscs were the most important taxa in all three rhodolith beds studied. Konar et al. (2006) reported that rhodolith beds in the North Pacific seem to be dominated by chitons instead of any other fauna described in previous studies. Gherardi (2004) determined the high contribution of carbonate sand from rhodolith beds in the Southwestern Atlantic in relation to the associated community structure. Perry (2005) has established the role of rhodolith beds in carbonate production in a reef environment at the Indian Ocean. Castriota et al. (2005) documented the dominance of molluscs in maërl beds in the Mediterranean, and Mathis et al. (2005) described the relevance of this habitat in Australian waters.

It has been demonstrated that maërl grounds provide refuge for juvenile life stages of commercially important shellfish species such as the queen scallop, *Aequipecten opercularis* (Kamenos et al. 2004a, b) and *Agropecten ventricosus* (Steller and Caceres 2009). Maërl has also been found to be important for the juvenile gadoid fish: Atlantic Cod (*Gadus morhua*), Saithe (*Pollachius virens*) and Pollack (*Pollachius pollachius*). These species have all been found to use maërl grounds during their juvenile life stages due to the availability of food and refuge from predators (Hall-Spencer et al. 2003; Kamenos et al. 2004c). Riosmena-Rodríguez and Medina (2011) have demonstrated that rhodolith beds are a relevant habitat for multispecies recruitment.

In the case of seaweeds associated with rhodoliths, a series of papers have been produced over the last 10 years that have shown the biodiversity, seasonal and spatial variation of the flora and their ecological interactions. Amado-Filho et al. (2007) compared the epibenthic macroalgal biomass and composition of rhodolith beds from zone 1 (4–18 m depth) surveyed in two seasons. They showed that samples from summer possessed up to twice as much biomass and species as those from winter. These results can be explained by the high instability of the beds along the southeastern Brazilian coast, caused by frequent storm disturbances, which are higher in winter than in summer (Soares-Gomes and Pires-Vanin 2003; Amado-Filho et al. 2007).

Peña and Bárbara (2008) developed a comprehensive literature review of all Galician studies and produced a list of 198 maërl-associated species (204 infraspecific taxa including life history stages) recorded from a total of 111 maërl beds. The survey (2003–2006) increased the Galician maërl epiflora records to 226 species, highlighting the occurrence of 10 non-native species. An a posteriori review was presented by Peña et al. (2014) in relation to the known seaweeds associated with maërl/rhodolith beds in the NE Atlantic from Svalbard to Portugal. A total of 350 macroalgal species has been recorded in this region, including 11 non-native species. The flora of maërl beds off Iceland and Norway is poorly known, but maërl beds off Britain, Ireland, France and Spain have been surveyed extensively and support several species that are maërl specialists (i.e. *Cruoria cruoriaeformis*, *Cladophora rhodolithicola*, *Gelidiella calcicola*).

Amado-Filho et al. (2010) studied seaweed populations growing on rhodoliths between 4 and 18 m water depth in the southern part of Espírito Santo State (Brazil).

They determined that 167 species are present in the flora with three new records for Brazil. Marked seasonal differences in fleshy algal species composition and abundance were related to seasonal instabilities caused by winter-storm disturbance over the rhodolith beds that were found to be supporting 25 % of the known macroalgal species richness along the Brazilian coast. Pascheli et al. (2013) showed that rhodolith density decreases with increasing depth and during winter. They found that turf-forming seaweeds accounted for 60 % of the biomass growing on rhodoliths and temporal variations were observed in species composition and rhodolith density.

In the case of invertebrates, Figueiredo et al. (2007) made an evaluation of rhodolith/maërl as living substrate for invertebrates comparing the proportion of invertebrates between live and dead at different seasons in a deep bank. They found a general seasonal trend of a reduction in invertebrate abundance from autumn/winter to spring. Amphipods were dominant in autumn (70 % of sample) but half as abundant in winter and spring when their population density was similar to that of polychaetes (49 %). Polychaete abundance was significantly lower in live than in dead transplanted rhodoliths, whereas amphipods showed no difference. Variation in infaunal abundances was related to variability in habitat heterogeneity due to changes in rhodolith size and shape resulting from changes in water turbulence.

In the case of amphipods, there are a series of papers dealing with new records (Bakalem et al. 2014, Algeria; Riera et al. 2014, Canary Islands) and new species (Carvalho et al. 2012, Portugal; Lowry and Stoddart 2012, Australia; Souto et al. 2014, Brazil; Sturaro and Guerra García 2012, Mediterranean).

In other areas of the world polychaetes contribute significantly to diversity present in rhodolith beds. Harvey and Bird (2008) found that polychaetes are dominant taxa in temperate Australian waters. Dos Santos et al. (2011) found four new records and a new species of the genus *Sabellaria* collected in rhodolith beds along the subtidal zone of the coast of João Pessoa and Cabedelo, Paraíba. Later, Berlandi et al. (2012) reviewed the polychaetes associated with several sites along the Brazilian coast. They found a significant difference between Espírito Santo and the Abrolhos, reflected in the polychaete assemblages, with 26 families identified, 4 exclusive to Espírito Santo and 9 in the Abrolhos. The most common families registered different species composition in both rhodolith beds, except for *Eunice multicylindris* and *Arabella mutans* found in both regions. Rhodoliths on the Abrolhos hosted polychaetes dominated by carnivorous or herbivorous families, such as the Syllidae, the most abundant and species-rich family. Rhodoliths found off Espírito Santo hosted polychaetes dominated by burrowers, such as the Lumbrineridae, which were favored by their boxwork structure, built of invertebrates and inorganic material.

In New Zealand, Dewas and O'Shea (2012) found that the large dog cockle, *Tucetona laticostata*, is widely but sporadically distributed throughout coastal waters of the Hauraki Gulf, where it resides partially buried in gravel and rhodoliths in shallow water, at 5–15 m depth. Avila et al. (2013) were the first to evaluate the relationship between sponges and the stability of rhodolith beds. This relationship has been documented for the Mediterranean at 88 m (Calcinai et al. 2013), and

Pereira-Filho et al. (2015) documented, in more detail, the close relationship between sponges and rhodolith/maërl beds in relation to reef formation in the Southwestern Atlantic.

Rhodolith beds are habitats in which new species have been discovered or where the distribution of poorly known species has been extended, such as for echinoderms (Gondim et al. 2014; Granja-Fernandez et al. 2014), octocorals (Bridge et al. 2011; Brice and Poliseno 2014), and sponges (Sitja and Maldonado 2014). Calcavanti et al. (2014) found differences in bacterial assemblages between rhodolith beds and other habitats.

More integrated studies, started by Foster et al. (2007), have documented the benthic macroalgae, benthic epifauna, fish fauna, and cryptofauna linked to a rhodolith bed mixed with *Sargassum* forest at Cabo Los Machos, where a total of 216 taxa were found. There were some large changes in abundance between sampling times, especially in *S. horridum* related to the season, encrusting nongeniculate coralline algae and *Helminthocladia australis* and an increase in "mixed red algal turf". Temporal variation in these invertebrates was characterized by reduction in taxa between March and October, with similar diversity as observed for macroalgae and fishes. In contrast to both macroalgae and benthic invertebrates, the average abundance/site in fish was highest in October, largely due to an increase in juveniles. The most abundant species (>4 individuals/200 m^2) in March were the Mexican barracuda (*Sphyraena lucasana*), spottail grunt (*Haemulon maculicauda*), Panamic sargent major (*Abudefduf troschelii*), and Pacific porgy (*Calamus brachysomus*). Cryptofaunal invertebrates living within large rhodoliths (more than 5 cm), greatly exceeded the number that was found in the smaller (less than 5 cm) size classes of rhodoliths: 114 total invertebrate taxa with an average of 40 taxa/individual large rhodolith, with polychaetes (52 taxa) and crustaceans (29 taxa) as the primary contributors, followed by molluscs (14 taxa) and echinoderms (10 taxa).

Riera et al. (2012) have studied the structure of the macrofaunal assemblages on different habitat types (*Cymodocea*, *Caulerpa*, sabellid field, garden eel and maërl beds) with samples taken from 14 to 46 m in depth. Significant differences were found in macrofaunal assemblages among seabed types, with highest abundances and lowest biodiversity in sabellid fields, where the sabellid *Bispira viola* dominated. The polychaetes *Aponuphis bilineata* and *Chone filicauda* and the mollusc *Turritella brochii* were the most abundant taxa on maërl beds in the Canary Islands. The mosaic of granulometric conditions would explain the associated macroinfaunal community structure and contribute to the creation of diversity on these relatively well preserved seabeds at a local scale.

In rhodolith beds from north New Zealand, Neill et al. (2014) found a highly diversified biota with 238 taxa: 197 invertebrates, 37 algae and 4 fish taxa. The highest abundance and taxon richness occurred in a bed in a high sedimentation rate. This finding points to the importance of sampling individual rhodolith beds and suggests that it would be risky to extrapolate findings to other beds, even those in close proximity. The findings of Neill et al. (2014) are well supported by Tamega et al. (2014), who described deep rhodolith beds dominated by bryozoans, cnidarians, brachiopods and sponges, a community composition structurally very different from shallow-water rhodolith beds. Sheehan et al. (2015) documented a similar situation comparing live and dead beds and highlighted the relevance of living material for the great abundance of epifauna that are not commonly found in dead areas.

1.4 Conservation and Protection Initiatives

During the 1990s and at the beginning of the 2000s, concerns about marine ecosystems around European waters resulted in the development of Special Areas for Conservation and Community Directives to protect ecosystems. The Decree 1626/94 of the European Union on conservation of Mediterranean living resources has a bearing on the conservation of maërl beds since it prohibits demersal trawling in the waters shallower than 50 m (Barbera et al. 2003). The Mediterranean Action Plan of the United Nations Environment Programme (UNEP/IUCN/GIS Posidonie 1990) considered 'les fonds de maërl' should be included in the Red Data Book of threatened marine vegetation. This document had no legislative standing but it served to highlight species and habitats that need protection (Barbera et al. 2003).

Birkett et al. (1998) developed the first evaluation of the maërl as a special habitat for conservation. BIOMAËRL (1999) was a project dedicated to understand the value of maërl/rhodolith beds in European waters (Atlantic and Mediterranean). Maërl is included as a key habitat for Ireland and the UK within the Annex I habitats, 'large shallow inlet and bays' and 'sand banks which are slightly covered by seawater at all times' such Special Areas of Conservation (SACs) designated under the Directive 92/43/EEC on the Conservation of Natural Habitats and Wild Fauna and Flora (Hall-Spencer et al. 2010). Costello (2000) developed the biodiversity plan in Ireland in which maërl was one of the main ecosystems considered for protection. Two maërl-forming species, *Lithothamnion corallioides* and *Phymatolithon calcareum*, are listed in Annex V of the Habitats Directive and in some locations maërl is also listed as a key (Hall-Spencer et al. 2010).

De Grave et al. (2000) evaluated the potential extraction of maërl beds in Ireland, concluding that a proposal to extract a live maërl bed or a maërl debris facies requires approval from the regulatory authorities and the application should be carefully scrutinized. They concluded that the inclusion of both the main maërl bed forming species in the EU Habitats Directive is desirable for an integral protection. In addition, any application to extract live or dead maërl should be required to submit an Environmental Impact Statement (EIS). However, there are no basic requirements in their EIS for maërl beds and De Grave et al. (2000) stated the need for generic guidelines dealing with the extraction of marine gravel deposits. No other country has regulated sand extraction for commercial purposes. Nicolleti et al. (2006) have documented how to monitor the impact of beach restoration on seagrasses and maërl in coastal areas where storms removed sand in nearshore areas. Lauwaert et al. (2009) have evaluated the impact of seabed extraction for beach replacement and construction, where maërl is one of the major components in the benthic environment, producing habitat fragmentation and degradation by the seabed extraction.

As a consequence of the concern about the state of the ecosystems in Europe, the EU developed a network of protected areas known as Natura 2000 sites. A series of publications on the conservation status of the maërl/rhodoliths in Atlantic and Mediterranean waters (Barbera et al. 2003), Brittany (Grall and Hall-Spencer 2003), Gulf of California (Steller et al. 2003) and their relationship with fisheries (Hall-Spencer et al. 2003) stated clearly that the health of rhodolith habitat in some areas of the world is decreasing, and there is an urgent need for management strategies. Bordehore et al. (2003) found that trawling activities that disturbed the sediments of fishing areas change the associated flora and fauna and thus the productivity of the area. The combination of the interest in developing rhodolith/maërl conservation in other countries, the decline of the Atlantic French maërl deposits, and the correlation of rhodolith/maërl presence in or near oil deposits has motivated the exploration of rhodoliths in other areas such as Brazil, México, Australia and New Zealand. Fredericq et al. (2014) have shown that rhodolith beds serve as seed banks for the fast recovery of the subtidal flora after the BP oil spill in the northern Gulf of México. It is necessary to develop monitoring methods for this habitat to understand potential impacts from other human activities.

Understanding is increasing about the ecological role of rhodoliths in nearshore environments worldwide, the biodiversity associated with rhodoliths, and how human activities are having an increasing impact. The importance of rhodolith beds as biodiversity centers has increased with the number of published papers and the growth in knowledge about the taxonomic status of the associated species.

Acknowledgements RRR acknowledges the support of CONACYT-SEMARNAT for the support to the current research throughout the grant 25343. Also, I thank the patience and the continuous review of this chapter by Julio Aguirre and Wendy Nelson to improve in many ways the present version. I deeply thank Mike Foster and Sung Min Boo for the access to some of their pictures for figures in this chapter.

References

Agegian C, Abbott I (1985) Deep water macroalgal communities: a comparison between Penguin Bank (Hawaii) and Johnston Atoll. In: Proceedings of the 5th international Coral Reef congress, 1985, pp 47–50

Aguirre J, Braga JC, Bassi D (this volume) Rhodoliths and rhodolith beds in the rock record. In: Riosmena-Rodríguez R., Nelson W., Aguirre J. (eds) Rhodolith/maerl beds: a global perspective. Springer, Berlin

Aguirre J, Peña V (2014) Advances in coralline algae research: insights from the IV International Rhodolith Workshop. Cryoptogamie Algol 35:3–5

Amado-Filho GM, Maneveldt G, Manso R, Marins-Rosa B, Pacheco M, Guimarães S (2007) Structure of rhodolith beds from 4 to 55 meters deep along the southern coast of Espírito Santo State, Brazil. Cienc Mar 33:399–410

Amado-Filho G, Maneveldt G, Pereira-Filho G, Manso R, Bahia R, Barros-Barreto M, Guimarães S (2010) Seaweed diversity associated with a Brazilian tropical rhodolith bed. Cienc Mar 36(4):371–391

Amado-Filho GM, Moura RL, Bastos AC, Salgado LT, Sumida PYG et al (2012a) Rhodolith beds are major $CaCO_3$ bio-factories in the Tropical South West Atlantic. PLoS ONE 7(4):e35171
Amado-Filho GM, Pereira-Filho GH, Bahia RG, Abrantes DP, Veras PC, Matheus Z (2012b) Occurrence and distribution of rhodolith beds on the Fernando de Noronha Archipelago of Brazil. Aquat Bot 101:41–45
Anon (1995) Biodiversity: the UK steering group report. HMSO, London, for Department of the Environment
Atabey N (1998) Facies characteristics, geographic distribution of rhodoliths, maërl (red algae) in southern shelf of the Sea of Marmara. Miner Res Explor Bull 120:55–61
Augier H, Boudouresque CF (1978) Vegetation marine de i'he de Port-Cros (Parc National). XVI: contribution a l'étude de l'épi fore du Déthtique Côtier. Tray Sci Parc Nation Port-Cros 4:101–125
Bakalem A, Dauvin JC, Grimes S (2014) New marine amphipod records on the Algerian coast. J Mar Biol Assoc U K 94:753–762
Ballesteros E (1989) Composición y estructura de los fondos de maërl de Tosa de Mar (Gerona, España). Collect Bot 17:161–182
Barbera C, Bordehore C, Borg JA, Glémarec M, Grall J, Hall-Spencer JM, de la Huz C, Lanfranco E, Lastra M, Moore PG, Mora J, Pita ME, Ramos-Esplá AA, Rizzo M, Sánchez-Mata A, Seva A, Schembri PJ, Valle C (2003) Conservation and management of northeast Atlantic and Mediterranean maerl beds. Aquat Conserv Mar Freshw Ecosyst 13:S65–S76
Basso D (1992) Le rodoficee calcaree dei fondi mobil circalitorali del Mar Tirreno: le rodoliti attuali in una prospettiva paleoecologica. Ph. D., thesis University of Milano, 139 pp
Basso D (1998) Deep rhodolith distribution in the Pontian Islands, Italy: a model for the paleoecology of a temperate sea. Palaeogeogr Palaeoclimatol Palaeoecol 137(1):173–187
Basso D, Brusoni F (2004) The molluscan assemblage of a transitional environment: the Mediterranean maërl from off the Elba Island (Tuscan Archipelago, Tyrrhenian Sea). Boll Mallacol 40:37–45
Berlandi RM, Figueiredo MAO, Paiva PC (2012) Rhodolith morphology and the diversity of polychaetes off the southeastern Brazilian coast. J Coast Res 28:280–287
BIOMAËRL (1999) Final report, BIOMAËRL project (Coordinator: Moore PG, University Marine Biological Station Millport, Scotland), EC Contract No. MAS3-CT95-0020, 1: 1–541, 2: 542–973 pp +Appendix
BIOMAËRL team (2003) Conservation and management of N.E Atlantic and Mediterranean maërl beds. Aquat Conserv Mar Freshw Ecosyst 13(S1):65–76
Birkett D, Maggs C, Dring MO (1998) Maerl, vol 5. An overview of dynamic and sensitivity characteristics for conservation management of SACs. Scottish Association for Marine Science
Blunden G, Farnham W, Jephson N, Fenn R, Plunkett B (1977) The composition of maerl from the Glenan Islands of Southern Brittany. Bot Mar 20(2):121–126
Blunden G, Farnham W, Jephson N, Barwell C, Fenn R, Plunkett B (1980) Composition of maerl beds of economic interest in northern Brittany, Cornwall and Ireland. In: Proceedings-international seaweed symposium
Blunden G, Campbell SA, Smith JR, Guiry MD, Hession CC, Griffin RL (1997) Chemical and physical characterization of calcified red algal deposits known as maërl. J Appl Phycol 9(1):11–17
Bordehore C, Ramos-Esplá AA, Riosmena-Rodríguez R (2003) Comparative study of two maerl beds with different otter trawling history, southeast Iberian Peninsula. Aquat Conserv Mar Freshw Ecosyst 13(S1):S43–S54
Bosence DWJ (1976) Ecological studies on two unattached coralline algae from western Ireland. Palaeontology 19:365–395
Bosence DWJ (1979) Live and dead faunas from coralline algal gravels, Co. Galway. Palaeontology 22:449–478
Bosence DWJ (1983a) Description and classification of rhodoliths (rhodoids, rhodolites). In: Peryt TM (ed) Coated grains. Springer, Berlin, pp 217–224

Bosence DWJ (1983b) The occurrence and ecology of recent rhodoliths – a review. In: Peryt TM (ed) Coated grains. Springer, Berlin, pp 225–242
Bosence DWJ, Wilson J (2003) Maerl growth, carbonate production rates and accumulation rates in the NE Atlantic. Aquat Conserv Mar Freshw Ecosyst 13(S1):S21–S31
Boudouresque C, Pergent G, Francour P, Harmelin-Vivien M, Jangoux M, Mazzella L, Panayotidis P, Pergent-Martini C, Ramos-Esplá AA, Romero J (1990) Le COST 647: posidonia project. Posidonia Newsl 3:27–34
Bressan G, Babbini L (2003) Biodiversita marina delle coste italiane: Corallinales del mar mediterraneo: guida alla determinazione. Soc Italina Biol Mar 10(2):1–237
Brice M, Poliseno A (2014) Two new records of octocorals (Anthozoa, Octocorallia) from northwest Australia. Rec West Aust Mus 29:159–168
Bridge TCL, Done TJ, Beaman RJ, Friedman A, Williams SB, Pizarro O, Webster JM (2011) Topography, substratum and benthic macrofaunal relationships on a tropical mesophotic shelf margin, central Great Barrier Reef, Australia. Coral Reefs 30(1):143–153
Cabioch J (1969) Les fonds de maerl de la baie de Morlaix et leur peuplement végétal. Cahiers de Biologie Marine
Calcavanti GS, Gregoracci GB, dos Santos EO, Silveira CB, Meirelles PM, Longo L, Gotoh K, Nakamura S, Iida T, Sawabe T, Rezende CE, Francini-Filho RB, Moura RL, Amado-Filho GM, Thompson FL (2014) Physiologic and metagenomic attributes of the rhodoliths forming the largest $CaCO_3$ bed in the South Atlantic Ocean. ISME J 8:52–62
Calcinai B, Moratti V, Martinelli M, Bavestrello G, Taviani M (2013) Uncommon sponges associated with deep coral bank and maerl habitats in the Strait of Sicily (Mediterranean Sea). Ital J Zool 80(3):412–423
Carvalho S, Cunha MR, Pereira F, Pousão-Ferreira P, Santos MN, Gaspar MB (2012) The effect of depth and sediment type on the spatial distribution of shallow soft-bottom amphipods along the southern Portuguese coast. Helgol Mar Res 66:489–501
Castriota L, Agamennone F, Sunseri G (2005) The mollusc community associated with maerl beds of Ustica Island (Tyrrhenian Sea). Cah Biol Mar 46(3):289
Cho TO, Riosmena-Rodriguez R (2008) *Ceramium periconicum* sp. nov. (Ceramiaceae, Rhodophyta): a new subtidal species from Baja California Sur, México. Bot Mar 51:307–312
Cho TO, Riosmena-Rodríguez R, Boo SM (2003) Fist record of *Ceramium giacconei* (Ceramiaceae, Rhdophyta) in the North Pacific: developmental morphology of vegetative and reproductive structures. Bot Mar 46:548–554
Clark RN (2000) The chiton fauna of the Gulf of California rhodolith beds (with the descriptions of four new species). Nemouria: Occasional Papers of the Delaware Museum of Natural History 43:1–18
Collins LB (1988) Sediments and history of the Rottnest Shelf, southwest Australia: a swell dominated, non-tropical carbonate margin. Sediment Geol 60:15–49
Conservation Do (1998) Kapiti marine reserve: conservation management plan Wellington conservancy paper presented at the conservation management planning series
Costello MJ (2000) A framework for an action plan on marine biodiversity in Ireland. Marine Institute, Ireland, 47 pp
Crouan PL, Crouan HM (1867) Florule du Finistère. Contenant les descriptions de 360 espèces nouvelles de sporogames, de nombreuses observations, et une synonomie des plantes cellulaires et vasculaires qui croissent spontanément dans ce département, etc. pp. [i]-x, [1]-262, frontisp., pi. 1–31, + 1 suppl. pl. Paris & Brest: Friedrich Klincksieck & J.B. et A. Lefournier
Dauvin JC (1997) Les biocénoses marines et littorales françaises des côtes atlantique, manche et mer du nord Synthèse, menaces et perspectives. Muséum National d'Hiostoire Naturelle
Dawson EY (1944) The marine algae of the Gulf of California. Allan Hancock pacific expeditions, 3:1–454
Dawson E (1960) Marine red algae of Pacific Mexico. Part 3. Cryptonemiales, Corallinaceae subfam. Melobesioideae. Pac Nat 3:3–125

De Grave S (1999) The influence of sedimentary heterogeneity on within maerl bed differences in infaunal crustacean community. Estuar Coast Shelf Sci 49(1):153–163

De Grave S, Whitaker A (1999) A census of maërl beds in Irish waters. Aquat Conserv Mar Freshw Ecosyst 9(3):303–311

De Grave S, Fazakerley H, Kelly L, Guiry M, Ryan M, Walshe J (2000) A study of selected maërl beds in Irish waters and their potential for sustainable extraction. Mar Resour Ser Mar Inst Dublin 10:1–44

Dewas S, O'Shea S (2012) The influence of *Tucetona laticostata* (Bivalvia: Glycymeridae) shells and rhodolith patches on benthic-invertebrate assemblages in Hauraki Gulf, New Zealand. N Z J Mar Freshw Res 46(1):47–56

Donze M (1968) The algal vegetation of the Ría de Arosa (NW. Spain). Blumea-Biodivers Evol Biogeogr Plants 16(1):159–192

Dos Santos AS, Riul P, Dos Santos Brasil AC, Christoffersen ML (2011) Encrusting Sabellariidae (Annelida: Polychaeta) in rhodolith beds, with description of a new species of Sabellaria from the Brazilian coast. J Mar Biol Assoc U K 91(02):425–438

Falconeti C (1970) Etude faunistique d'un faciès: "la Gravelette" ou maërl de Castiglione (Algérie). Tethys 1(4):1057–1096

Figueiredo MAO, Coutinho R, Villas-Boas AB, Tâmega FTS, Mariath R (2012) Deep-water rhodolith productivity and growth in the southwestern Atlantic. J Appl Phycol 24:487–493

Figueiredo MAO, Eide I, Reynier M, Villas-Bôas AB, Tâmega FTS, Ferreira CG, Nilssen I, Coutinho R, Johnsen S (2015) The effect of sediment mimicking drill cuttings on deep water rhodoliths in a flow-through system: experimental work and modeling. Mar Pollut Bull 95:81–88

Foster MS (2001) Rhodoliths: between rocks and soft places. J Phycol 37(5):659–667

Foster MS, Riosmena-Rodríguez R, Steller DL, Woelkerling WJ (1997) Living rhodolith beds in the Gulf of California and their implications. Pliocene carbonates and related facies flanking the Gulf of California, Baja California, Mexico 318. p 127

Foster M, McConnico L, Lundsten L, Wadsworth T, Kimball T, Brooks L, Medina-López M, Riosmena-Rodríguez R, Hernández-Carmona G, Vasquez-Elisando R (2007) The diversity and natural history of *Lithothamnion muelleri-Sargassum horridum* community in the Gulf of California. Cienc Mar 33(4):367–384

Fredericq S, Arakaki N, Camacho O, Gabriel D, Krayesky D, Self-Krayesky S, Rees G, Joseph Richards J, Sauvage T, Venera-Ponton D, Schmidth WE (2014) A dynamic approach to the study of rhodoliths a case study for the northwestern gulf of México. Cryptogam Algol 35:77–98

Freiwald A (1993) Coralline algal coral frameworks-Islands within the phaeophytic kelp belt. Facies 29(1):133–148

Freiwald A (1995) Sedimentological and biological aspects in the formation of branched rhodoliths in northern Norway. Beitr Paläontol Osterr 20:7–19

Freiwald A (1998) Modern nearshore cold-temperate calcareous sediments in the Troms District, northern Norway. J Sediment Res 68:763–776

Freiwald DGA, Henrich R, Schäfer P, Willkomm H (1991) The significance of high-boreal to subarctic maerl deposits in northern Norway to reconstruct Holocene climatic changes and sea level oscillations. Facies 25(1):315–339

Gardiner S (1920) The fauna and geography of the Maldive and Laccadive archipelagoes, vol 1. Cambridge University Press, Cambridge, pp 460–471

Gherardi DFM (2004) Community structure and carbonate production of a temperate rhodolith bank from Arvoredo Island, southern Brazil. Braz J Oceanogr 52:3–4

Gondim AI, Dias TLP, Duarte RCS, Riul P, Lacouth P, Christoffersen ML (2014) Filling a knowledge gap on the biodiversity of Rhodolith-associated Echinodermata from northeastern Brazil. Trop Conserv Sci 7(1):87–99

Grall J, Glémarec M (1997a) Using biotic indices to estimate macrobenthic community perturbations in the Bay of Brest. Estuar Coast Shelf Sci 44:43–53

Grall J, Glémarec M (1997b) Biodiversity of Maerl beds in Brittany: functional approach and anthropogenic impacts. Vie et Milieu (France) 47:339–349

Grall J, Hall-Spencer J (2003) Problems facing maerl conservation in Brittany. Aquat Conserv Mar Freshw Ecosyst 13(S1):S55–S64

Gran HH (1893) Algevegetationen i Tønsbergfjorden. Skrifter udgivne af Videnskabsselskabet i Christiania. I, Mathematisk-naturvidenskabelig klasse. 1893(7): 1–38, 2 tables

Granja-Fernandez R, Herrero-Pérezrul MD, López-Pérez RA, Hernández L, Fabián A. Rodríguez-Zaragoza FA, Wallace Jones R, Pineda-López R (2014) Ophiuroidea (Echinodermata) from coral reefs in the Mexican Pacific. Zootaxa, 406:101–145

Gray S, Kinross J, Read P, Marland A (2000) The nutrient assimilative capacity of maerl as a substrate in constructed wetland systems for waste treatment. Water Res 34(8):2183–2190

Hall-Spencer J (1998) Conservation issues relating to maerl beds as habitats for molluscs. J Conchol Spec Publ 2:271–286

Hall-Spencer J (2005) Ban on maerl extraction. Mar Pollut Bull 50:121–124

Hall-Spencer JM, Atkinson RJA (1999) *Upogebia deltaura* (Crustacea: Thalassinidea) in Clyde Sea maerl beds, Scotland. J Mar Biol Assoc U K 79:871–880

Hall-Spencer J, Moore PG (2000a) Effects of fishing on non-target species and habitats: biological, conservation and socio-economic issues

Hall-Spencer J, Moore PG (2000b) Scallop dredging has profound, long-term impacts on maerl habitats. ICES J Mar Sci J Cons 57(5):1407–1415

Hall-Spencer J, Rogers A, Davies J, Foggo A (2007) Deep-sea coral distribution on seamounts, oceanic islands, and continental slopes in the Northeast Atlantic. Bull Mar Sci 81(Supplement 1):135–146

Hall-Spencer J, Kelly J, Maggs CA (2010) Assessment of maerl beds in the OSPAR area and the development of a monitoring program. Department of Environment, Heritage and Local Government, Ireland

Hall-Spencer J, Grall J, Moore P, Atkinson R (2003) Bivalve fishing and maerl-bed conservation in France and the UK—retrospect and prospect. Aquat Conserv Mar Freshw Ecosyst 13(S1):S33–S41

Hardiman PA, Rolfe MS, White IC (1976). *Lithothamnium* studies off the southwest coast of England. ICES report no. CM 1979/K:9

Harvey A, Bird F (2008) Community structure of a rhodolith bed from cold-temperate waters (southern Australia). Aust J Bot 56(5):437–450

Hernández-Kantun J, Riosmena-Rodríguez R, López Vivas JM, Pacheco Ruiz I (2010) Range extension of *Kallymenia* spp. (Kallymeniaceae, Rhodophyta) associated with rhodolith beds, new records from the Gulf of California, México. Mar Biodivers Rec 3:1–5

Hinojosa-Arango G, Riosmena-Rodríguez R (2004) Influence of Rhodolith-forming species and growth-form on associated fauna of rhodolith beds in the central-west Gulf of California, México. Mar Ecol 25(2):109–127

IBAMA N° 08, 2 DE FEVEREIRO DE (1996) Regulatory document for fisheries in the Amazonas área. http://www.icmbio.gov.br/cepsul/images/stories/legislacao/Portaria/1996/p_ibama_08_1996_regulamentapescabaciahidroraficarioamazonas.pdf

Irvine L, Chamberlain Y (1994) Seaweeds of the British Isles. Volume 1. Rhodophyta. Part 2B. Corallinales, Hildenbrandiales. HMSO, London

Jacquotte R (1962) Etude des fonds des maerl de Méditerranée. Recl Trav Stn Mar Endoume 26:141–235

James D (2000) Diet, movement, and covering behavior of the sea urchin *Toxopneustes roseus* in rhodolith beds in the Gulf of California, México. Mar Biol 137(5–6):913–923

James NP, Collins LB, Bone Y, Hallock P (1999) Subtropical carbonates in a temperate realm: modern sediments on the southwest Australian shelf. J Sediment Petrol 69:1297–1321

Johnson ME, Baarli BG, Cachão M, da Silva CM, Ledesma-Vázquez J, Mayoral EJ, Ramalho RS, Santos A (2012) Rhodoliths, uniformitarianism, and Darwin: Pleistocene and recent carbonate

deposits in the Cape Verde and Canary archipelagos. Palaeogeogr Palaeoclimatol Palaeoecol 329:83–100
Kamenos N, Cusack M, Moore P (2008) Coralline algae are global palaeothermometers with bi-weekly resolution. Geochim Cosmochim Acta 72(3):771–779
Keegan B (1974) The macrofauna of maërl substrates of the West coast of Ireland. Cah Biol Mar 15(4):13–530
Kempf M (1970) Notes on the benthic bionomy of the N-NE Brazilian shelf. Mar Biol 5(3):213–224
Kempf M, Coutinho PN, Moñs JOS (1968) A plataforma continental do N-NE do Brasil. Nota preliminar sobre a natureza do fundo. D. H. N. Marinha do Brasil, Rio ~6, pp 579–600
Kendrick G, Brearley A (1997) Influence of *Sargassum* spp. attached to rhodoliths on sampling effort and demographic analyses of *Sargassum* spp. (Sargassaceae, Phaeophyta) attached to a reef. Bot Mar 40(1–6):517–522
Kjellman FR (1883) The algae of the arctic sea: a survey of the species, together with an exposition of the general characters and the development of the flora, vol 20. Kongliga svenska Vetenskaps – Akademiens Handlingar
Konar B, Riosmena-Rodríguez R, Iken K (2006) Rhodolith bed: a newly discovered habitat in the North Pacific Ocean. Bot Mar 49(4):355–359
Lauwaert B, Bekaert K, Berteloot M, De Backer A, Derweduwen J, Dujardin A, Fettweis M, Hillewaert H, Hoffman S, Hostens K, Ides S, Janssens J, Martens C, Michielsen T, Parmentier K, Van Hoey G, Verwaest T (2009) Synthesis report on the effects of dredged material disposal on the marine environment (licensing period 2008–2009). Report by BMM, ILVO, CD, aMT and WL BL/2009/01. 73pp
Lavradoi HP, Ignacio BL (2006) Biodiversidade bentónica da regiao central da Zona Economica Exclusiva Brasileira. Documentos REVIZEE Museo Nacional 389 pp
Lemoine M (1910) Repartition et mode de vie du maerl (*Lithothamnium calcareum*) aux environs de Concarneau (Finistère). Ann Inst Oceanogr Monaco 1(3):1–28
León-Cisneros K, Riosmena-Rodríguez R, Neto A, Hernández-Carmona G (2009) The red algal genus *Scinaia* (Nemaliales; Rhodophyta) on the Gulf of California, Mexico: a taxonomic account. Phycologia 48:186–210
Littler MM, Littler DS (1984) Models of tropical reef biogenesis: the contribution of algae. In: Round FE, Chapman VJ (eds) Progress in phycological research, vol 3. Biopress, London, pp 323–364
López-Benito M (1963) Estudio de la composición química del *Lithothamnion calcareum* (Aresch.) y su aplicación como corrector de cultivo. Investig Pesq 23:53–70
Lowry JK, Stoddart HE (2012) Australian and South African conicostomatine amphipods (Amphipoda: Lysianassoidea: Lysianassidae: Conicostomatinae subfam. nov.). Zootaxa 3248:43–65
Maggs C (1983) A phenological study of the epiflora of two maerl beds in Galway Bay. PhD thesis, National University of Ireland, Ireland
Martin S, Gattuso JP (2009) Response of Mediterranean coralline algae to ocean acidification and elevated temperature. Glob Chang Biol 15(8):2089–2100
Mathis BJ, Kohn AJ, Goldberg N (2005) Rhodoliths: the inside story. In: Wells FE, Walker DI, Kendrick (eds) The marine flora and fauna of esperance, Western Australia, 147.157 Wesern Australian Museum
Metri R (2006) Ecologia de un banco de algas calcareas de Reserva Biológica Marinha do Arvoredo, Santa Catarina, Brazil. Ph D thesis Universidad Federal Parana 110 p
Metri R, Rocha R (2008) Bancos de algas calcárias, um ecossistema rico a ser preservado. Nat Conserv Rev Bras Conserv Nat 6(1):8
Miranda F (1934) Materiales para una flora marina de las Rías Bajas gallegas. Bol R Soc Esp Hist Nat Secc Biol 34:165–180
Mora-Bermúdez J (1980) Poblaciones bentónicas de la Ría de Arosa. Ph.D. thesis, The University of Santiago de Compostela, Spain

Myers A, McGrath D (1980) A new species of *Stenothoe* Dana (Amphipoda, Gammaridea) from maërl deposits in Kilkieran Bay. J Life Sci (Dublin) 2:15–18
Myers A, McGrath D (1983) The genus *Listriella* (Crustacea: Amphipoda) in British and Irish waters, with the description of a new species. J Marine Biol Assoc U K 63(02):347–353
Neill KF, Nelson WA, D'Archino R, Leduc D, Farr TJ (2014) Northern New Zealand rhodoliths: assessing faunal and floral diversity in physically contrasting beds. Mar Biodivers. doi:10.1007/s12526-014-0229-0
Nelson W (2009) Calcified macroalgae–critical to coastal ecosystems and vulnerable to change: a review. Mar Freshw Res 60(8):787–801
Nicoletti L, Paganelli D, Gabellini M (2006) Environmental aspects of relict sand dredging for beach nourishment: proposal for a monitoring protocol. Quaderno ICRAM n. 5. 155 pp
Nunn J (1993) The molluscan fauna associated with maerl. Conchologists Newsl 125:161–167
Pascheli C, Riul P, Riosmena-Rodríguez R, Scherner F, Nunes M, Hall-Spencer JM, Cabral de Oliveira E, Horta P (2013) Seasonal and depth-driven changes in rhodolith bed structure and associated macroalgae off Arvoredo island (southeastern Brazil). Aquat Bot 111:62–65
Peña V, Bárbara I (2008) Biological importance of an Atlantic European maerl bed off Benencia Island (northwest Iberian Peninsula). Bot Mar 51:493–505
Peña V, De Clerck O, Afonso-Carrillo J, Ballesteros E, Bárbara I, Barreiro R, Le Gall L et al (2014) An integrative systematic approach to species diversity and distribution in the genus *Mesophyllum* (Corallinales, Rhodophyta) in Atlantic and Mediterranean Europe. Eur J Phycol 50(1):1–17
Pérès JM, Piccard J (1955) Biotopes et biocoenoses de la Méditerranée occidentale comparées à ceux de la manche et de l'Atlantique nord-oriental. Arch Zool Exp Géné 92(1):1–71
Perry CT (2005) Morphology and occurrence of rhodoliths in siliciclastic, intertidal environments from a high latitude reef setting, southern Mozambique. Mar Geol 214:143–161
Pruvot G (1897) Essai sur les fonds et la faune de la Manche occidentale comparés à ceux du Golfe du Lion.-,irch. zooz. exp. gén., 30 sér., 5, 511–617
Ramos-Esplá AA (1992) Ascidias litorales del Mediterráneo Ibérico: Faunística, Ecología y Biogeografía. Unpublished Ph. D. thesis, Universidad de Alicante 440 pp
Reyes-Bonilla et al (1998) Reyes-Bonilla H, Riosmena-Rodríguez R, Foster MS 1997. Hermatipyc corals associated to rhodolith beds in the Gulf of California, México. Pac Sci 51(3):328–337
Riera R, Delgado JD, Rodríguez M, Monterroso O, Ramos E (2012) Macrofaunal communities of threatened subtidal maërl seabeds on Tenerife (Canary Islands, north-east Atlantic Ocean) in summer. Acta Oceanol Sin 31:98–105
Riera R, Ramos E, Herrera R, Moro L (2014) *Jassa marmorata* (Holmes, 1905) and *Monocorophium acherusicum* (Costa, 1853) New Amphipods (Crustacea: Amphipoda) to the Canary Islands. Rev Acad Canaria Cienc XXVI:27–31
Riosmena-Rodríguez R, Medina-Lopez M (2011) The role of rhodolith beds in the recruitment of invertebrate species in the Southwestern Gulf of Mexico. All Flesh is Green, Plant-Animal Interrelationships Cellular Origin, Life in Extreme Habitats and Astrobiology. Springer, New York, 16:417–428
Riosmena-Rodríguez R, Steller DS, Foster MS (2007) Prefacio a trabajos selectos sobre rodolitos. Cienc Mar 33(4):i–iii
Riosmena-Rodríguez R, Steller D, Hinojosa-Arango G, Foster M (2010) Reefs that rock and roll: biology and conservation of rhodolith beds in the Gulf of California. In: Marine biodiversity and conservation in the Gulf of California. University of Arizona Press and Sonoran Desert Museum, Tuscon
Riosmena-Rodríguez R, Mandenvelt G, Horta P, Figueiredo M (2015) Systematics and biogeography of the subclass Corallinophycideae (Rhodophyta) from the Atlantic Ocean. Phytotaxa 190:1–2

Riul P, Targino CH, Farias JDN, Visscher PT, Horta PA (2008) Decrease in *Lithothamnion* sp. (Rhodophyta) primary production due to the deposition of a thin sediment layer. J Mar Biol Assoc UK 88(01):17–19

Rivera M, Riosmena-Rodríguez R, Foster M (2004) Age and growth of *Lithothamnion muelleri* (Corallinales, Rhodophyta) in the southwestern Gulf of California, Mexico. Cienc Mar 30(1B):235–249

Rocha RM, Metri R, Omuro J (2006) Spatial distribution and abundance of ascidians in a bank of coralline algae at Porto Norte, Arvoredo Island, Santa Catarina. J Coast Res 39:1676–1679

Rowe GA, Sheader M, Jensen AC (1990) The infauna of handfast point maërl bed. Nature Conservancy Council, Peterborough, 57 pp

Sala E, Aburto-Oropeza O, Paredes G, Parra I, Barrera JC, Dayton PK (2002) A general model for designing networks of marine reserves. Science 298(5600):1991–1993

Sánchez-Ibarra C, Bermúdez-García DM, Bezaury-Creel JE, Lasch-Thaler C, Rodríguez-Dowdell N, Cárdenas-Torres N, Rojas-González de Castilla S, Gondor A (2013). Plan de acción para la conservación y aprovechamiento sustentable de la biodiversidad terrestre y marina de la región Golfo de California y Pacífico Sudcaliforniano. Comisión Nacional de Áreas Naturales Protegidas (CONANP), The Nature Conservancy (TNC), Fondo Mexicano para la Conservación de la Naturaleza, A.C., 294 pp, México

Segawa S (1959) Coralline algae. Bull Jpn Soc Phycol 2(2):33–39

Sewell AA, Johnson ME, Backus DH, Ledesma-Vázquez J (2007) Rhodolith detritus impounded by a coastal dune on Isla Coronados. Gulf Calif Cienc Mar 33(4):483–494

Sheehan EV, Bridger D, Attrill MJ (2015) The ecosystem service value of living versus dead biogenic reef. Estuar Coast Shelf Sci 154:248–254

Siesser WG (1972) Relict algal nodules (rhodolites) from the South African continental shelf. J Geol 80:611–616

Sitja C, Maldonado M (2014) New and rare sponges from the deep shelf of the Alboran Island (Alboran Sea, Western Mediterranean). Zootaxa 3760(2):141–179

Soares-Gomes A, Pires-Vanin AMS (2003) Padrões de abundância, riqueza e diversidade de moluscos bivalves na plataforma continental ao largo de Ubatuba, São Paulo, Brasil: uma comparação metodológica. Rev Bras Zool 20:717–772

Steller DL, Cáceres-Martínez C (2009) Coralline algal rhodoliths enhance larval settlement and early growth of the Pacific calico scallop *Argopecten ventricosus*. Mar Ecol Prog Ser 396:49–60

Steller DL, Foster MS (1995) Environmental factors influencing distribution and morphology of rhodoliths in Bahía Concepción, BCS, México. J Exp Mar Biol Ecol 194(2):201–212

Steller DL, Riosmena-Rodríguez R, Foster M, Roberts C (2003) Rhodolith bed diversity in the Gulf of California: the importance of rhodolith structure and consequences of disturbance. Aquat Conserv 13:S5–S20

Steneck RS (1986) The ecology of coralline algal crusts: convergent patterns and adaptative strategies. Annu Rev Ecol Syst 17:273–303

Sturaro N, Guerra-García JM (2012) A new species of *Caprella* (Crustacea: Amphipoda) from the Mediterranean Sea. Helgol Mar Res 66(1):33–42

Tâmega FTS, Riosmena-Rodríguez R, Mariath R, Figueiredo MAO (2014) Nongeniculate coralline red algae (Rhodophyta: Corallinales) in coral reefs from Northeastern Brazil and a description of *Neogoniolithon atlanticum* sp. nov. Phytotaxa 190(1):277–298, 28 fig., 4 tables

Teichert S (2013) Rhodoliths (Corallinales, Rhodophyta) as a Biosedimentary System in Arctic Environments(Svalbard Archipelago, Norway). PhD Thesis, Erlangen University: Germany. 160 pp

UNEP/IUCN/GIS Posidonie (1990) Livre rouge 'Gerard Vuignier' des végétaux, peuplements et paysages marins menacés de Mediterranée. MAP Technical Report Series 43; United Nations Environment Programme, Athens, ! 250 pp

Veale L, Thompson R, Bates M (1999). Isle of Man sublittoral survey: 1994–1997. In: Davies H (ed) Irish sea forum. Isle of Man, pp 35–49

Villas-Bôas AB, Tâmega FTDS, Andrade M, Coutinho R, Figueiredo MADO (2014) Experimental effects of sediment burial and light attenuation on two coralline algae of a deep water rhodolith bed in Rio de Janeiro, Brazil. Cryptogam Algol 35(1):67–76
Weber-van Bosse A (1913) Marine algae, Rhodophyceae, of the "Sealark" Expedition, collected by Mr. J Stanley Gardiner. Botany 8:105–142
Weber-van Bosse A, Foslie M (1904) The Corallinaceae of the Siboga-expedition. Siboge Expeditie 61:1–110
Wilson S, Blake C, Berges JA, Maggs CA (2004) Environmental tolerances of free-living coralline algae (maerl): implications for European marine conservation. Biol Conserv 120(2):279–289
Woelkerling WJ, Irvine L, Harvey AS (1993) Growth-forms in non-geniculate coralline red algae (Corallinales, Rhodophyta). Aust Syst Bot 6(4):277–293

Chapter 2
Coralline Algae as Recorders of Past Climatic and Environmental Conditions

Nicholas A. Kamenos, Heidi L. Burdett, and Nicolas Darrenougue

Abstract Calcifying marine organisms can be used as recorders, or proxies, of past environmental conditions if they lock physical or chemical signals within their skeletal material. Coralline algae lay down regular growth bands and the study of their structure and composition has gained increasing attention as a technique for reconstructing past environments in tropical, temperate and polar regions. Structurally, growth band width and percentage calcification have been used as records of historic light availability (e.g. due to cloud cover and sea ice extent). The chemical composition of their high Mg calcite skeleton has received significantly more attention, being used to reconstruct temperature, salinity, dissolved inorganic carbon, upwelling patterns and wider climate indices. At the ecosystem level, such reconstructions have been used to shed light on the drivers of past changes in marine productivity. Against a backdrop of projected ocean acidification coralline algae show significant potential for reconstructing historic changes in ocean acidification-driven marine carbonate chemistry. Due to their global distribution, coralline algae are becoming a regularly used tool for understanding environmental and ecosystem change, particularly in areas where other proxies are not available or instrumental records are sparse.

N.A. Kamenos (✉)
School of Geographical and Earth Sciences, University of Glasgow,
Glasgow G12 8QQ, Scotland
e-mail: nick.kamenos@glasgow.ac.uk

H.L. Burdett
Scottish Oceans Institute, University of St Andrews, St Andrews KY16 8LB, Scotland

Department of Earth and Environmental Sciences, University of St Andrews,
St Andrews KY16 9AL, Scotland

N. Darrenougue
Earth Environment Group, Research School of Earth Sciences, The Australian National University, Canberra ACT 0200, Australia

R. Riosmena-Rodríguez et al. (eds.), *Rhodolith/Maërl Beds: A Global Perspective*, Coastal Research Library 15, DOI 10.1007/978-3-319-29315-8_2

2.1 Overview

The study of skeletal banding associated with the growth of terrestrial and marine organisms has been crucial in the development of organisms as recorders, or proxies, of past environmental conditions. As organisms grow, they record the physical or chemical signatures of their surrounding environment within growth banding patterns. By accessing those signatures, we can determine the nature of past environments. Dendrochronology, the study of growth banding patterns, is the most developed field of organismal proxy reconstruction, where bands are used to understand past environments (Fritts 1991; Hughes 2002). For example, the width of tree rings has been linearly related to climatic variables such as temperature and precipitation (Hughes 2002). Tree rings also allow exact dating of particular climatic events by ring counting (Fritts 1991). Sclerochronology, the study of growth banding in marine organisms, is also advancing via the use of corals (e.g. Lough and Barnes 2000; De'ath et al. 2009) and the long-lived bivalve *Arctica islandica* L. (e.g. Schöne et al. 2005; Scourse et al. 2006; Wanamaker et al. 2011b).

More recently, algochronology (Kamenos and Law 2010), the study of growth patterns in marine coralline algae, has gained increasing attention as a technique for reconstructing past environments. The growing number of published studies has already stimulated the discussion of coralline algal palaeoenvironemantal proxies within wider-reaching reviews on coralline algae (e.g. Nelson 2009) and marine calcifying proxies (Wanamaker et al. 2011a). Here we expand on these previous mini-reviews. Proxy research has been conducted on both free-living coralline algae (rhodoliths or maerl) (e.g. Kamenos and Law 2010), and encrusting growth-forms (e.g. Williams et al. 2011). In some cases, the same species can form both morphotypes (e.g. *Lithothamnion glaciale* Kjellman). Therefore, we discuss proxy development based on investigations into both free-living and encrusting growth forms.

Coralline algae have several characteristics that make them ideal palaeoenvironmental proxies:

1. They lay down annual and sub-annual calcium carbonate growth bands composed of high-Mg calcite (Henrich et al. 1996; Foster 2001; Kamenos et al. 2008).
2. As new thalli grow at the surface of a rhodolith bed, older thalli become covered by sediment and form part of the dead deposit. Individual coralline thalli can grow for between 700 (Frantz et al. 2005) to ~850 (Halfar et al. 2007) years, with accumulations of rhodolith deposits spanning the last 20,000 years (Bosence 1983).
3. They have a worldwide distribution from polar (Schwarz et al. 2005) to tropical (Littler et al. 1991) shallow seas making them useful tools for reconstructing environmental conditions in areas where few other proxies exist (e.g. mid and high latitudes) and for high-resolution comparisons in areas where other proxies are available (low latitudes).
4. Despite being composed of high-Mg calcite, there is mixed evidence regarding post-mortem diagenetic effects, which can skew environmental reconstructions.

In some cases, lack of diagenesis may be due to the presence of a membrane covering the carbonate skeleton (Alexandersson 1974). In others, diagenesis begins post-mortem (Moberly 1970).

5. Little evidence exists for coralline algal stress-induced geochemical deviation in elemental composition (e.g. Kamenos et al. 2008). While in some cases there is an offset from isotopic equilibrium (~3.5‰) (e.g. Halfar et al. 2000; Lee and Carpenter 2001; Williams et al. 2011), in other cases there is no evidence of isotope disequilibrium (Rahimpour-Bonab et al. 1997).

To date, coralline algae have been used to understand: in situ temperature (the water surrounding the algae) (e.g. Kamenos 2010; Darrenougue 2013), sea surface temperature (SST) (e.g. Halfar et al. 2000, 2008), atmospheric air temperature (Hetzinger et al. 2009), freshwater/glacial runoff (Kamenos et al. 2012), salinity (Chan et al. 2011), atmospheric cloud cover (Burdett et al. 2011), upwelling (Caragnano et al. 2014), sea ice extent (Halfar et al. 2013), ^{14}C (Frantz et al. 2000), ^{13}C and dissolved inorganic carbon (Williams et al. 2011), Aleutian Low Pressure index (Halfar et al. 2011b), North Atlantic Oscillation index (Hetzinger et al. 2012), Atlantic Multidecadal Oscillation index (Kamenos 2010), Decadal Sea Level pressure (Hetzinger et al. 2012) and, indirectly, changes in copepod (Kamenos 2010) and salmon (Halfar et al. 2011b) abundance.

2.2 Implementation of Coralline Algae as Proxies

During the development of coralline algae as environmental recorders, they have gone through three stages:

1. **Calibration**: laboratory and field experiments were conducted specifically to understand how the 'signals' recorded by the algae, for example Mg content, correspond to their environmental surroundings (e.g. Moberly 1968; Halfar et al. 2000; Kamenos et al. 2008). These experiments allowed determination of the relationships between environmental variables and the chemical or physical proxy locked within the algae.
2. **Validation**: the process of ensuring the relationships observed during calibration are (i) directly related, and (ii) do not co-vary with a third (or more) variable. The absence of a direct relationship may introduce temporal non-linearity between the proxy-variable relationship, or non-continuous coherence. For example, using a synchrotron light source, Mg has been confirmed as being part of the calcite lattice of *Lithothamnion glaciale*, rather than being associated with organic matter (Kamenos et al. 2009). This is important as Mg thermometry assumes replacement of the Ca^{2+} ions by Mg^{2+} ions within the calcite lattice at higher temperatures (Oomori et al. 1987).
3. **Environmental reconstruction**: by applying the defined calibration, coralline algae may be used in palaeoenvironmental reconstruction using both chemical (elemental and isotopic composition) and skeletal structure components.

2.3 Geographical Coverage of Proxies

The global distribution of coralline algae has the potential to yield more information than other biogenic climate proxies such as corals, which are restricted to the low-latitude oceans (<30°). However, the distribution of coralline algal proxy records is sparse compared to the global distribution of known rhodolith/maërl beds (Fig. 2.1). The majority of coralline algae used in proxy studies have originated from the North Atlantic and North Pacific Oceans, reflecting the geographical location of research groups working in this field of research. Only one record is currently available for the Southern Hemisphere (20°S) (Darrenougue et al. 2013). As such, proxy records focus primarily on temperate, sub-polar and, to a lesser extent, tropical climates, specifically targeting large-scale climatic phenomenon such as the North Atlantic Oscillation (Kamenos 2010), the Atlantic Multi-decadal Oscillation (Kamenos 2010; Halfar et al. 2011a), the Pacific Decadal Oscillation (Halfar et al. 2007) and the El Niño Southern Oscillation (Darrenougue 2013).

To date, seven species of red coralline algae have been successfully used in palaeoclimate reconstructions (Table 2.1): *Clathromorphum nereostratum* Lebednik, *C. compactum* Kjellman, *Lithothamnion crassiusculum* Mason, *L. glaciale*, *Lithophyllum kotschyanum* Unger, *Phymatolithon calcareum* Adey and McKibbin and *Sporolithon durum* Townsend and Woelkerling. The choice of species has been determined by a number of factors:

Fig. 2.1 Geographical distribution of previously reported rhodolith/maerl beds (*black dots*, From: Burdett 2013) and the origin of coralline algae thalli used in proxy reconstructions to date (*red dots*)

Table 2.1 Geographical distribution of currently published coralline algal-derived environmental reconstructions, ordered by descending length of record

Location	Species	Depth (m)	Reconstruction	Method	Time period	Record length (y)	Resolution	Reference
West coast of Scotland	*Lithothamnion glaciale*	6	IST, Copepod abundance	Mg/Ca	1353–2006	653	Fortnightly	Kamenos (2010)
Canadian Arctic	*Clathromorphum compactum*	15–17	Sea ice cover	Mg/Ca, growth increment widths	1365–2009	646	2 monthly	Halfar et al. (2013)
Aleutian archipelago	*Clathromorphum nereostratum*	10–33	Aleutian low pressure index Salmon abundance	Growth increment widths	1783–2008	225	Annual	Halfar et al. (2011b)
Aleutian Islands	*Clathromorphum nereostratum*	25	Decadal sea-level pressure, SST, NAO	Mg/Ca	1818–1967	149	Monthly	Hetzinger et al. (2012)
Aleutian archipelago	*Clathromorphum nereostratum*	n/a	SST/salinity	$\delta^{18}O$	1886–2003	117	Annually resolved, decadally smoothed	Halfar et al. (2007)
Aleutian archipelago	*Clathromorphum nereostratum*	10–25	Dissolved inorganic carbon/SST	$\delta^{18}O$ Mg/Ca ratios	1887–2003	116	Annual	Williams et al. (2011)
Newfoundland, Canada	*Clathromorphum compactum*	10	SST	Mg/Ca	1890–2006	116	Annual	Gamboa et al. (2010)
Newfoundland and Quebec	*Clathromorphum compactum*	3–10	SST	Average growth increment widths	1893–2007	115	Annually resolved, decadally smoothed	Halfar et al. (2011a)

(continued)

Table 2.1 (continued)

Location	Species	Depth (m)	Reconstruction	Method	Time period	Record length (y)	Resolution	Reference
Loch Sween, west coast of Scotland	*Lithothamnion glaciale*	7	Cloud cover and SST	Cell calcification	1910–2006	96	Seasonal	Burdett et al. (2011)
Newfoundland, Canada	*Clathromorphum compactum*	10–25	SST	Mg/Ca	1938–2007	69	Seasonal	Hetzinger et al. (2011)
Aleutian Islands	*Clathromorphum nereostratum*			U/Ca				
Newfoundland, Canada	*Clathromorphum compactum*	10–25	Upwelling	Ba/Ca	1938–2007	69	Seasonal	Hetzinger et al. (2011)
Aleutian Islands	*Clathromorphum nereostratum*							
Aleutian Islands	*Clathromorphum compactum*	10	SST	Mg/Ca	1902–1976	65	Seasonal	Hetzinger et al. (2009)
Søndre Strømfjord, Western Greenland	*Lithothamnion glaciale*	5	Glacial run off (temperature and salinity)	Mg/Ca	1939–2002	63	Fortnightly – 5 yearly	Kamenos et al. (2012)
				$\delta^{18}O$				
Akun Bay, Alaska	*Clathromorphum nereostratum*	10	SSS/upwelling	Ba/Ca	1945–2005	60	Sub-annual	Chan et al. (2011)
Newfoundland, Canada	*Clathromorphum compactum*	10	SSS	Ba/Ca	1950–2006	56	Annual – sub-annual	Hetzinger et al. (2013)
Nouméa, New Caledonia	*Sporolithon durum*	4–5	IST	Mg/Ca	1964–2008	44	Annual – sub-annual	Darrenougue (2013)
Newfoundland	*Lithothamnion glaciale*	Na	SST	$\delta^{18}O$	1957–1995	38	Sub-annual	Halfar et al. (2000)
				Mg/Ca				
Gulf of Maine	*Clathromorphum compactum*	Na	SST	$\delta^{18}O$	1971–2001	30	Sub-annual	Halfar et al. (2008)

Loch Sween, west coast of Scotland	*Lithothamnion glaciale*	7	SST, IST	$MgCO_3$ mol% $SrCO_3$ mol%	2000–2007	7	Fortnightly	Kamenos et al. (2008)
Gulf of Aden, Yemen	*Lithophyllum kotschyanum*	6–8	SST	Mg/Ca	2004–2008	4	Seasonal	Caragnano et al. (2014)
Gulf of Aden, Yemen	*Lithophyllum kotschyanum*	6–8	SST	Li/Ca	2004–2008	4	Seasonal	Caragnano et al. (2014)
Gulf of Aden, Yemen	*Lithophyllum kotschyanum*	6–8	Upwelling	Ba/Ca	2004–2008	4	Seasonal	Caragnano et al. (2014)
Isle of Arran, west coast of Scotland	*Phymatolithon calcareum*	13	SST, IST	$MgCO_3$ mol% $SrCO_3$ mol%	2005–2007	1.5	Fortnightly	Kamenos et al. (2008)
La Paz, Gulf of California	*Lithothamnion crassiusculum*	1–5	SST	$\delta^{18}O$ Mg/Ca	n/a	1	Monthly	Halfar et al. (2000)

SST sea surface temperature, *IST* in situ temperature, *SSS* sea surface salinity, *NAO* North Atlantic Oscillation, *n/a* not available

1. Requirement for regular banding patterns (annual banding is the most common), limiting the species that are suitable.
2. Restricted distribution range of specific species (primarily driven by defined temperature tolerances), limiting inter-species comparisons.
3. Ease of access to samples, usually via SCUBA diving or snorkelling, limiting the depth range of proxies (maximum proxy depth is currently 25 m, compared to >250 m depth recorded for live rhodoliths).

These factors result in a sparse distribution of coralline algal proxy information, despite their superior performance compared to other biogenic proxies such as corals or bivalves in terms of data resolution.

2.4 Structurally-Derived Coralline Algal Proxies

2.4.1 Environmental Drivers of Structure

The overall shape of free-living coralline algal thalli may be determined by, amongst other factors, water conditions and depth (Peña and Bárbara 2008). Highly branched thalli may form in slow moving (Peña and Bárbara 2008) or shallow waters (Steller et al. 2003). Discoidal (i.e. flat) forms may be more abundant in deep waters where downward growth is unfavourable, whilst spherical and ellipsoidal forms may occur in shallower water (Peña and Bárbara 2008). Similarly, a number of environmental factors can affect sub-organismal growth patterns within coralline algae (Foster 2001; Kamenos et al. 2008), including:

- The slowing of growth following burial, in response to a reduction in light availability and water movement.
- Summer/winter growth patterns in response to cyclical patterns of light availability and water temperature.
- Monthly/lunar growth influences, affecting tidal patterns.
- Possible El Niño effect, affecting local climatic regimes and prevailing environmental conditions.

2.4.2 The Cellular Structure of Coralline Algae

The cellular structure of red coralline algae is well defined; cell morphology has historically been used in the taxonomic identification of coralline algal species. The branches of free-living thalli generally have a radial core, where the cells divide simultaneously to form bands of cells which can be visible without a microscope

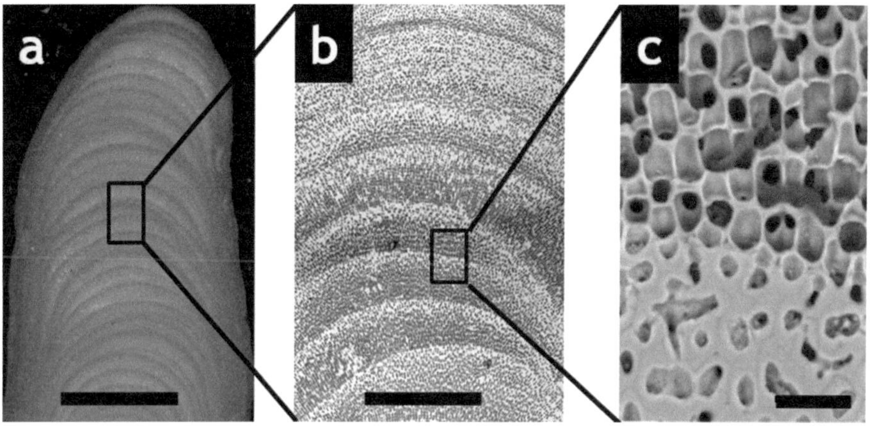

Fig. 2.2 Longitudinal cross-section of a branch of *Lithothamnion glaciale*. Growth bands are clearly visible in (**a**) [scale bar = 1 mm] and (**b**) [scale bar = 500 µm]. Darker areas (less calcified) represent summer growth and lighter areas (more calcified) represent winter growth. Individual cell growth is visible in (**c**), showing the less calcified summer cells (*top*) and the highly calcified winter cells (*bottom*) within one annual growth band [scale bar = 20 µm] (Images: N. Kamenos)

(Fig. 2.2) (Foster 2001). As the algae grow, a high-magnesium (high-Mg) calcite skeleton is deposited within each cell's wall (Chave 1984) (Fig. 2.2). Growth is generally faster in the summer, leading to larger, less calcified cells, whilst winter growth is slow, with smaller, more heavily calcified cells (Figs. 2.2 and 2.3) (Kamenos and Law 2010). This creates internal growth bands that may form on an annual basis (Figs. 2.2 and 2.3), in a manner similar to tree rings.

Four orders of coralline algal banding have been proposed (Fig. 2.3) (Bosence 1976; Freiwald and Henrich 1994; Foster 2001)

- 1st order bands: single row of cells.
- 2nd order bands: rows of cells separated by sutures – organically enriched 'joins' that have been observed in *Lithothamnion glaciale.*
- 3rd order bands: numerous rows of first level bands formed in summer (larger, less calcified cells) or winter (smaller, more heavily calcified cells).
- 4th order bands: full band growth (may be formed annually).

2.4.3 Structure and Environmental Proxies

This regular growth pattern, which is determined by prevailing environmental conditions, allows the structure of coralline algae to be used as a proxy for environmental conditions. Whilst the resolution of structurally-based records is lower than chemical analyses (seasonal – annual), the methods for preparation and analysis are less time-consuming and considerably less expensive – sections may be imaged with a light microscope, and measurements can be semi-automated using computer

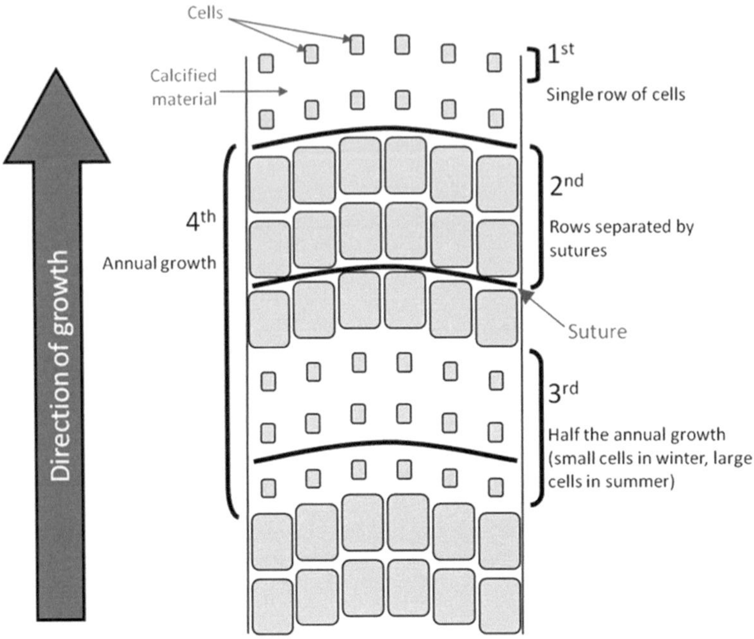

Fig. 2.3 Diagrammatic representation of the proposed growth patterns of free-living coralline algae, from 1st order (one row of cells) to 4th order (annual growth) bands (From: Burdett (2013))

programs. However, banding periodicity is not consistent across all genera, affecting reconstructions – *Lithothamnion glaciale* (Kamenos et al. 2008), *L. muelleri* Lenormand ex Rosanoff (Rivera et al. 2004) and *Clathromorphum compactum* (Halfar et al. 2008) exhibit annual banding, whilst other species such as *Phymatolithon calcareum* and *Sporolithon durum* have sub-annual banding (Blake and Maggs 2003; Darrenougue 2013). Thus, it is necessary to understand the banding patterns of all species selected for structurally-based climate proxy construction.

2.4.3.1 Band Width

When coralline algae are longitudinally sectioned and polished, internal growth bands are clearly visible. In reproductive specimens, conceptacles (circular cavities which harbour reproductive sporangia) may be formed annually, providing another method of identifying growth bands (Fig. 2.4).

The width of annual growth bands of encrusting *Clathromorphum compactum* has been used to reconstruct a 225-year (1782–2007) proxy record of the Aleutian Low climate patterns using 29 algal samples from throughout the Aleutian archipelago (Fig. 2.5). A negative relationship was observed between band width and the Aleutian Low Pressure (ALP) Index, which in turn is related to cloudiness (and thus

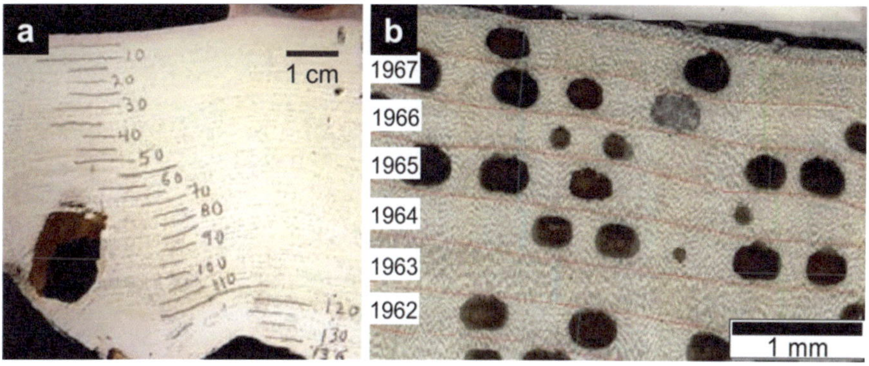

Fig. 2.4 Cross-section of *Clathromorphum nereostratum* from the Aleutian Islands. (**a**) Annual growth increments; this particular sample was 136 years old (newest growth at the *top*). Detail of growth bands is shown in (**b**) – large circular structures are conceptacles, which, in this species, develop annually (From: Halfar et al. (2011a))

light availability) in the subarctic North Pacific and the Bering Sea (Fig. 2.5). Halfar et al. (2011b) also highlighted a statistically significant relationship between biologically-driven light availability and the growth band width of *C. compactum*; band width was reduced by ~100 μm $year^{-1}$ in areas with 100% kelp cover compared to areas with no kelp cover. Consequently, a widespread decline in kelp cover since the 1990s has limited this proxy's capacity to reconstruct the recent ALP index (Halfar et al. 2011b).

In contrast to *C. compactum*, the annual band widths of free-living *L. glaciale* is not consistently related to in situ water temperature (Kamenos 2010), the Atlantic Multi-decadal Oscillation (AMO) (Kamenos 2010), sea surface temperature (Burdett et al. 2011), nor seasonal cloud cover (Burdett et al. 2011). This has been attributed to a sample preparation artefact: since free-living coralline algal branches do not grow in straight lines, it is impossible to section a branch consistently parallel to the maximum width of each growth band.

2.4.3.2 Percentage Calcification

Densitometric algochronology uses the ratio between calcified and un-calcified areas within growth bands as an environmental proxy. High-resolution images of sectioned and polished algal branches may be overlain with digital 'quadrats' to act as sampling areas (Fig. 2.6). The small size of the quadrats (typically 2,500 μm^2) enables the percentage calcification of 3rd-order bands (i.e. summer/winter) to be determined (Fig. 2.6). Using this method, a record with 6-monthly resolution may be quickly and easily achieved with minimal instrumentation.

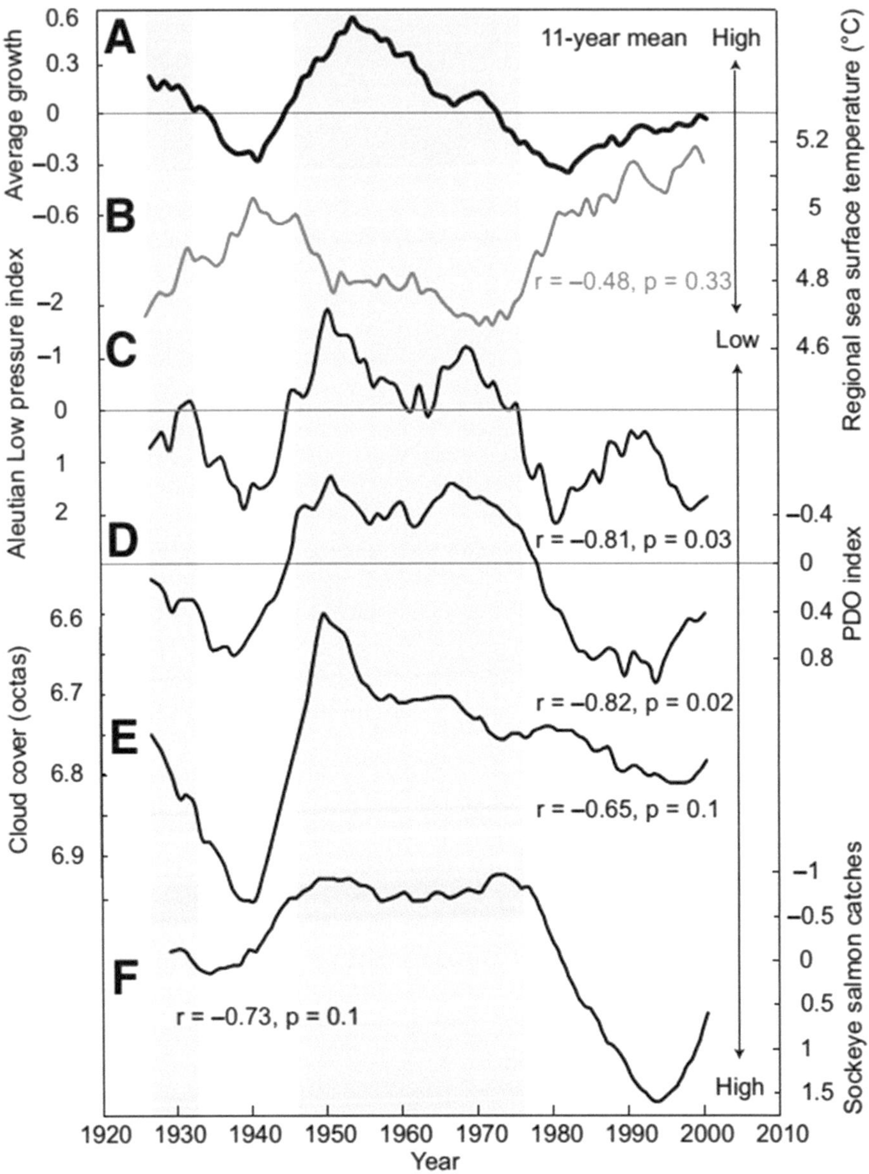

Fig. 2.5 (**a**) *C. compactum* algal chronology compared to (**b**) sea surface temperature, (**c**) Aleutian Low Pressure Index, (**d**) Pacific Decadal Oscillation Index, (**e**) regional cloud cover and (**f**) sockeye salmon catch (From: Halfar et al. (2011a))

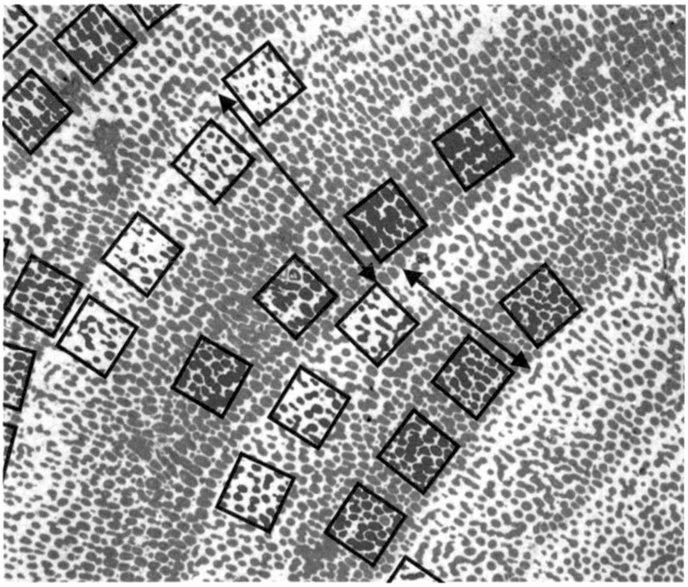

Fig. 2.6 Example microscope image of a longitudinal cross-section of a free-living *Lithothamnion glaciale* specimen. Distinct 3rd level (seasonal) growth bands are visible, with highly calcified winter cells (*lighter*) and less calcified summer cells (*darker*). *Black overlays* indicate the analyses undertaken for the calculation of maximum band width (*double headed arrows*) and percentage calcification using multiple quadrats within each 3rd level band (*black squares*). Quadrats are 50 μm wide (Image: N. Kamenos)

A linear relationship between summer, but not winter, calcification, and in situ temperature has been defined for laboratory maintained specimens of *L. glaciale* (Kamenos and Law 2010). A weaker relationship was observed between percentage calcification (both summer and winter bands) and the AMO Index. However, the AMO Index primarily reflects sea-surface, rather than in situ, temperature, and may be confounded by other environmental parameters such as light (Kamenos and Law 2010). Indeed, cloud cover (which will directly impact light availability) is negatively correlated to both summer and winter calcification of *L. glaciale* (Fig. 2.7) (Burdett et al. 2011). This allowed for the reconstruction of a 96-year (1910–2006) cloud cover record for the west coast of Scotland, demonstrating an average 12-year cyclicity in cloud cover, and a general increase from ~1930, coinciding with an increase in sea-surface temperature (Fig. 2.7) (Burdett et al. 2011).

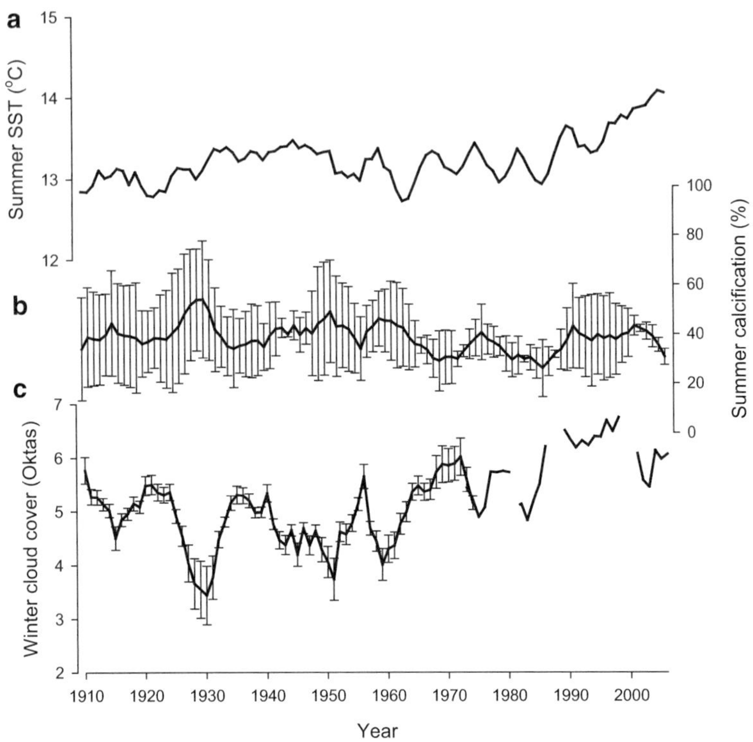

Fig. 2.7 Time series from 1910 to 2006 for the west coast of Scotland of (**a**) summer sea-surface temperature, (**b**) summer calcification of *Lithothamnion glaciale* (mean±SD) and (**c**) hindcast cloud cover (1910–1972, *solid line*, mean ± SE), calculated from the relationship between summer calcification and observed cloud cover (1972–2006, *dashed line*) (From: Burdett et al. (2011))

2.5 Elemental and Isotopic Proxies

The elemental and isotopic composition of the high-Mg calcite deposited within coralline algal growth bands has been used to produce palaeoenvironmental reconstructions with up to fortnightly resolution. Environmental reconstructions from coralline algae generated using elemental and isotopic proxies are more widespread than structural proxies but require significantly more instrumental investment. These techniques depend on geochemical analytical instrumentation and generally use elemental microprobe and inductively-coupled plasma mass-spectrometry for elemental reconstructions, as well as isotope-ratio mass-spectrometry and accelerator mass-spectrometry for isotopic reconstructions. In addition, significant sample preparation and instrumental calibration is required. However, elemental- and isotopic-based reconstructions are generally of higher temporal resolution (up to fortnightly) than structural reconstruction (up to seasonal).

2.5.1 *Magnesium*

Initial observations indicated that Mg concentrations within coralline algae varied directly (Chave and Wheeler 1965), and indirectly (via the influence of growth rate/light) (Moberly 1968), with marine temperatures. More recently, dedicated experiments have quantified some of the variability and determined the relationships that exist. Mg-palaeothermometry theory assumes Ca^{2+} ion substitution by Mg^{2+} ions within the calcite lattice at higher temperatures, thus a positive relationship between Mg and temperature is expected (Oomori et al. 1987). Mg occurs solely in the calcite lattice of some species (e.g. *Lithothamnion glaciale*, Kamenos et al. 2009) and, while being found in the lattice of other species, it may also occur as a non-lattice component (e.g. in *Porolithon* sp.) (Milliman et al. 1971). Mg concentrations in coralline algae range from 7.7 to 28.8 mol% $MgCO_3$ (Chave 1954b), and are controlled by seawater temperature (Chave 1954a; Henrich et al. 1996; Halfar et al. 2000; Kamenos et al. 2008). At geological time scales, secular changes in the Mg content of seawater (Ries 2006) also control coralline algal Mg content. For every 1°C increase there is ~1–1.5 mol% $MgCO_3$ increase within the calcite skeleton, depending on the species and whether in situ temperature or sea-surface temperature is being considered (Henrich et al. 1996; Halfar et al. 2008; Kamenos et al. 2008).

Mg concentrations within coralline algae have been observed to represent in situ seawater temperature (e.g. Fig. 2.8) (Kamenos et al. 2008, 2012; Kamenos 2010; Darrenougue 2013), sea-surface temperature (Halfar et al. 2000; Hetzinger et al. 2009, 2012; Gamboa et al. 2010; Caragnano et al. 2014) and atmospheric temperature (Hetzinger et al. 2009). The longest coralline algal palaeoclimatic reconstruction is a 651 year-long (1350–2006), fortnightly resolution, Mg-derived in situ temperature record from the eastern Atlantic (Kamenos 2010) (Fig. 2.8).

The Mg/Ca ratios deposited by foraminifera and sea stars may be controlled by both temperature and salinity (Borremans et al. 2009; Dissard et al. 2010). While light intensity and physiological cycles have also been suggested as possible drivers of Mg concentrations in coralline algae (Moberly 1968), recent reconstructions in areas with large freshwater inputs and variable light intensity suggest that the Mg/Ca ratios deposited by coralline algae are most highly correlated to instrumental temperature (e.g. Halfar et al. 2000; Kamenos et al. 2008; Williams et al. 2011).

2.5.2 *Strontium*

Strontium substitution in calcium carbonate is expected (thermodynamically) to have a negative temperature dependence due to the negative enthalpy of Ca^{2+} substitution by Sr^{2+} (Kinsman and Holland 1969). While Sr-palaeothermometry has been successfully used in aragonitic corals (e.g. Cohen et al. 2001), in coralline algae there appears to be strong organism effects leading to kinetic incorporation of Sr^{2+} and Mg^{2+} into the calcite lattice, creating an unexpected positive Sr-temperature

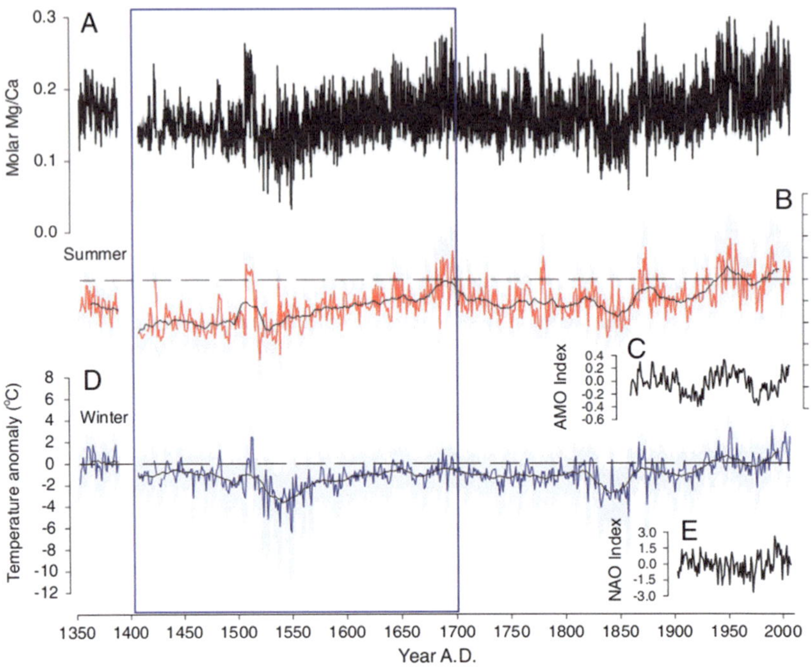

Fig. 2.8 North East Atlantic summer and winter algal-derived in situ temperature anomaly. (**a**) Molar Mg/Ca (*black line*) extracted from *Lithothamnion glaciale* from the west coast of Scotland. Peaks represent maximum summer temperature, troughs represent minimum winter temperature (*grey shading* indicates standard deviation of replicate measurements). (**b**) Summer (*red line*) and (**d**) winter (*blue line*) extreme temperature anomalies, extracted from the fortnightly resolution, algal-derived time series. *Black line* is the 21-year moving average; *grey shading* indicates standard error (measurement and regression). *Blue box* indicates the temporal extent of the Little Ice Age. For comparison, winter Atlantic Multidecadal Oscillation (AMO) index (**c**) and winter (December–March) North Atlantic Oscillation (NAO) index (**e**) are given (From: Kamenos (2010))

relationship in coralline algae. This minimises the value of Sr as a direct temperature proxy (Kamenos et al. 2008; Hetzinger et al. 2011; Darrenougue 2013).

2.5.3 Barium

Ba^{2+}, which can be been substituted for Ca^{2+} in calcium carbonate skeletons, is representative of seawater Ba concentrations, which is controlled by seasonal upwelling patterns (Lea et al. 1989; Shen and Dunbar 1995) and riverine runoff (McCulloch et al. 2003). Coralline algal Ba/Ca ratios have shown significant promise for reconstructing historic changes in sea-surface salinity due to upwelling in the Canadian Atlantic (Hetzinger et al. 2013), Pacific (Hetzinger et al. 2011) (Fig. 2.9) and in the Gulf of Aden, Yemen (Caragnano et al. 2014) as well as Ba-enriched freshwater runoff from the land (Chan et al. 2011).

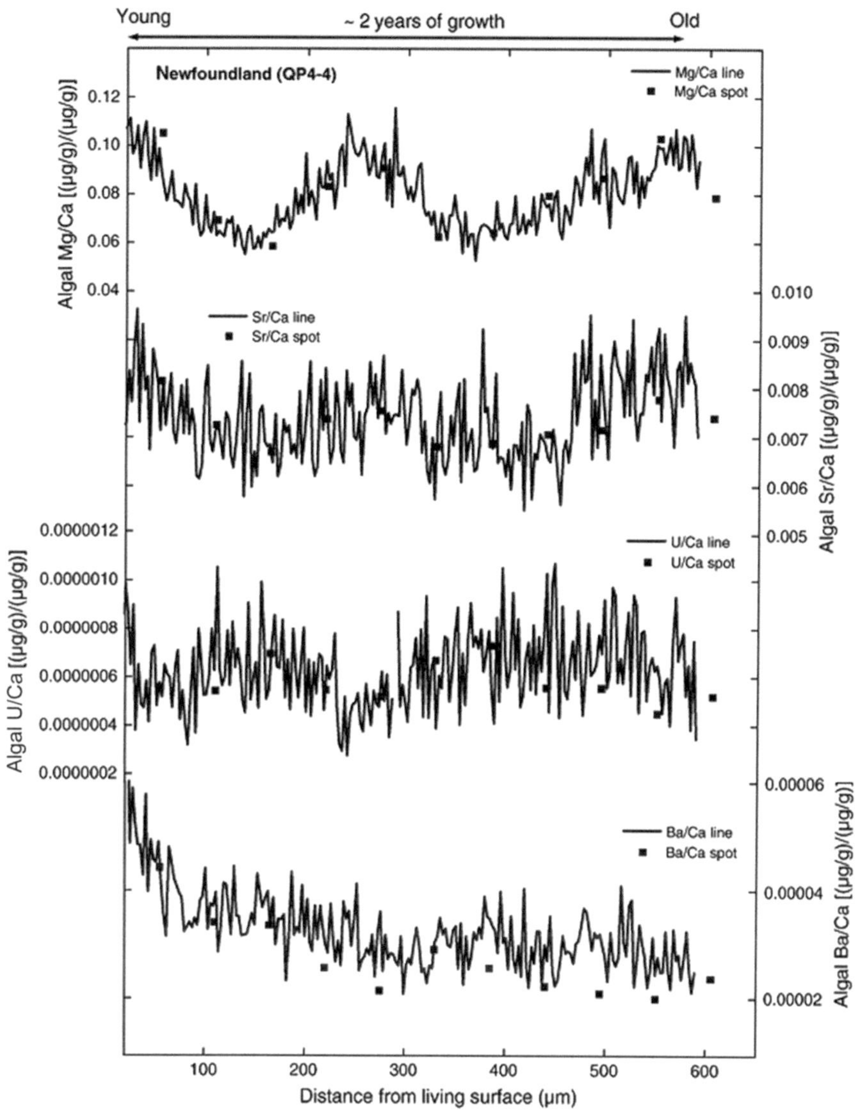

Fig. 2.9 Mg/Ca, Sr/Ca, U/Ca and Ba/Ca from Newfoundland *Clathromorphum compactum* over a 2-year growth period, determined using laser-ablation inductively-coupled plasma mass-spectrometry (From: Hetzinger et al. (2011))

2.5.4 Lithium

Li concentrations in the ocean are controlled by riverine inflow (Huh et al. 1998) and hydrothermal activity (Stoffyn-Egli and Mackenzie 1984). However, the incorporation of Li into biomineralisers remains poorly constrained (Rollion-Bard et al. 2009). Li/Ca have been used to reconstruct temperature in corals and brachiopods (Delaney et al. 1985) while in foraminifera no relationship has been observed (Rollion-Bard et al. 2009). In coralline algae, Li has recently been used to reconstruct SST in the Gulf of Aden at sub-annual resolution over a 5-year growth period (Caragnano et al. 2014) (Fig. 2.10).

2.5.5 Uranium

It is uncertain whether U is part of the crystal lattice in calcitic and aragonitic species, thus there is uncertainty around its role as a Ca replacement proxy (Shen and Dunbar 1995). However, there is evidence of a U-temperature relationship in corals which may also be controlled by seawater alkalinity (indicating changes in salinity) (e.g. Shen and Dunbar 1995). However, U/Ca ratios in encrusting *Clathromorphum* sp. (Hetzinger et al. 2011) do not show significant relationships to temperature in the Atlantic and Pacific oceans (Fig. 2.9). This is most likely due to the low U concentrations present in seawater, and thus within the coralline algal structure, making it difficult to measure with current instrumentation (Hetzinger et al. 2011).

2.5.6 Oxygen Isotopes

Measurement of oxygen isotopic composition ($\delta^{18}O$) is commonly used as a proxy for palaeoenvironmental reconstructions in a wide range of marine biogenic carbonates including foraminifera (e.g. Elderfield and Ganssen 2000) and corals (e.g. Tudhope et al. 1995). The use of oxygen isotopes is based on the thermodynamic fractionation between ^{18}O and ^{16}O during calcium carbonate precipitation (Urey 1947). The $\delta^{18}O$ of marine biogenic carbonates is determined by the temperature and salinity of the seawater (Schmidt et al. 2001), thus meteoric changes in seawater salinity need to be considered when using $\delta^{18}O$ as a palaeotemperature proxy (Lea 2006).

$\delta^{18}O$ composition of red coralline algae can show a negative offset from the $\delta^{18}O$ equilibrium, which was attributed to vital effects (e.g. Milliman 1977). Offsets of up to 4.8 ‰ (Wefer and Berger 1991) and more often 3.5 ‰ (Halfar et al. 2000, 2008; Lee and Carpenter 2001; Williams et al. 2011) have been observed. However, Rahimpour-Bonab et al. (1997) observed higher latitude coralline algae are sometimes not significantly depleted in ^{18}O since their $\delta^{18}O$ composition was near equilibrium.

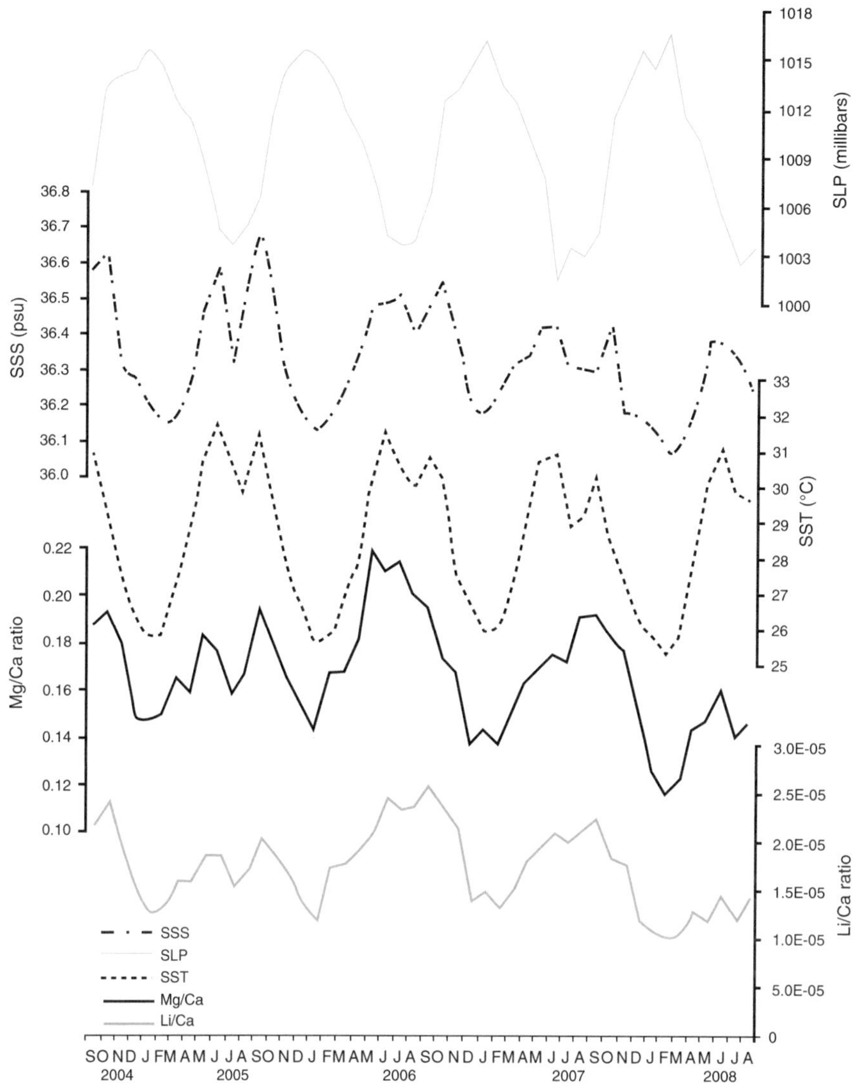

Fig. 2.10 Comparison between Mg/Ca cycles (*black line*), reconstructed sea-surface temperature (SST, *dotted line*), sea level pressure (SLP, *thin line*), sea surface salinity (SSS, *dashed-dotted line*), and Li/Ca cycles (*grey line*) in *Lithophyllum kotschyanum* from the Gulf of Aden, Yemen (From: Caragnano et al. (2014))

$\delta^{18}O$ composition has been used to reconstruct sea-surface temperature in *Lithothamnion crassiusculum* (subtropical) and *Lithothamnion glaciale* (subarctic) (Halfar et al. 2000), *Clathromorphum nereostratum* (Arctic) (Halfar et al. 2007; Hetzinger et al. 2009), *Sporolithon durum* (tropical) (Darrenougue 2013) and *C. compactum* (subarctic) (Halfar et al. 2008) (Fig. 2.11). In situ temperature

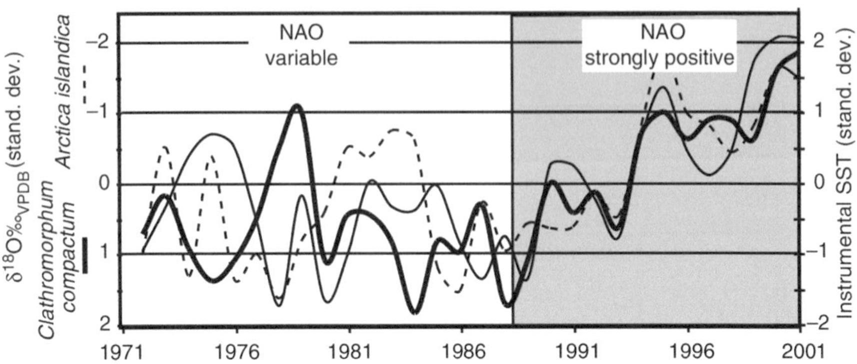

Fig. 2.11 30 year, annually averaged, $\delta^{18}O$ anomalies derived from *Clathromorphum compactum* (*thick black line*), compared to Boothbay Harbor (Gulf of Maine) instrumental sea-surface temperatures (SST, *thin black line*) and a $\delta^{18}O$ record of *Arctica islandica* from central Gulf of Maine (*dashed line*) (From: Halfar et al. (2008))

has also been reconstructed using *L. glaciale* (Arctic) (Kamenos et al. 2012). Consequently, large-scale climatic phenomenon characterised by changes in water temperature may also be reconstructed using the $\delta^{18}O$ signature of coralline algal skeletons, including the El Niño Southern Oscillation and Pacific Decadal Oscillation (*C. nereostratum*, Halfar et al. 2007) and the North Atlantic Oscillation (*C. compactum*, Halfar et al. 2008). By taking into account the Mg/Ca ratio, $\delta^{18}O$ has also been used to reconstruct in situ seawater salinity from *L. glaciale* (Arctic) (Kamenos et al. 2012), and sea-surface salinity from *S. durum* (tropical) (Darrenougue 2013).

2.5.7 Carbon Isotopes

The carbon isotope ratio ($^{13}C/^{12}C$) of seawater dissolved inorganic carbon (DIC), can be used to understand anthropogenic CO_2 dissolution into surface waters (Quay et al. 1992). $\delta^{13}C$ locked in the skeleton of *Clathromorphum nereostratum*, representative of dissolved inorganic carbon, has been used as a water mass tracker, providing insight into upwelling variability in the North Pacific and the Bering Sea. An increase in the transport of deeper water toward the surface has been linked to the intensity of the Aleutian Low pressure system (Williams et al. 2011).

Additionally, ^{14}C measurements may be used as a tracer to determine growth rates in carbonate depositing organisms such as coralline algae (Frantz et al. 2005; Darrenougue et al. 2013). Anthropogenic activities have contributed significantly to the dilution, or addition, of natural atmospheric ^{14}C since the Industrial Revolution, allowing determination of organism age (Levin et al. 1980); coralline algal ^{14}C is also related to atmospheric ^{14}C concentrations (Frantz et al. 2000, 2005). The ^{14}C

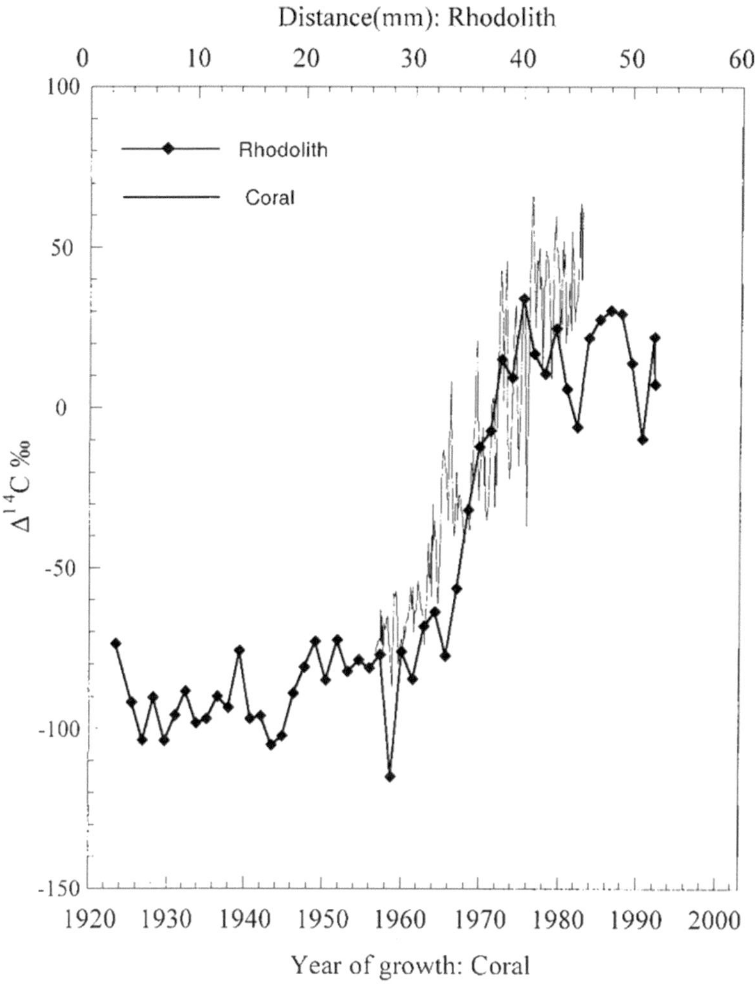

Fig. 2.12 ^{14}C record from a rhodolith collected near La Paz, Baja California Sur (*solid diamonds*) and a $\delta^{14}C$ record from a Galapagos Island coral with independent age chronology (*thin line*). The increase in ^{14}C is due to an increase in nuclear weapons testing from the 1960s. *Top x-axis* represents sample interval (mm) (From: Frantz et al. (2000))

concentration of coralline algae is critical for determining long-term growth rates and/or age, and for dating the exact timing of climatic events recorded by the algal growth band structure or chemistry (Fig. 2.12). ^{14}C dating of multiple rhodoliths from a deposit, in conjunction with dendrochronology techniques on the growth bands, has also been used to reconstruct an age model for rhodolith accumulation creating a 651 year-long master chronology (Kamenos 2010).

2.5.8 *Magnesium Isotopes*

Magnesium is a major element in the oceans, playing key roles in biogeochemical cycles (Hippler et al. 2009). While investigating the factors controlling Mg isotope fractionation, Hippler et al. (2009) observed that $\delta^{26}Mg$ in coralline algae did not appear to be affected by sea-surface temperature or salinity (Hippler et al. 2009). However, given that no metabolic influences were observed, it was suggested that coralline algal $\delta^{26}Mg$ may help in the reconstruction of seawater chemistry changes over geological timescales (Hippler et al. 2009).

2.6 Understanding Regional Productivity

Despite the increased use of coralline algae in environmental reconstruction, they have received very little utility in understanding historic or future productivity.

2.6.1 *Copepod Abundance*

In the North Atlantic, Mg-temperature reconstructions using *Lithothamnion glaciale* were negatively related to the abundance of the copepod *Calanus finmarchicus* Gunnerus, determined using a continuous plankton recorder. Using calculated copepod-Mg temperature relationships, together with future temperature projections for the North Atlantic from the ECHAM4 climate model, copepod abundances were projected to fall over the next four decades (Kamenos 2010). This may have a significant knock-on effect for North Atlantic cod stocks, whose preferred food source is *C. finmarchicus* (Kamenos 2010).

2.6.2 *Salmon Abundance*

In the Bering Sea, growth increments in *Clathromorphum compactum* were used to reconstruct the Aleutian Low, a dominant climate pattern in the subarctic North Pacific (Halfar et al. 2011b). Via the Aleutian Low's influence on Bering Sea cloudiness, the Aleutian Low was observed to force regional plankton productivity and historic landings in Sockeye salmon, a major US fishery (Halfar et al. 2011b).

2.7 Future Prospects for the Use of Coralline Algae as Environmental Proxies

The structural and chemical composition of coralline algae has been used to reconstruct components of the marine environment and associated atmospheric climatic drivers. However, the records held by coralline algal deposits may be at risk from global change. Specifically, continued increases in atmospheric CO_2 have led to a measurable decline in ocean pH over the past 25 years (IPCC 2013). With ocean pH projected to continue to decline by up to 0.5 units by 2100 (Caldeira and Wickett 2005), a process know as ocean acidification (OA) (See Chap. 3). In a palaeoenvironmental reconstruction context, recent studies have suggested that, under low pH/high CO_2 conditions, (1) the cellular structure of coralline algae may be compromised (Burdett et al. 2012; Ragazzola et al. 2012), (2) calcification may be reduced (Martin and Gattuso 2009; Budenbender et al. 2011) or increased (over-compensation for dissolution) (Kamenos et al. 2013; Martin et al. 2013), (3) that dead thalli, such as those that comprise a deposit, show greater dissolution than live thalli in water at ambient CO_2 concentrations (Kamenos et al. 2013), and (4) while the Mg content of dead rhodoliths appears to remain unchanged (Kamenos et al. 2013), live coralline algae appear to reduce their Mg content under OA to reduce their skeletal reactivity to the low pH/high CO_2 conditions (Ries 2011; Kamenos et al. 2013).

These findings present various possible benefits and disadvantages for the use of coralline algae as palaeoenvironmental reconstructors under OA: (1) for species that show increased calcification rates, this may provide more skeletal carbonate material from which to extract environmental signals. However, for species that dissolve, or for dissolving dead thalli, this will limit the skeletal carbonate material available for reconstructions, and (2) OA may directly impact growth banding, changing the (currently known) growth/calcification responses of coralline algae to other environmental parameters such as temperature or light, necessitating a re-calibration. With these points in mind, efforts should be made to monitor changing relationships in proxy calibrations with the aim of enabling the use of multiple calibration relationships depending on the temporal period being reconstructed.

Coralline algae could themselves prove useful in reconstructing historic OA. As OA is projected to be most intense at high latitudes (IPCC 2013), the distribution of coralline algae at high latitudes suggests they may contain information on historic acidification events. Recently published studies indicate coralline algae have scope in understanding historic OA by reconstructing pCO_2 using skeletal bonding strength (Pauly et al. 2015) and palaeo pH using boron isotopes locked within their growth bands (Fietzke et al. 2015).

Thus, against a backdrop of projected OA, it is likely that, at least for the next 50–100 years, coralline algae will continue to be used as high-resolution palaeoenvironmental proxies. Given the recent development of many of the techniques, several of which are easily accessible, not requiring expensive instrumentation nor specialist technical support, it is likely that they will receive wider application from the increasing number of research teams addressing environmental questions at low, mid and high latitudes.

References

Alexandersson T (1974) Carbonate cementation in coralline algal nodules in Skagerrak, North-Sea – biochemical precipitation in undersaturated waters. J Sediment Petrol 44:7–26
Blake C, Maggs CA (2003) Comparative growth rates and internal banding periodicity of maerl species (Corallinales, Rhodophyta) from northern Europe. Phycologia 42:606–612
Borremans C, Hermans J, Baillon S et al (2009) Salinity effects on the Mg/Ca and Sr/Ca in starfish skeletons and the echinoderm relevance for paleoenvironmental reconstructions. Geology 37:351–354
Bosence D (1976) Ecological studies on two unattached coralline algae from western Ireland. Palaeontology 19:365–395
Bosence DWJ (1983) The occurrence and ecology of recent rhodoliths-a review. In: Peryt TM (ed) Coated grains. Springer, Berlin
Budenbender J, Riebesell U, Form A (2011) Calcification of the Arctic coralline red algae *Lithothamnion glaciale* in response to elevated CO2. Mar Ecol Prog Ser 441:79–87
Burdett HL (2013) DMSP dynamics in coralline algal habitats. PhD thesis, School of Geographical and Earth Sciences. University of Glasgow, Glasgow, 300 pp
Burdett HL, Kamenos NA, Law A (2011) Using coralline algae to understand historic marine cloud cover. Palaeogeogr Palaeoclimatol Palaeoecol 302:65–70
Burdett HL, Aloisio E, Calosi P et al (2012) The effect of chronic and acute low pH on the intracellular DMSP production and epithelial cell morphology of red coralline algae. Mar Biol Res 8:756–763
Caldeira K, Wickett ME (2005) Ocean model predictions of chemistry changes from carbon dioxide emissions to the atmosphere and ocean. J Geophys Res Oceans 110: C09S04. doi:10.1029/2004jc002671
Caragnano A, Basso D, Jacob DE et al (2014) The coralline red alga *Lithophyllum kotschyanum f. affine* as proxy of climate variability in the Yemen coast, Gulf of Aden (NW Indian Ocean). Geochim Cosmochim Acta 124:1–17
Chan P, Halfar J, Williams B et al (2011) Freshening of the Alaska coastal current recorded by coralline algal Ba/Ca ratios. J Geophys Res Biogeosci 116:G01032
Chave KE (1954) Aspects of biogeochemistry of magnesium. 1: calcareous marine organisms. J Geol 62:266–283
Chave KE (1984) The physics and chemistry of biomineralisation. Annu Rev Earth Planet Sci 112:293–305
Chave KE, Wheeler BD (1965) Mineralogic changes during growth in red algae, *Clathromorphum compactum*. Science 147:621
Cohen AL, Layne G, Hart S et al (2001) Kinetic control of skeletal Sr/Ca in a symbiotic coral: implications for the paleotemperature proxy. Paleoceanography 16:20–26
Darrenougue N (2013) Rhodoliths as environmental archives in the tropics. The Australian National University, Canberra, 244 pp
Darrenougue N, De Deckker P, Payri C et al (2013) Growth and chronology of the rhodolith-forming, coralline red alga *Sporolithon durum*. Mar Ecol Prog Ser 474:105–119
De'ath G, Lough JM, Fabricius KE (2009) Declining coral calcification on the great barrier reef. Science 323:116–119
Delaney ML, Be AWH, Boyle EA (1985) Li, Sr, Mg, and Na in foraminiferal calcite shells from laboratory culture, sediment traps, and sediment cores. Geochim Cosmochim Acta 49:1327–1341
Dissard D, Nehrke G, Reichart GJ et al (2010) The impact of salinity on the Mg/Ca and Sr/Ca ratio in the benthic foraminifera *Ammonia tepida*: results from culture experiments. Geochim Cosmochim Acta 74:928–940
Elderfield H, Ganssen G (2000) Past temperature and delta O-18 of surface ocean waters inferred from for aminiferal Mg/Ca ratios. Nature 405:442–445

Fietzke J, Ragazzola F, Halfar J, Dietze H, Foster LC, Hansteen TH, Eisenhauer A, Steneck RS (2015) Century-scale trends and seasonality in pH and temperature for shallow zones of the Bering Sea. Proc Natl Acad Sci USA 112:2960–2965

Foster MS (2001) Rhodoliths: between rocks and soft places. J Phycol 37:659–667

Frantz BR, Kashgarian M, Coale KH et al (2000) Growth rate and potential climate record from a rhodolith using C-14 accelerator mass spectrometry. Limnol Oceanogr 45:1773–1777

Frantz BR, Foster MS, Riosmena-Rodríguez R (2005) *Clathromorphum nereostratum* (Corallinales, Rhodophyta): the oldest alga? J Phycol 41:770–773

Freiwald A, Henrich R (1994) Reefal coralline algal build-ups within the Arctic circle: morphology and sedimentary dynamics under extreme environmental seasonality. Sedimentology 41:963–984

Fritts HC (1991) Reconstructing large-scale climatic patterns from tree-ring analysis. The University of Arizona Press, Tuscon

Gamboa G, Halfar J, Hetzinger S et al (2010) Mg/Ca ratios in coralline algae record northwest Atlantic temperature variations and North Atlantic Oscillation relationships. J Geophys Res Oceans 115:C12044

Halfar J, Zack T, Kronz A et al (2000) Growth and high-resolution paleoenvironmental signals of rhodoliths (coralline red algae): a new biogenic archive. J Geophys Res Oceans 105:22107–22116

Halfar J, Steneck R, Schöne B et al (2007) Coralline alga reveals first marine record of subarctic North Pacific climate change. Geophys Res Lett 34:L07702

Halfar J, Steneck RS, Joachimski M et al (2008) Coralline red algae as high-resolution climate recorders. Geology 36:463–466

Halfar J, Hetzinger S, Adey WH et al (2011a) Coralline algal growth-increment widths archive North Atlantic climate variability. Palaeogeogr Palaeoclimatol Palaeoecol 302:71–80

Halfar J, Williams B, Hetzinger S et al (2011b) 225 years of Bering Sea climate and ecosystem dynamics revealed by coralline algal growth-increment widths. Geology 39:579–582

Halfar J, Adey WH, Kronz A et al (2013) Arctic sea-ice decline archived by multicentury annual-resolution record from crustose coralline algal proxy. Proc Natl Acad Sci 110:19737–19741

Henrich R, Freiwald A, Wehrmann A et al (1996) Nordic-cold water carbonates: occurrence and controls. In: Reitner J, Neuweiler F, Gunkel F (eds) Global and regional controls on biogenic sedimentation. Gottinger Arbeiten Geol. Palaonton, Gottingen

Hetzinger S, Halfar J, Kronz A et al (2009) High-resolution Mg/Ca ratios in a coralline red alga as a proxy for Bering Sea temperature variations from 1902 to 1967. Palaios 24:406–412

Hetzinger S, Halfar J, Zack T et al (2011) High-resolution analysis of trace elements in crustose coralline algae from the North Atlantic and North Pacific by laser ablation ICP-MS. Palaeogeogr Palaeoclimatol Palaeoecol 302:81–94

Hetzinger S, Halfar J, Mecking JV et al (2012) Marine proxy evidence linking decadal North Pacific and Atlantic climate. Clim Dyn 39:1447–1455

Hetzinger S, Halfar J, Zack T et al (2013) Coralline algal Barium as indicator for 20th century northwestern North Atlantic surface ocean freshwater variability. Sci Rep 3:1761

Hippler D, Buhl D, Witbaard R et al (2009) Towards a better understanding of magnesium-isotope ratios from marine skeletal carbonates. Geochim Cosmochim Acta 73:6134–6146

Hughes MK (2002) Dendrochronology in climatology-the state of the art. Dendrochronologia 20:95–116

Huh Y, Chan LH, Zhang L et al (1998) Lithium and its isotopes in major world rivers: implications for weathering and the oceanic budget. Geochim Cosmochim Acta 62:2039–2051

IPCC (2013) Summary for policymakers: the physical science basis. In: Working Group I contribution to the IPCC fifth assessment report

Kamenos NA (2010) North Atlantic summers have warmed more than winters since 1353, and the response of marine zooplankton. Proc Natl Acad Sci U S A 107:22442–22447

Kamenos NA, Law A (2010) Temperature controls on coralline algal skeletal growth. J Phycol 46:331–335

Kamenos NA, Cusack M, Moore PG (2008) Coralline algae are global palaeothermometers with bi-weekly resolution. Geochim Cosmochim Acta 72:771–779
Kamenos NA, Cusack M, Huthwelker T et al (2009) Mg-lattice associations in red coralline algae. Geochim Cosmochim Acta 73:1901–1907
Kamenos NA, Hoey T, Nienow P et al (2012) Reconstructing Greenland Ice sheet runoff using coralline algae. Geology 40:1095–1098
Kamenos NA, Burdett HL, Aloisio E et al (2013) Coralline algal structure is more sensitive to rate, rather than the magnitude, of ocean acidification. Glob Chang Biol 19:3621–3628
Kinsman DJJ, Holland HD (1969) Co-precipitation of cations with $CaCO_3$. 4: Co-precipitation of Sr^{2+} with aragonite between 16–96°C. Geochim Cosmochim Acta 33:1–17
Lea DW (2006) Elemental and Isotopic proxies of past ocean temperatures. In: Elderfield H (ed) The oceans and marine geochemistry. Elsevier, Amsterdam
Lea DW, Shen GT, Boyle EA (1989) Coralline barium records temporal variability in equatorial Pacific upwelling. Nature 340:373–376
Lee D, Carpenter SJ (2001) Isotopic disequilibrium in marine calcareous algae. Chem Geol 172:307–329
Levin I, Munnich KO, Weiss W (1980) The effect of anthropogenic CO_2 and C-14 sources on the distribution of C-14 in the atmosphere. Radiocarbon 22:379–391
Littler MM, Littler DS, Hanisak MD (1991) Deep-water rhodolith distribution, productivity, and growth history at sites of formation and subsequent degradation. J Exp Mar Biol Ecol 150:163–182
Lough JM, Barnes DJ (2000) Environmental controls on growth of the massive coral *Porites*. J Exp Mar Biol Ecol 245:225–243
Martin S, Gattuso JP (2009) Response of Mediterranean coralline algae to ocean acidification and elevated temperature. Glob Chang Biol 15:2089–2100
Martin S, Cohu S, Vignot C et al (2013) One-year experiment on the physiological response of the Mediterranean crustose coralline alga, *Lithophyllum cabiochae*, to elevated pCO2 and temperature. Ecol Evol 3:676–693
McCulloch M, Fallon S, Wyndham T et al (2003) Coral record of increased sediment flux to the inner Great Barrier Reef since European settlement. Nature 421:727–730
Milliman JD (1977) Role of calcareous algae in Atlantic continental margin segmentation. In: Flugel E (ed) Fossil algae. Springer, Berlin
Milliman JD, Gastner M, Müller J (1971) Utilization of magnesium in coralline algae. Geol Soc Am Bull 82:573–580
Moberly RJ (1968) Composition of magnesian calcites of algal and pelcypods by electron microprobe analysis. Sedimentology 11:61–82
Moberly R (1970) Microprobe study of diagenesis in calcareous algae. Sedimentology 14:113–123
Nelson WA (2009) Calcified macroalgae – critical to coastal ecosystems and vulnerable to change: a review. Mar Freshw Res 60:787–801
Oomori T, Kaneshima H, Maezato Y et al (1987) Distribution coefficient of Mg2+ ions between calcite and solution at 10–50-Degrees-C. Mar Chem 20:327–336
Pauly M, Kamenos NA, Donohue P, LeDrew E (2015) Coralline algal Mg-O bond strength as a marine pCO2 proxy. Geology 43:267–270
Peña V, Bárbara I (2008) Biological importance of an Atlantic European maërl bed off Benencia Island (northwest Iberian Peninsula). Bot Mar 51:493–505
Quay PD, Tilbrook B, Wong CS (1992) Oceanic uptake of fossil-fuel CO_2 – C-13 evidence. Science 256:74–79
Ragazzola F, Foster LC, Form A et al (2012) Ocean acidification weakens the structural integrity of coralline algae. Glob Chang Biol 18:2804–2812
Rahimpour-Bonab H, Bone Y, Moussavi-Harami R et al (1997) Geochemical comparisons of modern cool-water calcareous biota, Lacepede Shelf, south Australia. Soc Sediment Geol 56:77–92

Ries JB (2006) Mg fractionation in crustose coralline algae: geochemical, biological, and sedimentological implications of secular variation in the Mg/Ca ratio of seawater. Geochim Cosmochim Acta 70:891–900

Ries JB (2011) Skeletal mineralogy in a high-CO_2 world. J Exp Mar Biol Ecol 403:54–64

Rivera MG, Riosmena-Rodríguez R, Foster MS (2004) Age and growth or *Lithothamnion muelleri* (Corallinales, Rhodophyta) in the southwestern Gulf of California, Mexico. Ciecias Mar 30:235–249

Rollion-Bard C, Vigier N, Meibom A et al (2009) Effect of environmental conditions and skeletal ultrastructure on the Li isotopic composition of scleractinian corals. Earth Planet Sci Lett 286:63–70

Schmidt GA, Hoffamnn G, Thresher D (2001) Isotopic tracers in coupled models: a new paleo-tool. PAGES News 9:10–11

Schöne BR, Fiebig J, Pfeiffer M et al (2005) Climate records from a bivalved Methuselah (Arctica islandica, Mollusca; Iceland). Palaeogeogr Palaeoclimatol Palaeoecol 228:130–148

Schwarz AM, Hawes I, Andrew N et al (2005) Primary production potential of non-geniculate coralline algae at Cape Evans, Ross Sea, Antarctica. Mar Ecol Prog Ser 294:131–140

Scourse J, Richardson C, Forsythe G et al (2006) First cross-matched floating chronology from the marine fossil record: data from growth lines of the long-lived bivalve mollusc *Arctica islandica*. The Holocene 16:967–974

Shen GT, Dunbar RB (1995) Environmental controls on uranium in reef corals. Geochim Cosmochim Acta 59:2009–2024

Steller DL, Riosmena-Rodríguez R, Foster MS et al (2003) Rhodolith bed diversity in the Gulf of California: the importance of rhodolith structure and consequences of disturbance. Aquat Conserv Mar Freshwat Ecosyst 13:S5–S20

Stoffyn-Egli P, Mackenzie FT (1984) Mass balance of dissolved lithium in the oceans. Geochim Cosmochim Acta 48:859–872

Tudhope AW, Shimmield GB, Chilcott CP et al (1995) Recent changes in climate in the far western equatorial Pacific and their relationship to the Southern Oscillation; oxygen isotope records from massive corals, Papua New Guinea. Earth Planet Sci Lett 136:575–590

Urey HC (1947) The thermodynamic properties of isotopic substances. J Chem Soc 562–581

Wanamaker AD Jr, Hetzinger S, Halfar J (2011a) Reconstructing mid- to high-latitude marine climate and ocean variability using bivalves, coralline algae, and marine sediment cores from the Northern Hemisphere. Palaeogeogr Palaeoclimatol Palaeoecol 302:1–9

Wanamaker AD Jr, Kreutz KJ, Schöne BR et al (2011b) Gulf of Maine shells reveal changes in seawater temperature seasonality during the medieval climate anomaly and the little Ice Age. Palaeogeogr Palaeoclimatol Palaeoecol 302:43–51

Wefer G, Berger WH (1991) Isotope paleontology – growth and composition of extant calcareous species. Mar Geol 100:207–248

Williams B, Halfar J, Steneck RS et al (2011) Twentieth century delta C-13 variability in surface water dissolved inorganic carbon recorded by coralline algae in the northern North Pacific Ocean and the Bering Sea. Biogeosciences 8:165–174

Chapter 3
Effects of Ocean Warming and Acidification on Rhodolith/Maërl Beds

Sophie Martin and Jason M. Hall-Spencer

Abstract Coralline algae are expected to be adversely impacted by global warming and ocean acidification, although there has been no synthesis of these effects on habitat-forming species. We compiled published responses of maërl and rhodolith-forming species to ocean acidification and warming. Although the response is variable among species, their recruitment, growth, health and survival are usually negatively affected under elevated CO_2. Most studies show that coralline algal calcification is adversely affected under near-future ocean acidification scenarios and that in combination with a 1–3 °C increase in seawater temperature this has an even larger impact. Most research has involved relatively short-term experiments on single species, which makes it difficult to predict long-term effects at the ecosystem level because the impact of global changes on coralline algal habitats will depend on the direct impacts on individual species and the indirect effects of altered interspecific interactions. Studies in areas with naturally high CO_2 levels show that coralline algae are adversely affected by long-term acidification through increased competition from non-calcified competitors. Coralline algal habitats such as vermetid reefs, coralligene and beds of rhodoliths or maerl are likely to decline in the near future as higher CO_2 levels benefit fleshy algae and corrosive waters reduce calcareous habitat complexity and associated biodiversity.

Ocean warming and acidification are currently under scrutiny since in combination they are expected to reduce marine biodiversity and profoundly alter ecosystem function. Habitats where coralline algae play an important ecological role, such as

---

S. Martin (✉)
CNRS, UMR 7144, Laboratoire Adaptation et Diversité en Milieu Marin, Station Biologique de Roscoff, Place Georges Teissier, 29688 Roscoff Cedex, France

Sorbonne Universités, UPMC Université de Paris 06, Station Biologique de Roscoff, Place Georges Teissier, 29688 Roscoff Cedex, France
e-mail: sophie.martin@sb-roscoff.fr

J.M. Hall-Spencer
Marine Biology and Ecology Research Centre, Plymouth University, Plymouth PL4 8AA, UK

Shimonda Marine Research Centre, University of Tsukuba, Japan

R. Riosmena-Rodríguez et al. (eds.), *Rhodolith/Maërl Beds: A Global Perspective*, Coastal Research Library 15, DOI 10.1007/978-3-319-29315-8_3

rhodolith and maerl beds, are threatened by dissolution due to the high solubility of their high-magnesium calcite skeletons. Those in the Arctic, where carbonate saturation levels are already low and falling, are of particular concern. As well as direct corrosive effects on coralline algae, rising CO_2 emissions are expected to have knock-on effects on ecosystems because reduced seabed habitat complexity causes a reduction in biodiversity and simplifies food webs. We anticipate that degradation of coastal calcareous habitats due to ocean acidification will facilitate a proliferation of fleshy algae that may benefit from an increase in dissolved inorganic carbon, to the detriment of calcified algae. It is not all doom and gloom: coralline algae have survived previous mass extinctions and many species tolerate highly variable CO_2 levels. Which species survive the Anthropocene will depend upon their ability to acclimate and adapt. These topics warrant further research since coralline algae provide highly biodiverse habitats that benefit commercially important species of fish and molluscs. Data so far suggest that (1) this important algal group is especially vulnerable to ocean acidification and warming and (2) protecting these habitats has long-term benefits, not least because coralline algae provide habitat for species of commercial importance.

3.1 Climate Change and Ocean Acidification

Anthropogenic emissions have increased the atmospheric carbon dioxide (CO_2) concentration from 280 ppm prior to the beginning of the industrial revolution (1750) to more than 400 ppm in 2015; this is *ca.* 100 ppm higher than at any time in the past 740,000 years. The present rate of increase in atmospheric CO_2 and temperature is unprecedented in recent Earth history (Hoegh-Guldberg et al. 2007).

Approximately 25 % of CO_2 emissions dissolve into the ocean, increasing pCO_2 in the surface water and altering seawater carbonate chemistry (Canadell et al. 2007). This CO_2 reacts with the water to form carbonic acid most of which dissociates into ions of hydrogen (H^+) and bicarbonate (CO_3^{2-}) (Fig. 3.1). The increased concentration of H^+ reduces pH ($pH = -\log_{10}[H^+]$) and carbonate ion (CO_3^{2-}) concentration and increases the concentration of HCO_3^-.

During the twentieth century, increasing atmospheric CO_2 caused the surface ocean to warm by 0.7 °C and to acidify by 0.1 pH units (Hoegh-Guldberg et al. 2007). The atmospheric CO_2 concentration is expected to reach between 421 and 936 ppm by 2100 causing a global mean surface temperature increase of between 0.3 and 4.8 °C and a surface ocean pH decrease of 0.06–0.32 units relative to the period 1986–2005 (IPCC 2013; Fig. 3.2). Ocean acidification and warming will not be regionally uniform; marine organisms in the Arctic region are thought to face more rapid and stronger warming and acidification than the global mean (Fig. 3.3).

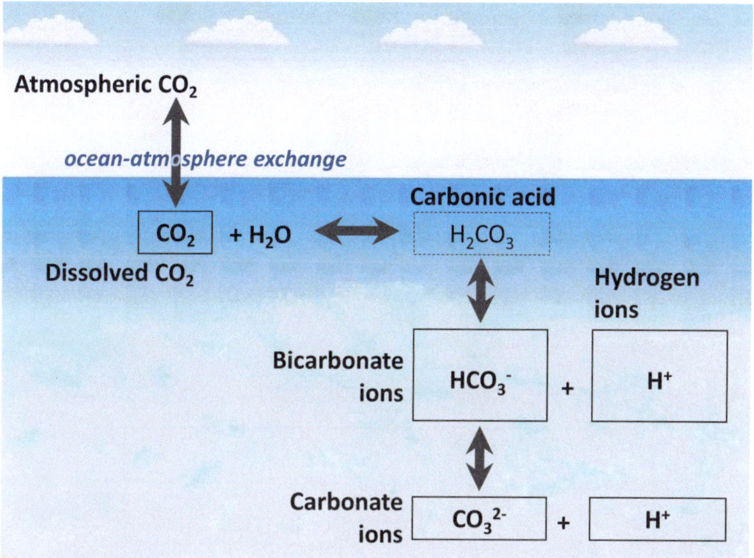

Fig. 3.1 Equilibrium between the three different species of dissolved inorganic carbon in seawater. Atmospheric CO_2 is absorbed at the surface of the oceans where it combines with the water molecules to form carbonic acid, which dissociates into bicarbonate, carbonate, and hydrogen ions. Different forms of dissolved inorganic carbon $\left(DIC = [CO_2] + [HCO_3^-] + [CO_3^{2-}]\right)$ follow thermodynamic equilibria in seawater for the following reactions: $CO_2 + H_2O \leftrightarrow H_2CO_3 \leftrightarrow HCO_3^- \leftrightarrow H^+ + CO_3^{2-} + 2H^+$

3.2 Effects of Ocean Warming and Acidification on Seaweeds

Global warming and ocean acidification will alter biochemical and physiological processes in seaweeds, causing changes in their ecological interactions (Koch et al. 2012; Harley et al. 2012). Here we base predictions about the fate of rhodolith/maerl beds upon a growing body of information about the effects of warming and acidification on seaweeds in general and on coralline algae in particular. While some seaweed taxa seem likely to benefit from ocean warming and acidification, coralline algae are expected to be adversely impacted, putting the habitats they form at risk. This is a concern since beds of rhodoliths and maerl form highly biodiverse but very slow-growing habitats (Peña et al. 2014).

3.2.1 Effects of Ocean Warming on Seaweeds

Temperature directly affects enzymatic processes and so alters rates of photosynthesis and growth; it also determines whether seaweeds can survive and reproduce and sets limits to their geographic distribution (Lüning 1990). Seaweeds, as with other

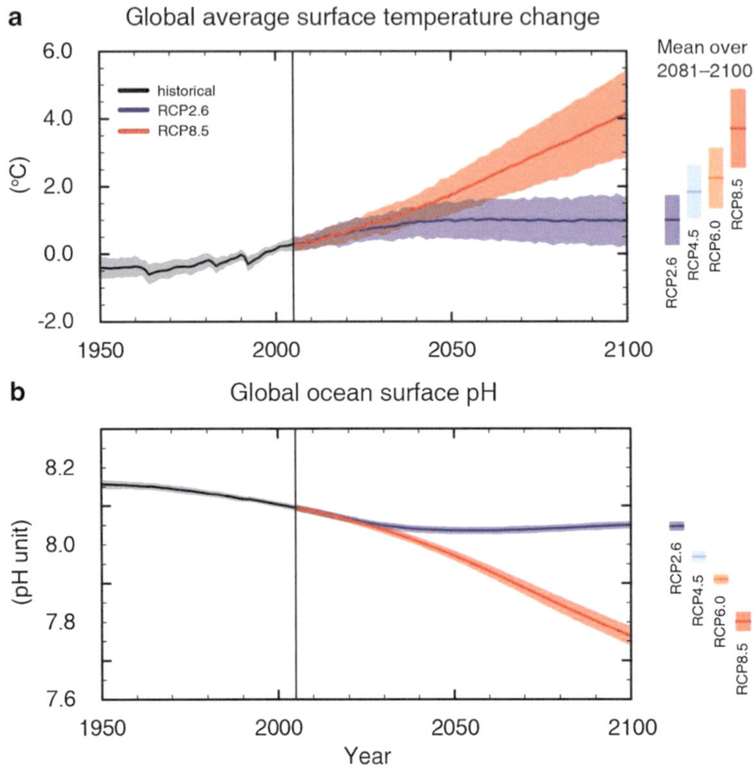

Fig. 3.2 Projected change in (**a**) global annual mean surface temperature and (**b**) global mean ocean surface pH, relative to 1986–2005 for Representative Concentration Pathways (RCPs) RCP2.6 and RCP8.5 (Source: IPCC 2013) RCP 2.6 assumes that emissions (measured in CO_2-equivalents) peak between 2010 and 2020 and decline substantially thereafter. In RCP 8.5, emissions continue to rise throughout the twenty-first century

organisms, have thermal performance curves that show beneficial effects of increased temperature at suboptimal temperature but detrimental effects above a threshold (Lüning 1990). In order for some individual species of seaweed to survive rising sea surface temperature they will need to acclimate (on short timescales), adapt (on medium and long timescales) or migrate (by dispersion) (Wiencke and Bischof 2012). Seaweeds that are intertidal or have a broad biogeographic range are more tolerant of temperature extremes than subtidal species and those with narrower temperature ranges. The changes in temperature we see today are driving a poleward shift in seaweed biogeographical regions (Wernberg et al. 2011). The tropical region is widening polewards, to the detriment of organisms that occupy warm-temperate regions and the cold-temperate regions are shrinking. Arctic seaweeds are at particular risk due to warming winter temperatures (Wiencke and Bischof 2012). In coralline algae, a small rise in temperature, within the range of temperature experienced in natural habitats, can increase growth, photosynthesis and calcification in both temperate and tropical species (Martin et al. 2006; Steller et al. 2007) but this has limits and rising temperature above these levels is

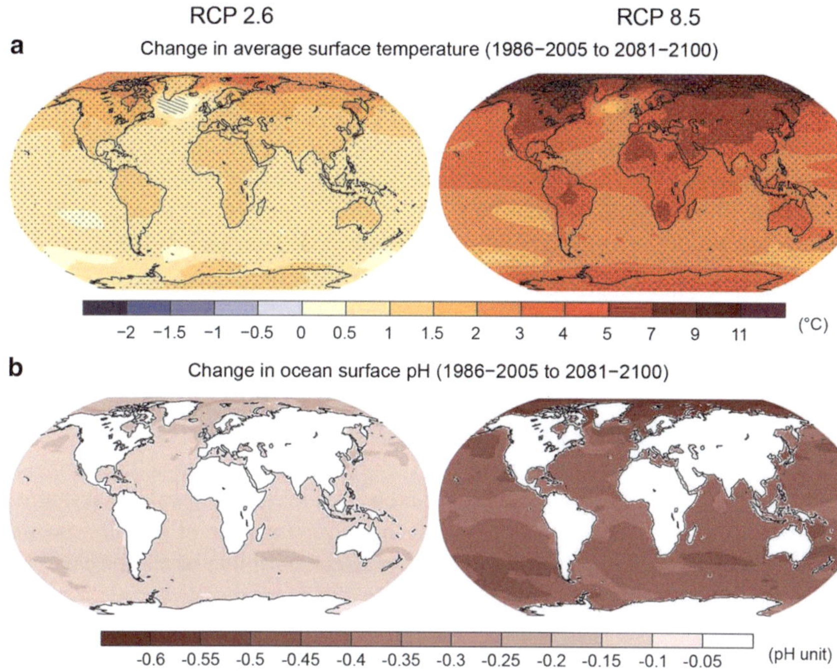

Fig. 3.3 Maps of projected changes in (**a**) annual mean surface temperature and (**b**) ocean surface pH for the scenarios RCP2.6 and RCP8.5 in 2081–2100 relative to 1986–2005 (Source: IPCC 2013)

detrimental (Table 3.1). For example, Agegian (1985) showed that growth of tropical *Porolithon gardineri* slowed dramatically at temperatures above 29–30 °C. An increase of +3 °C above that normally experienced by coralline algae causes bleaching and adversely affects health, survival, and the rates of photosynthesis and calcification in both tropical and warm-temperate coralline algae (Anthony et al. 2008; Martin and Gattuso 2009; Diaz-Pulido et al. 2011; Martin et al. 2013; Table 3.1).

3.2.2 Impact of Ocean Acidification on Seaweeds

Decreasing pH in the surface ocean will cause major shifts in seawater chemistry over the course of this century that are likely to affect photosynthesis and calcification since these processes use dissolved inorganic carbon (DIC: HCO_3^-, CO_3^{2-} and CO_2) as substrate

1. Photosynthesis (left to right) and respiration (right to left) processes:

$$CO_2 + H_2O \leftrightarrow CH_2O + O_2$$
$$HCO_3^- + H_2O \leftrightarrow CH_2O + O_2 + OH^-$$

2. Gross calcification (left to right) and dissolution (right to left) processes:

$$CO_3^{2-} + Ca^{2+} \leftrightarrow CaCO_3$$
$$2\,HCO_3^{-} + Ca^{2+} \leftrightarrow CaCO_3 + H_2O + CO_2$$

Algae can use dissolved CO_2 entering the cell by diffusion as the carbon source for photosynthesis but most of them have carbon concentrating mechanisms which actively take up HCO_3^- which is then converted to CO_2 in the cells (Raven et al. 2002; Hepburn et al. 2011; Cornwall et al. 2012). This mechanism counteracts the limited availability of CO_2 in seawater (Raven and Beardall 2003). An increase in seawater pCO_2 is expected to enhance photosynthesis in primary producers that rely exclusively on CO_2 diffusion (Kübler et al. 1991) and may be favourable to algae that can down-regulate their carbon concentrating mechanisms to save energy (Hepburn et al. 2011; Cornwall et al. 2012; Raven et al. 2012).

Coralline algae are thought to be one of the groups of species most vulnerable to ocean acidification due to the solubility of their high magnesium-calcite skeletons. Ocean acidification is causing a decrease in the saturation state of calcium carbonate ($\Omega = [Ca^{2+}] \times [CO_3^{2-}] / K_{sp}$) which is likely to affect the ability of marine calcifiers to form their carbonate skeleton or shells by a decline in calcification rates (Kroeker et al. 2010). Although the physiological response in terms of calcification is variable among taxa and species (Ries et al. 2009), seawater acidification is related to reduced growth rates in calcified macroalgae (Kroeker et al. 2013). The recruitment and growth of coralline algae are usually negatively affected under elevated pCO_2 (Table 3.1). Reductions in calcification rate at elevated pCO_2 have been demonstrated for most coralline algae (Harley et al. 2012) but this response is variable among species (Fig. 3.4 and Table 3.1).

3.2.3 Interactive Effects of Ocean Warming and Acidification on Seaweeds

While the effects of climate change and ocean acidification on seaweeds have mostly been studied in isolation from each other, there is a lack of studies how these factors interact. The combination of these two factors could however have a larger impact than alone. Recent studies have effectively shown that the negative effect of ocean acidification on coralline algal calcification is exacerbated by further ocean warming (Anthony et al. 2008). Some authors reported a significant pCO_2 effect on coralline calcification, health and survival only in combination with increased temperature, suggesting that elevated pCO_2 aggravates the sensitivity of coralline algae to temperature (Martin and Gattuso 2009).

Table 3.1 Compilation of published responses of coralline algae to ocean acidification (elevated pCO_2) and warming (elevated temperature), alone or in combination

Effects on:	Species	Type of study	Experimental conditions		Response to:			References
			pCO_2 (μatm)	T (°C)	↑ T	↑ pCO_2	T×pCO_2	
Recruitment								
	Temperate							
	Cold-temperate							
	Lithophyllum sp. (CCA)	Lab	210/300[b]	17		↓		Russell et al. (2009)
	Lithophyllum sp. (CCA)	Out	277–438/377–930	14		↓		Russell et al. (2011)
	Phymatolithon lenormandii (CCA)	Lab	398/1,261	18		↓		Bradassi et al. (2013)
	Arthrocardia corymbosa (ACA)	Lab	145–415/410-1130	11		—		Cornwall et al. (2013a)
	Warm-temperate							
	Corallina sp. (ACA); *Hydrolithon farinosum* (CCA)	Vent	334–957[d]	14–27		↓		Porzio et al. (2013)
	unclassified CCA	Vent	440–567/1,075–1,581	17.5–24		—		Kroeker et al. (2013)
	Tropical and sub-tropical							
	Porolithon gardineri (CCA)[ξ]	Out	pH 7.5–8.5[c]	19.5–30		↓		Agegian (1985)
	Lithophyllum cf. *pallescens*, *Hydrolithon* sp. and *Porolithon* sp. (CCA)[a]	Out	365/765	23–29		↓		Jokiel et al. (2008); Kuffner et al. (2008)

(continued)

Table 3.1 (continued)

Effects on:	Species	Type of study	Experimental conditions		Response to:			References
			pCO_2 (μatm)	T (°C)	↑ T	↑ pCO_2	T×pCO_2	
Cover								
	Temperate							
	Cold-temperate							
	Corallina vancouveriensis (ACA)	Field	pH 8.0–8.4[c]	–		—		Wootton et al. (2008)
	Warm-temperate							
	Amphiroa, Corallina, Jania, Lithophyllum, Mesophyllum, and *Peyssonnelia* spp. (ACA & CCA)	Vent	296–953[d]	13–25		↓		Hall-Spencer et al. (2008)
	Hydrolithon & *Pneophyllum* sp. (CCA)	Vent	306–1,564	18–19		↓		Martin et al. (2008)
	unclassified CCA	Vent	440–567/1,075–1,581	17.5–24		↓		Kroeker et al. (2013b)
	Tropical and sub-tropical							
	Lithophyllum, Hydrolithon and *Porolithon* sp. (CCA)	Out	365/765	23–29		↓		Kuffner et al. (2008)
	Hydrolithon, Porolithon, Titanoderma, Lithophyllum, Mesophyllim, Neogoniolithon, Sporolithon, and *Porolithon*(CCA)	Out	401/807/1,299	26		— or ↓		Doropoulos et al. (2012)
	Crustose coralline algae (CCA)	Vent	296–953[d]	–		↓		Fabricius et al. (2011)

Growth								
	Polar and sub-polar							
	Lithothamnion glaciale (M)	Lab [L]	422/589/755/1018	8		↓		Ragazzola et al. (2012)
	Lithothamnion glaciale (M)	Lab [L]	422/589/755/1018	8		↓		Ragazzola et al. (2013)
	Temperate							
	Cold-temperate							
	Phymatolithon calcareum (M)	Lab [L]	–	10/14/18	—			Blake and Maggs (2003)
	Lithothamnion corallioides (M)	Lab [L]	–	10/14/18	∩			Blake and Maggs (2003)
	Lithophyllum sp. (CCA)	Lab [W]	210/300[b]	17		↓		Russell et al. (2009)
	Lithophyllum sp. (CCA)	Out [W]	277–438/377–930	14		↓		Russell et al. (2011)
	Corallina officinalis (ACA)	Lab [L]	385/665/1486	–		↓		Hofmann et al. (2012a)
	Corallina officinalis (ACA)	Lab [W]	384/1313/1935	15		↓		Hofmann et al. (2012b)
	Arthrocardia corymbosa (ACA)	Lab [W]	145–415/410–1130	11		↓		Cornwall et al. (2013a)
	Arthrocardia corymbosa (ACA)	Lab [L, W]	475/820/1500	15		↓		McGraw et al. (2010)
	Tropical and sub-tropical							
	Porolithon gardineri (CCA)	Out [L]	pH 7.5–8.5[c]	19.5–30	∩	↓		Agegian (1985)
	Lithophyllum margaritae (M)	Field [L]	–	18–30.5	↑			Steller et al. (2007)
	Hydrolithon samoense (CCA)	Lab [W]	263/399/960	27		↓		Kato et al. (2013)

(continued)

Table 3.1 (continued)

Effects on:	Species	Type of study	Experimental conditions		Response to:			References
			pCO_2 (µatm)	T (°C)	↑ T	↑ pCO_2	T×pCO_2	
	Lithophyllum kotschyanum (CCA)	Lab [W]	263/399/960	27		↓		Kato et al. (2013)
Health and survival								
	Temperate							
	Cold-temperate							
	Arthrocardia corymbosa (ACA)	Lab[∂]	145–415/410–1130	11		—		Cornwall et al. (2013a)
	Lithophyllum incrustans (CCA)	Lab	450/637/790/1,002	16		↓		Noisette et al. (2013a)
	Warm-temperate							
	Lithophyllum cabiochae (CCA)	Lab	417/728	13–22/+3 °C	↓	—	↓↓	Martin and Gattuso (2009)
	Tropical and sub-tropical							
	Porolithon onkodes (CCA)	Out	298/613/1,185	25–26/+3 °C	↓	↓	↓↓	Anthony et al. (2008)
	Porolithon onkodes (CCA)	Out	298/613/1,185	25–26/+3 °C	↓	↓	↓↓	Diaz-Pulido et al. (2012)
Respiration								
	Polar and sub-polar							
	Clathromorphum circumscriptum (CCA)	Lab	–	0–20	↑			Adey (1973)
	Temperate							
	Cold-temperate							
	Clathromorphum circumscriptum (CCA)	Field	–	7–19	↑			Digby (1977)
	Corallina officinalis (ACA)	Field	–	10–19	—			Digby (1977)

	Lithophyllum yessoense (CCA)	Lab	–	5–30	↑			Ichiki et al. (2001)
	Lithothamnion corallioides (R)	Field	324–517[pers]	13–17	↑			Martin et al. (2006)
	Corallina officinalis (ACA)	Lab	385/665/1,486	–		—		Hofmann et al. (2012a)
	Corallina officinalis (ACA)	Lab	384/1,313/1,935	15		∩		Hofmann et al. (2012b)
	Ellisolandia elongata (ACA)	Lab	362/547/770/977	18		—		Egilsdottir et al. (2013)
	Lithothamnion glaciale (M)	Lab	498/1081/2,278	10		—		Kamenos et al. (2013)
	Ellisolandia elongata (ACA)	Lab	450/637/790/1,002	16		—		Noisette et al. (2013a)
	Lithophyllum incrustans (CCA)	Lab	450/637/790/1,002	16		↑		Noisette et al. (2013a)
	Lithothamnion corallioides (M)	Lab	450/637/790/1,002	16		—		Noisette et al. (2013a)
	Lithothamnion corallioides (M)	Lab	380/550/750/1,000	13–19		—		Noisette et al. (2013b)
	Warm-temperate							
	Lithophyllum cabiochae (CCA)	Lab	417/728	13–22/+3 °C	↑ or —	—	— or ↓	Martin et al. (2013)
	Tropical and sub-tropical							
	Lithophyllum margaritae (M)	Lab	–	10–30	↑			Steller et al. (2007)
	Hydrolithon sp. (R)	Lab	400/900	25		—		Semesi et al. (2009)

(continued)

Table 3.1 (continued)

Effects on:	Species	Type of study	Experimental conditions		Response to:			References
			pCO_2 (μatm)	T (°C)	↑ T	↑ pCO_2	T×pCO_2	
Photosynthesis								
	Polar and sub-polar							
	Clathromorphum circumscriptum (CCA)	Lab	–	0–20	—			Adey (1973)
	Temperate							
	Cold-temperate							
	Clathromorphum circumscriptum (CCA)	Field	–	7–19	↑			Digby (1977)
	Corallina officinalis (ACA)	Field	–	7–19	↑			Digby (1977)
	Lithophyllum yessoense (CCA)	Lab	–	5–30	∩			Ichiki et al. (2001)
	Lithothamnion corallioides (M)	Field	324–517[pers]	10–17	↑			Martin et al. (2006)
	Corallina officinalis (ACA)	Lab	759/1,866/2,014[b]	12		—		Cornwall et al. (2012)
	Corallina officinalis (ACA)	Lab	385/665/1486	–		—		Hofmann et al. (2012a)
	Corallina officinalis (ACA)	Lab	384/1,313/1,935	15		↓		Hofmann et al. (2012b)
	Arthrocardia corymbosa (ACA)	Lab	145–415/410–1,130	11		—		Cornwall et al. (2013a)
	Ellisolandia elongata (ACA)	Lab	362/547/770/977	18		—		Egilsdottir et al. (2013)
	Lithothamnion glaciale (M)	Lab	498/1081/2,278	10		—		Kamenos et al. (2013)

	Ellisolandia elongata (ACA)	Lab	450/637/790/1,002	16		—		Noisette et al. (2013a)
	Lithophyllum incrustans (CCA)	Lab	450/637/790/1,002	16		—		Noisette et al. (2013a)
	Lithothamnion corallioides (M)	Lab	450/637/790/1,002	16		—		Noisette et al. (2013a)
	Lithothamnion corallioides (M)	Lab	380/550/750/1,000	13–19		∩		Noisette et al. (2013b)
	Warm-temperate							
	Lithophyllum cabiochae (CCA)	Lab	417/728	13–22/+3 °C	— or ↑	↓	—	Martin et al. (2013)
	Tropical and sub-tropical							
	Amphiroa foliacea (ACA)	Lab	pH 7–9[c]	28		↑		Borowitzka (1981)
	Amphiroa anceps (ACA)	Lab	pH 7–9[c]	18/22		↑		Borowitzka (1981)
	Lithophyllum margaritae (M)	Lab	–	10–30	∩			Steller et al. (2007)
	Porolithon onkodes (CCA)	Out	298/613/1,185	25–26/+3 °C	↓	↓	↓↓	Anthony et al. (2008)
	Hydrolithon sp. (R)	Lab	400/900	25		↑		Semesi et al. (2009)
			1–1400			↑		
	Corallina sessilis (ACA)	Lab	380/1000	23–27		↓		Gao and Zheng (2010)
Photosystem II efficiency (effective quantum yield) and relative electron transport rate								
	Temperate							
	Cold-temperate							
	Phymatolithon calcareum (R)	Lab $^{Fv/Fm}$	–	9–25	—			Wilson et al. (2004)
	Lithophyllum sp. (CCA)	Lab $^{Fv/Fm}$	210/300[c]	17		↓		Russell et al. (2009)
	Lithophyllum sp. (CCA)	Out ^{rETR}	277–438/377–930	14		↓		Russell et al. (2011)

(continued)

Table 3.1 (continued)

Effects on:	Species	Type of study	Experimental conditions		Response to:			References
			pCO_2 (μatm)	T (°C)	↑ T	↑ pCO_2	T×pCO_2	
	Corallina officinalis (ACA)	Lab rETR	384/1313/1935	15		—		Hofmann et al. (2012b)
	Arthrocardia corymbosa (ACA)	Lab Fv/Fm	145–415/410–1,130	11		—		Cornwall et al. (2013a)
Calcification								
	Polar and sub-polar							
	Lithothamnion glaciale (M)	Lab TA	387/819/974/1,568	7–9	↑	↓	—	Büdenbender et al. (2011)
	Lithothamnion glaciale (M)	Lab SEM	422/589/755/1,018	8		↓		Ragazzola et al. (2012)
	Lithothamnion glaciale (M)	Lab SEM	422/589/755/1,018	8		—		Ragazzola et al. (2013)
	Temperate							
	Cold-temperate							
	Bossiella orbiniana (ACA)	Lab 45Ca	170–2400	19		∩		Smith and Roth (1979)
	Phymatolithon calcareum (M)	Lab 45Ca	–	0–20	↑ or ∩			King and Schramm (1982)
	Corallina pilulifera (ACA)	Lab Ca	350/1600	20		↓		Gao et al. (1993)
	Lithothamnion corallioides (M)	Field TA	324–517 pers	10–17	↑			Martin et al. (2006)
	Corallina officinalis (ACA)	Lab TA	385/665/1,486	–		∩		Hofmann et al. (2012a)
	Ellisolandia elongata (ACA)	Lab TA	362/547/770/977	18		—		Egilsdottir et al. (2013)
	Lithothamnion glaciale (M)	Lab TA	498/1,081/2,278	10		↑		Kamenos et al. (2013)

	Ellisolandia elongata (ACA)	Lab TA	450/637/790/1,002	16		—		Noisette et al. (2013a)
	Lithophyllum incrustans (CCA)	Lab TA	450/637/790/1,002	16		↓		Noisette et al. (2013a)
	Lithothamnion corallioides (M)	Lab TA	450/637/790/1,002	16		↓		Noisette et al. (2013a)
	Lithothamnion corallioides (M)	Lab TA	380/550/750/1,000	13–19		↓		Noisette et al. (2013b)
	Warm-temperate							
	Lithophyllum cabiochae (CCA)	Lab BW	417/728	13–22/+3 °C	↑	—	↓	Martin and Gattuso (2009)
	Lithophyllum cabiochae (CCA)	Lab TA	417/728	13–22/+3 °C	↑ or —	↑ or —	↓ or —	Martin et al. (2013)
	Tropical and sub-tropical							
	Amphiroa foliacea (ACA)	Lab 14C	pH 7.2/8.3 [c]	28		—		Borowitzka (1981)
	Lithophyllum margaritae (M)	Lab AT, BW	–	10–30	↑			Steller et al. (2007)
	Porolithon onkodes (CCA)	Out BW	298/613/1,185	25–26/+3 °C	—	↓	↓↓	Anthony et al. (2008)
	Lithophyllum cf. *pallescens, Hydrolithon* sp. and, *Porolithon* sp. (R)a	Out BW	365/765	23–29		↓		Jokiel et al. (2008)
	Neogoniolithon sp. (CCA)	Lab BW	409/606/903/2,856	25		∩		Ries et al. (2009)
	Hydrolithon sp. (R)	Lab TA	400/900	25		↓		Semesi et al. (2009)
			1–1,400			↓		
	Corallina sessilis (ACA)	Lab TA	380/1000	23–27		↓		Gao and Zheng (2010)
	Hydrolithon onkodes (CCA)	Lab BW, TA	100–2,200	27		↓		Comeau et al. (2012)

(continued)

Table 3.1 (continued)

Effects on:	Species	Type of study	Experimental conditions		Response to:			References
			pCO_2 (μatm)	T (°C)	↑ T	↑ pCO_2	T×pCO_2	
	Hydrolithon onkodes (CCA)	Lab [BW]	420/530/830	26/+3 °C	—	↓	—	Johnson and Carpenter (2012)
	Hydrolithon onkodes (CCA)	Lab [BW]	262–2,141	27		↓		Comeau et al. (2013)
	Hydrolithon samoense (CCA)	Lab [SEM]	263/399/960	27		↓		Kato et al. (2013)
	Lithophyllum kotschyanum (CCA)	Lab [SEM]	263/399/960	27		↓		Kato et al. (2013)

Coralline algae are classified as articulated coralline algae (ACA), crustose coralline algae (CCA), maerl (M), and rhodoliths (R)
Studies are classified as laboratory-based experiments and benthic mesocosm studies (Lab), outdoor mesocosm studies under a natural light regime (Out), field experiments (Field) and Vent studies (Vent)
[pers], Personal data; –, information not provided
Responses are: ↑, increase; ↓, decrease; ↓↓ more-pronounced decrease; —, no effect; ∩, parabolic response
Health was determined from bleaching and/or necroses or from the ratio of variable (Fv) to maximal (Fm) quantum yield of photosystem II (Fv/Fm) greater than 0.57[d] being indicative of photosynthetically healthy coralline algae
Growth rates were determined from variations in wet or dry weight [W] and variations in length [L] (linear extension beyond red alizarin stain or difference between initial and final length)
Respiration and photosynthesis rates presented here were determined from measurements of oxygen and C_T exchanges. Photophysiology was also assessed through measurements of the optimal quantum yield of photosystem II (the ratio of variable to maximal fluorescence, Fv/Fm) [Fv/Fm] or the relative electron transport rate (rETR)[rETR] by using a pulse amplitude modulated (PAM) fluorometer
Calcification rates were determined from alkalinity anomaly[TA] or buoyant weight[BW] techniques as well as changes in Ca^{2+}concentration in the medium[Ca] or ^{45}Ca incorporation[45Ca] or ^{14}C incorporation[14C] in the thallus. Data on the degree of calcification of cell walls are also given in this section as the amount of $CaCO_3$ deposition within the cell wall (Ragazzola et al. 2012) or percent of calcified cell wall per whole cell (Kato et al. 2013) by using scanning electron microscope (SEM)[SEM]
[a]Rhodoliths consisting of a mixed CCA community including *Lithophyllum* cf. *pallescens*, *Hydrolithon* sp. and, *Porolithon* sp.
[b]When pCO_2 was not provided in the publication, it was estimated from pH and total alkalinity data
[c]Otherwise a pH range was provided
[d]The vent studies show significant decreases in crustose coralline algae from pH_T (on the total scale) 8.1–7.8 (corresponding to pCO_2 from *ca.* 300 to 950 *μ*atm). Lower pH_T were not considered here

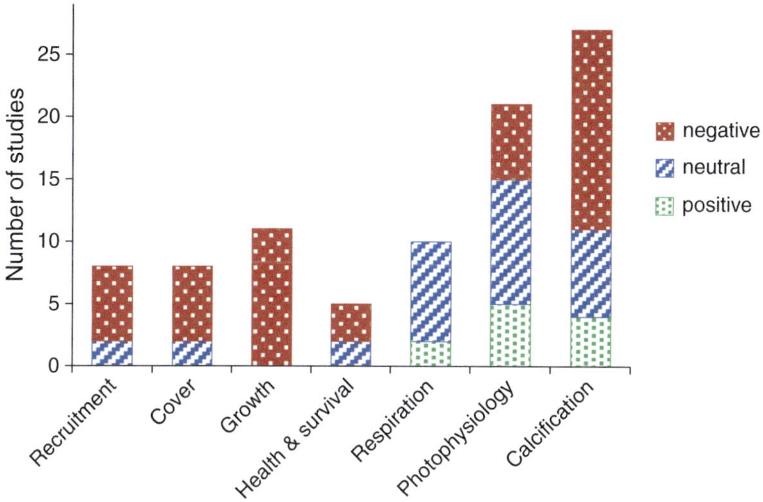

Fig. 3.4 Summary of the expected impacts of ocean acidification on coralline algae this century (based on studies in Table 3.1). Negative, neutral and positive effects correspond to decreases, no change, and increases in the processes, respectively. Photophysiology includes both photosynthesis and photosystem II efficiency and relative electron transport rate

3.2.4 Physiological Response of Coralline Algae to Climate Change and Ocean Acidification

Photosynthesis, respiration, and calcification in coralline algae are usually determined from measurements of oxygen, DIC and alkalinity in incubation chambers (Fig. 3.5). Most of these studies show that coralline algal calcification is negatively affected under elevated pCO_2 (Table 3.1) and that this effect is exacerbated by warming (Anthony et al. 2008). However, some work only shows a significant pCO_2 effect on calcification when this is combined with an increase in temperature (Martin and Gattuso 2009) and some experiments have shown a positive effect of moderate increases in pCO_2 (Smith and Roth 1979; Ries et al. 2009; Martin et al. 2013; Kamenos et al. 2013). Such responses may be related to the ability of the algae to maintain an elevated pH at the site of calcification despite reduced external pH to facilitate $CaCO_3$ precipitation (Borowitzka 1987; Ries et al. 2009; Hurd et al. 2011; Cornwall et al. 2013b).

The response of coralline photosynthesis to increased pCO_2 is also variable among species with negative, neutral, positive and parabolic responses (Table 3.1) that can depend on the light levels used (Martin et al. 2013). Studies investigating the effect of increased pCO_2 on respiration showed no response for most species (Fig. 3.4). Very few studies of coralline algae have investigated photosynthesis, respiration, and calcification all together yet we know that these processes are complex and tightly linked. Photosynthesis affects calcification through the formation of the fibrous organic matrix of the cell walls where the nucleation of calcite crystals is thought to occur (Borowitzka 1981). In addition, both photosynthesis and

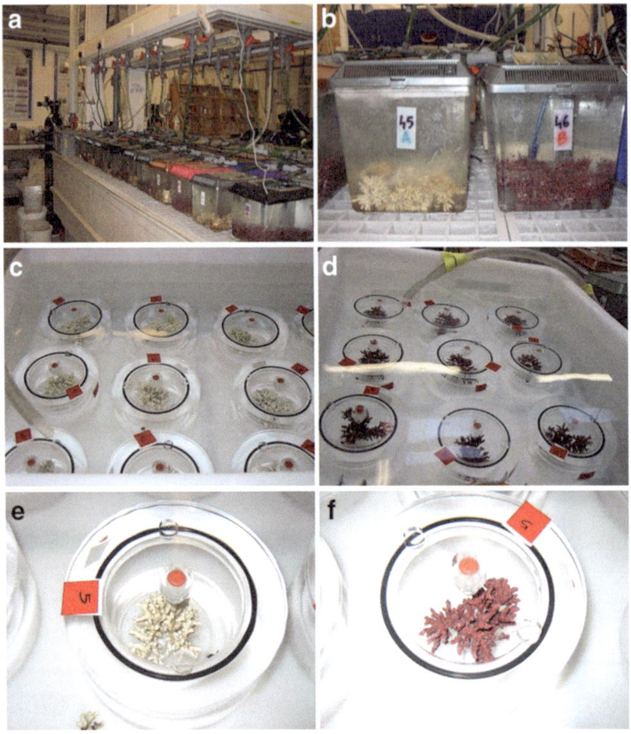

Fig. 3.5 Experimental set-up (**a**) with replicated aquaria (**b**) supplied with seawater at ambient or elevated pCO_2 containing alive or dead *Lithothamnion glaciale* maerl that were incubated in Perspex chambers for metabolic fluxes analyses (**c**, **e**, dead maerl and **d**, **f**, live maerl) (Photo S. Martin (Experiments performed at Plymouth Marine Laboratory))

respiration affect calcification through changes in pH that occur in the cell walls at the site of calcification (Smith and Roth 1979; Gao et al. 1993) but also in the diffusion boundary layer between the algal surface and external seawater (Hurd et al. 2011). Photosynthesis (or respiration) increases (or decreases) pH and thereby increases (or decreases) $CaCO_3$ saturation state, promoting (or hindering) the precipitation of $CaCO_3$. Coralline algae are able to maintain calcification in the dark even at the relatively low pH values generated by respiration. Digby (1977) and Hofmann et al. (2012b) postulate that carbonic anhydrase may also play a role in the calcification of coralline algae by catalysing the conversion of CO_2 into HCO_3^- and then CO_3^{2-}. The stimulation of carbonic anhydrase activity could help prevent a decrease in calcification at elevated pCO_2 as reported for the Mediterranean crustose coralline alga *Lithophyllum cabiochae* (Martin et al. 2013). However, carbonic anhydrase is also used by photosynthesis to convert HCO_3^- to CO_2. The maintenance or enhancement of calcification rates under elevated pCO_2 in *L. cabiochae* may thus be detrimental to photosynthesis, as indicated by reduced photosynthesis under elevated pCO_2 (Martin et al. 2013).

The physiological response of maerl to warming and ocean acidification has been investigated in *Lithothamnion corallioides* which is a temperate species (Noisette et al. 2013b) and *Lithothamnion glaciale* which is a cold-water species (Büdenbender et al. 2011; Kamenos et al. 2013). In *L. corallioides* diel (24 h) calcification decreased by 50 % at 750 µatm and 80 % at 1,000 µatm, and in *L. glaciale* the thalli dissolve under elevated pCO_2 (Büdenbender et al. 2011). Kamenos et al. (2013) found that at night *L. glaciale* calcified in the control treatment but dissolved in the low pH treatment. In both species calcification is less affected by pCO_2 in the light that in the dark, although in *L. glaciale* upregulated calcification occurs at low pH cf. control treatments during the day (Noisette et al. 2013b; Kamenos et al. 2013). High latitude maerl beds are at risk as surface waters are becoming more corrosive which is expected to cause dissolution of these habitats (Büdenbender et al. 2011). Tropical rhodolith beds may also be affected since in multispecies rhodoliths, made up of *Lithophyllum*, *Hydrolithon* and *Porolithon* spp., there is a decrease calcification by 20–250 % between control and acidified conditions (Jokiel et al. 2008; Semesi et al. 2009).

3.2.5 Response of Early Life Stages to Ocean Acidification and Warming

Although fragmentation is the main source of new thalli in beds of rhodoliths and maerl, they are initiated through recruitment from spores (Foster 2001). Unfortunately, there is mounting evidence that ocean acidification and warming will have negative impacts upon the recruitment and growth of early life history stages of coralline algae (Agegian 1985; Jokiel et al. 2008; Kuffner et al. 2008; Russell et al. 2009; Porzio et al. 2013; Bradassi et al. 2013; Table 3.1). Kuffner et al. (2008) found that impacts on settlement led to 90 % lower tropical crustose coralline cover at pH 7.9 than at pH 8.2. Some coralline algal species show a reduction in reproductive structures in areas with naturally high pCO_2 conditions (Porzio et al. 2011) and germination of spores in the laboratory reveal developmental abnormalities and increased mortality in acidified conditions. Bradassi et al. (2013) found that in acidified conditions the germlings of an intertidal species of coralline algae were able to fight dissolution by up-regulating their rates of calcification; this must increase energy costs and helps explain why coralline algae are replaced by fleshy algae in naturally high pCO_2 conditions (Hall-Spencer et al. 2008; Porzio et al. 2013).

3.2.6 Ecosystem Level Responses

Most research into the effects of ocean warming and acidification has involved relatively short-term (1 year or less) experiments on single species which makes it difficult to scale-up and predict long-term effects at the community and ecosystem levels (Russell et al. 2013). To tackle this, areas with naturally high CO_2 (and/or low pH and low calcium carbonate saturation states) are being used to investigate which organisms can tolerate the long-term consequences of ocean acidification and reveal how communities of primary producers respond (Hall-Spencer et al. 2008; Johnson et al. 2012; Inoue et al. 2013). As with laboratory and mesocosm experiments, the vent systems cannot accurately mimic future ocean conditions; acidified areas are open systems so corallines can recruit from unaffected habitats and grazing fish can swim in and out, these systems typically have large variations in carbonate chemistry, and in some cases there are confounding factors that may mask or amplify the effects of CO_2. Some efforts have been made to determine the combined effects of warming and acidification at such sites, although such work has so far been restricted to corals, molluscs and bryozoans (Rodolfo-Metalpa et al. 2011). Some volcanic vents are proving to be particularly useful 'natural laboratories' for the study of ocean acidification as they reveal tipping points in recruitment, growth, survival and species interactions along pCO_2 gradients (Porzio et al. 2011, 2013). Many species of microalgae, macroalgae and seagrasses are remarkably tolerant of long-term exposures to high and variable carbon dioxide levels at tropical and temperate CO_2 seeps (Johnson et al. 2012, 2013; Russell et al. 2013). That they tolerate these conditions does not mean that they will necessarily thrive; seagrasses for example lose the ability to defend themselves against herbivores and become over-run by competing species of invasive seaweeds (Arnold et al. 2012).

One clear pattern in studies at multiple natural CO_2 gradients is that the Corallinales are adversely affected (Porzio et al. 2011; Fabricius et al. 2011) which is a particular concern since this group of organisms underpin maerl, rhodolith habitat production (Hall-Spencer et al. 2010). At Mediterranean CO_2 seeps the Corallinales dissolve at low carbonate saturation states (Martin et al. 2008) and are outcompeted by heterokont algae as CO_2 levels increase (Fig. 3.6). Epilithic diatoms, Dictyotales (e.g. *Dictyota* spp., *Padina* spp.) and Fucales (e.g. *Cystoseira* spp., *Sargassum* sp.) thrive as CO_2 levels increase although *Padina* spp. loose calcification (Porzio et al. 2011, 2013; Johnson et al. 2012). At the ecosystem level the coralligenous habitats loose biodiversity as carbonate levels fall along transects of increasing levels of CO_2; benthic recruitment is disrupted in calcified organisms (Cigliano et al. 2010); some organisms such as soft corals and anemones are tolerant but many are physiologically unable to cope with the increased CO_2 levels (Suggett et al. 2012; Calosi et al. 2013; Inoue et al. 2013). Reefs formed by an association between coralline algae and vermetid molluscs are expected to become extinct this century unless CO_2 emissions are reduced and emergency conservation measures are taken (Milazzo et al. 2014).

Additional insights into the community-level effects of warming and ocean acidification are beginning to come from longer-term multispecies laboratory experiments (Hale et al. 2011). These also demonstrate that future increases in CO_2 are likely to have strong negative effects on coralline algae and positive or neutral effects on noncalcified seaweeds both directly via improved growth and indirectly via reduced consumption by calcified herbivores such as sea urchins (Connell and Russell 2010; Diaz-Pulido et al. 2011; Cornwall et al. 2012; Olischläger et al. 2012;

a

b

Fig. 3.6 Sea urchin and coralline algae dominated rocky shore at (a) ambient CO_2 (Photograph by David Liittschwager, National Geographic) and (b) the loss of coralline algae and proliferation of diatoms and phaeophytes at a CO_2 seep off Ischia, Italy (Photograph by Luca Tiberti, Associazione Nemo)

Roleda et al. 2012; Olabarria et al. 2013). It is clear that the impact of climate change and ocean acidification on rhodolith/maerl beds will depend on the combined influence of direct environmental impacts on individual species and indirect effects mediated by changes in interspecific interactions (Harley et al. 2012). For example, ocean acidification may disrupt invertebrate recruitment to maerl/rhodolith beds by altering chemical settlement cues associated with crustose coralline algae, as shown with coral recruits (Doropoulos and Diaz-Pulido 2013).

Ocean warming and acidification degrade the habitat complexity of coralline algal sediments with profound effects since these algae induce settlement and recruitment of numerous invertebrates. The rapid rate of physicochemical change is predicted to cause a shift from calcareous to fleshy algal-dominated coastal ecosystems (Hall-Spencer et al. 2008; Hoegh-Guldberg et al. 2007; Anthony et al. 2011; Diaz-Pulido et al. 2011). Given that rhodolith/maerl habitats usually only form in waters that have high carbonate saturation states, the spread of low saturation state waters in upwelling areas, lowered salinity water masses and in polar waters, are likely to reduce their habitat complexity and thus cause biodiversity loss. Beds of unattached coralline algae provide nursery grounds and habitat for numerous commercial species of invertebrates and fishes (Kamenos et al. 2004a, b) so a decline in these habitats due to dissolution is also likely to have serious consequences for society and economy through the impact on fisheries.

3.2.7 Impact on Global C and $CaCO_3$ Budget

Rhodolith/maerl beds are a significant component of carbon and carbonate cycles in shallow coastal ecosystems, being major contributors to CO_2 fluxes through high community photosynthesis and respiration (Martin et al. 2005, 2007) and through high $CaCO_3$ production and dissolution (Martin et al. 2006, 2007). The habitats formed by coralline algae are expected to be degraded by ocean acidification and warming as they have slow growth rates and are easily corroded due to their soluble high Mg-calcite skeletons. Any major decline in coralline algae would have dramatic consequences since they cover vast areas of rock and can form very extensive beds on sediments (Amado-Filho et al. 2012). Changes in the balance between algal carbonate production and dissolution induced by elevated pCO_2 and temperature in maerl and rhodolith beds may have major implications for carbon dynamics in coastal systems by affecting the carbonate chemistry of the water column and the ability of the oceans to take up atmospheric CO_2 (Andersson et al. 2005).

In temperate maerl beds, *L. corallioides* calcification is expected to decrease by up to 80 % at 1,000 µatm pCO_2, relative to present day conditions of pCO_2 (Noisette et al. 2013b). This would cause a reduction in $CaCO_3$ precipitation from a current value of *ca.* 500 g $CaCO_3$ m^{-2} year^{-1} (Martin et al. 2007) to less than 100 g $CaCO_3$ m^{-2} year^{-1} in the near future under 1,000 µatm pCO_2 (Noisette et al. 2013b). The net calcification in maerl/rhodolith beds could even become negative because the production of $CaCO_3$ will be exceeded by its dissolution (Martin and Gattuso 2009).

Arctic maerl beds are at particular at risk due to seawater becoming corrosive to algal carbonate since live specimens of polar *L. glaciale* are corroded under elevated pCO_2 (Büdenbender et al. 2011). Dissolution also occurs in the dark in live maerl under elevated pCO_2 in temperate maerl species (Noisette et al. 2013b; Kamenos et al. 2013). As maerl beds are also composed of dead thalli, dissolution is likely to strongly increase; dissolution of dead *L. glaciale* thalli increases by more than ten-fold from ambient condition to elevated pCO_2 of *ca.* 1,100 µatm (Kamenos et al. 2013). Precipitation and dissolution of $CaCO_3$ contribute to the global CO_2 balance through shifts in the seawater carbonate equilibrium. The precipitation of one mole of $CaCO_3$ ($Ca^{2+} + 2HCO_3^- \leftrightarrow CaCO_3 + H_2O + CO_2$) releases *ca.* 0.6 mol of CO_2 in seawater (Ware et al. 1992). In that way, the CO_2 released by calcification in maerl/rhodolith beds will also be reduced. These changes will thus have major implications for both carbon and carbonate budgets in coastal systems.

3.2.8 Acclimation and Adaptation

The coralline fossil record for the past 300 Million years shows they have been able to survive past mass extinctions including periods of very high CO_2 (Wood 1999). However, the present rate of ocean change may be too rapid for genetic adaptation of habitat-forming coralline algae. Work at CO_2 vents that are 100s of years old indicates that corallines have a limited capacity to acclimate to ocean acidification, since most species are intolerant of chronic exposures to increases in pCO_2 levels predicted this century and those that can survive are outcompeted by fleshy algae (Martin et al. 2008; Porzio et al. 2011; Fabricius et al. 2011). Localised seeps with high CO_2 levels are not well suited to the study of coralline algal adaptation since they are open to colonization by algal spores from outside the acidified areas and so presently we have no knowledge about the ability of rhodolith/maerl species to adapt to present day rates of warming and acidification. We find it striking, however, that rhodolith/maerl beds are common along the Atlantic seaboard of the Americas, where carbonate saturation states are high, but they are scarce along the Pacific seaboard where carbonate saturation states are low.

Adaptation is the evolutionary response of a population over multiple generations to environmental changes but, according the current magnitude and rate of ocean warming and acidification, the potential for evolutionary adaptation is limited in organisms with long generation times such as coralline algae. However, coralline algae may have the potential to adjust to modified environment within their lifetime (acclimation). In particular, the ability of organisms to tolerate significant temperature and/or pCO_2 fluctuations may be a result of adaptation (a genetic trait shared by the population) and/or acclimation (owing to phenotypic plasticity of the individual). Organisms surviving in highly variable environments are likely to be more robust. For instance, Egilsdottir et al. (2013) reported that coralline algae inhabiting variable environments where pH/pCO_2 fluctuates naturally are likely to exhibit fewer negative responses to elevated pCO_2 than those inhabiting relatively stable environments, supporting the assumption of a greater resilience of organisms acclimated and/or adapted to highly variable pH/pCO_2 environments of future ocean acidification.

Coralline algae may be able to tolerate ocean acidification through changes in the composition of their skeletons (Agegian 1985; Ries et al. 2009; Egilsdottir et al. 2013). A decrease in mMg/Ca ratio would confer resilience to elevated pCO_2 as this would lower the solubility of their skeletons. However, it is questionable whether mineralogical plasticity associated with a decrease in Mg incorporation will help since warming simultaneously increases Mg incorporation (Agegian 1985). Dolomite ($MgCO_3$) rich crustose coralline algae have six to tenfold lower rates of dissolution than predominantly Mg-calcite species (Nash et al. 2013). Dolomite-rich crustose coralline algae are widespread in shallow wave-exposed habitats in the tropics but have not been recorded in cooler waters suggesting that its formation may be constrained by temperature. Thus dolomite intracellular calcification may confer an advantage on tropical corallines but not those found at higher latitudes (Nash et al. 2011). It is clear that elevated pCO_2 weakens the skeletal structure in the high latitude maerl species *L. glaciale* (Ragazzola et al. 2012).

Society can also adapt help mitigate adverse effects of warming and acidification. In the NE Atlantic steps have been taken to remove damaging dredging, fishing and aquaculture practices to protect maerl beds in a network of protected areas (Chap. 10). If this approach is adopted world-wide the reduced pressure on coralline algal systems would help increase their resilience to the adverse effects of acidification and warming.

3.3 Conclusions

Beds of unattached coralline algae are scarce in waters with naturally low saturation states of carbonate and are likely to decline in the near future as corrosive waters spread due to ocean acidification. This will lead to a degradation of these ecosystems, reducing habitat complexity and their associated biodiversity. Although responses to ocean warming and acidification are variable among species, beds of maerl and rhodoliths are at risk at high latitudes where seawater is becoming corrosive to their high Mg-calcite skeletons. Loss of both live and dead maerl is expected to spread to lower latitudes over the century. Work on the effects of ocean warming and acidification indicates that certain fleshy algae and non-calcified animals may benefit and outcompete corallines and those calcified animals that characterise these habitats (Fig. 3.7).

Given the importance of coralline algae, and the scale of the repercussions if they are adversely impacted by rising CO_2 levels, concerted efforts are required to test whether our concerns are justified. More detailed examinations of the physiology/ecology of corallines need to be undertaken as we have an incomplete understanding of how ocean acidification and temperature influence gross calcification, growth, dissolution and competitive interactions between seaweeds. Given that society

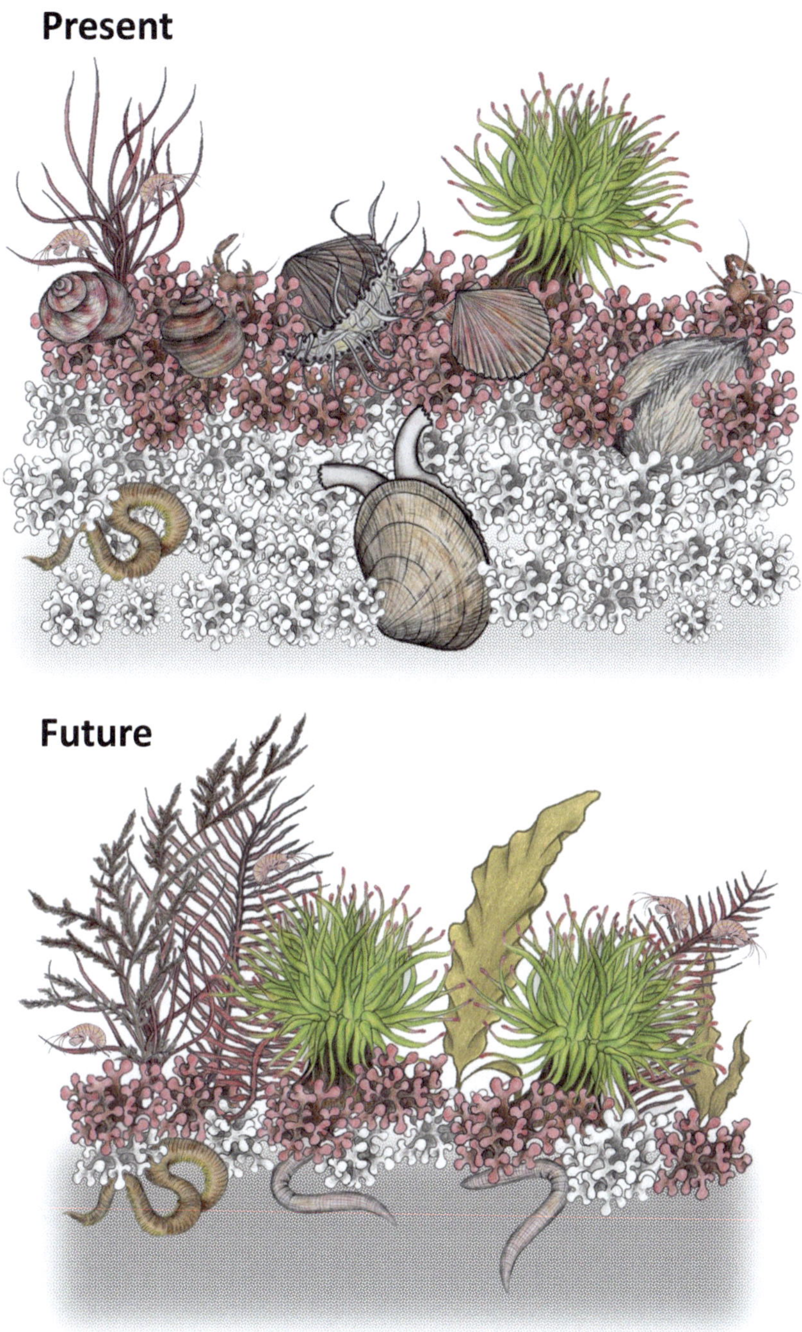

Fig. 3.7 Predicted changes to a typical maerl bed with rising temperature and CO_2 levels. Present day maerl beds with alive (*above*) and dead (*below*) unattached corallines are inhabited by a rich diversity of calcifying (e.g. gastropods, bivalves, decapods, echinoids) and non-calcifying species. As waters become corrosive to carbonate future maerl beds are expected to be degraded, with the loss of habitat complexity and biodiversity, although certain fleshy macroalgae and non-calcified fauna (e.g. anemones, polychaetes) are expected to proliferate (Drawing by Sophie Martin)

relies upon coralline algae for ecosystems services, such as providing settlement cues or providing habitat for commercially important species, we recommend that biological monitoring programmes begin as soon as possible since changes may be occurring at a rate that will exceed the environmental niches of numerous coralline alga taxa, testing their capacities for acclimation and genetic adaptation.

3.3.1 Credits

IPCC 2013: Summary for Policymakers. In Climate Change 2013: The Physical Science Basis. Working Group I Contribution to the Fifth Assessment Report of the Intergovernmental Panel on Climate Change, Figure SPM.7 (a) and (c); Figure SPM.8 (a) and (d). [Cambridge University Press (2013)].

Acknowledgements The authors are grateful to the reviewer for valuable comments and suggestions on a previous version of this manuscript. This work is a contribution to the "European Project on Ocean Acidification" (EPOCA) and EU'Mediterranean Sea Acidification under a changing climate' project (MedSeA) which received funding from the European Community (grant agreements 211384 and 265103).

References

Adey WH (1973) Temperature control of reproduction and productivity in a subarctic coralline alga. Phycologia 12:111–118

Agegian CR (1985) The biogeochemical ecology of *Porolithon gardineri* (Foslie). PhD thesis, University of Hawaii, Honolulu

Amado-Filho GM, Moura RL, Bastos AC, Salgado LT, Sumida PY, Guth AZ, Francini-Filho RB, Pereira-Filho GH, Abrantes DP, Brasileiro PS, Bahia RG, Leal RN, Kaufman L, Kleypas JA, Farina M, Thompson FL (2012) Rhodolith beds are major $CaCO_3$ bio-factories in the tropical South West Atlantic. PLoS ONE 7:e35171

Andersson AJ, MacKenzie FT, Lerman A (2005) Coastal ocean carbonate ecosystems in the high CO_2 world of the Anthropocene. Am J Sci 305:875–918

Anthony KRN, Kline DI, Diaz-Pulido G, Dove S, Hoegh-Guldberg O (2008) Ocean acidification causes bleaching and productivity loss in coral reef builders. Proc Natl Acad Sci U S A 105:17442–17446

Anthony KRN, Kleypas JA, Gattuso J-P (2011) Coral reefs modify their seawater carbon chemistry – implications for impacts of ocean acidification. Glob Chang Biol 17:3655–3666

Arnold T, Mealey C, Leahey H et al (2012) Ocean acidification and the loss of protective phenolics in seagrasses. PLoS One 7(4):e35107

Blake C, Maggs CA (2003) Comparative growth rates and internal banding periodicity of maerl species (Corallinales, Rhodophyta) from northern Europe. Phycologia 42:606–612

Borowitzka MA (1981) Photosynthesis and calcification in the articulated coralline red algae *Amphiroa anceps* and *Amphiroa foliacea*. Mar Biol 62:17–23

Borowitzka MA (1987) Calcification in algae: mechanism and the role of metabolism. Crit Rev Plant Sci 6:1–45

Bradassi F, Cumani F, Bressan G, Dupont S (2013) Early reproductive stages in the crustose coralline alga *Phymatolithon lenormandii* are strongly affected by mild ocean acidification. Mar Biol 160:2261–2269

Büdenbender J, Riebesell U, Form A (2011) Calcification of the Arctic coralline red algae *Lithothamnion glaciale* in response to elevated CO_2. Mar Ecol Prog Ser 441:79–87

Calosi P, Rastrick SPS, Graziano M, Thomas SC, Baggini C, Carter HA, Hall-Spencer JM, Milazzo M, Spicer JI (2013) Distribution of sea urchins living near shallow water CO_2 vents is dependent upon species acid–base and ion-regulatory abilities. Mar Pollut Bull 73:470–484

Canadell JG, Le Quéré C, Raupach MR, Field CB, Buitenhuis ET, Ciais P, Conway TJ, Gillett NP, Houghton RA, Marland G (2007) Contributions to accelerating atmospheric CO_2 growth from economic activity, carbon intensity, and efficiency of natural sinks. Proc Natl Acad Sci U S A 104:18866–18870

Cigliano M, Gambi MC, Rodolfo Metalpa R, Patti FP, Hall-Spencer JM (2010) Effects of ocean acidification on invertebrate settlement at volcanic CO_2 vents. Mar Biol 157:2489–2502

Comeau S, Carpenter RC, Edmunds PJ (2012) Coral reef calcifiers buffer their response to ocean acidification using both bicarbonate and carbonate. Proc R Soc B 280:20122374

Comeau S, Edmunds PJ, Spindel NB, Carpenter RC (2013) The responses of eight coral reef calcifiers to increasing partial pressure of CO_2 do not exhibit a tipping point. Limnol Oceanogr 58:388–398

Connell SD, Russell BD (2010) The direct effects of increasing CO_2 and temperature on non-calcifying organisms: increasing the potential for phase shifts in kelp forests. Proc R Soc Lond B Biol Sci 277:1409–1415

Cornwall CE, Hepburn CD, Pritchard D, Currie KI, McGraw CM, Hunter KA, Hurd CL (2012) Carbon-use strategies in macroalgae: differential responses to lowered pH and implications for ocean acidification. J Phycol 48:137–144

Cornwall CE, Hepburn CD, McGraw CM, Currie KI, Pilditch CA, Hunter KA, Boyd PW, Hurd CL (2013a) Diurnal fluctuations in seawater pH influence the response of a calcifying macroalga to ocean acidification. Proc R Soc B 280:20122201

Cornwall CE, Hepburn CD, Pilditch CA, Hurd CL (2013b) Concentration boundary layers around complex assemblages of macroalgae: implications for the effects of ocean acidification on understory coralline algae. Limnol Oceanogr 58:121–130

Diaz-Pulido G, Gouzezo M, Tilbrook B, Dove S, Anthony K (2011) High CO_2 enhances the competitive strength of seaweeds over corals. Ecol Lett 14:156–162

Diaz-Pulido G, Anthony KRN, Kline DI, Dove S, Hoegh-Guldberg O (2012) Interactions between ocean acidification and warming on the mortality and dissolution of coralline algae. J Phycol 48:32–39

Digby P (1977) Photosynthesis and respiration in the coralline algae, *Clathromorphum circumscriptum* and *Corallina officinalis* and the metabolic basis of calcification. J Mar Biol Assoc U K 57:1111–1124

Doropoulos C, Ward S, Diaz-Pulido G, Hoegh-Guldberg O, Mumby PJ (2012) Ocean acidification reduces coral recruitment by disrupting intimate larval-algal settlement interactions. Ecol Lett 15:338–346

Doropoulos C, Diaz-Pulido G (2013) High CO_2 reduces the settlement of a spawning coral on three common species of crustose coralline algae. Mar Ecol Prog Ser 475:93–99

Egilsdottir H, Noisette F, Noel LMLJ, Olafsson J, Martin S (2013) Effects of pCO_2 on physiology and skeletal mineralogy in a tidal pool coralline alga *Corallina elongata*. Mar Biol 160:2103–2112

Fabricius KE, Langdon C, Uthicke S, Humphrey C, Noonan S, De'ath G, Okazaki R, Muehllehner N, Glas MS, Lough JM (2011) Losers and winners in coral reefs acclimatized to elevated carbon dioxide concentrations. Nat Clim Chang 1:165–169

Foster MS (2001) Rhodoliths: between rocks and soft places. J Phycol 37:659–667

Gao K, Aruga Y, Asada K, Ishihara T, Akano T, Kiyohara M (1993) Calcification on the articulated coralline alga *Corallina pilulifera*, with special reference to the effect of elevated CO_2 concentration. Mar Biol 117:129–132

Gao KS, Zheng YQ (2010) Combined effects of ocean acidification and solar UV radiation on photosynthesis, growth, pigmentation and calcification of the coralline alga *Corallina sessilis* (Rhodophyta). Glob Chang Biol 16:2388–2398

Hale R, Calosi P, McNeill L, Mieszkowska N, Widdicombe S (2011) Predicted levels of future ocean acidification and temperature rise could alter community structure and biodiversity in marine benthic communities. Oikos 120:661–674

Hall-Spencer JM, Rodolfo-Metalpa R, Martin S, Ransome E, Fine M, Turner SM, Rowley SJ, Tedesco D, Buia MC (2008) Volcanic carbon dioxide vents show ecosystem effects of ocean acidification. Nature 454:96–99

Hall-Spencer JM, Kelly J, Maggs CA (2010) Background document for maerl beds. OSPAR Commission, London, Publication 491/2010 36 pp. ISBN 978-1-907390-32-6

Harley CDG, Anderson KM, Demes KW, Jorve JP, Kordas RL, Coyle TA, Graham MH (2012) Effects of climate change on global seaweed communities. J Phycol 48:1064–1078

Hoegh-Guldberg O, Mumby PJ, Hooten AJ, Steneck RS, Greenfield P, Gomez E, Harvell CD, Sale PF, Edwards AJ, Caldeira K, Knowlton N, Eakin CM, Iglesias-Prieto R, Muthiga N, Bradbury RH, Dubi A, Hatziolos ME (2007) Coral reefs under rapid climate change and ocean acidification. Science 318:1737–1742

Hepburn CD, Pritchard DW, Cornwall CE, McLeod RJ, Beardall J, Raven JA, Hurd CL (2011) Diversity of carbon use strategies in a kelp forest community: implications for a high CO_2 ocean. Glob Chang Biol 17:2488–2497

Hofmann LC, Straub S, Bischof K (2012a) Competition between calcifying and noncalcifying temperate marine macroalgae under elevated CO_2 levels. Mar Ecol Prog Ser 464:89–105

Hofmann LC, Yildiz G, Hanelt D, Bischof K (2012b) Physiological responses of the calcifying rhodophyte, *Corallina officinalis* (L.), to future CO_2 levels. Mar Biol 159:783–792

Hurd CL, Cornwall CE, Currie KI, Hepburn CD, McGraw CM, Hunter KA, Boyd P (2011) Metabolically-induced pH fluctuations by some coastal calcifiers exceed projected 22nd century ocean acidification: a mechanism for differential susceptibility? Glob Chang Biol 17:3254–3262

Ichiki S, Mizuta H, Yasui H, Yamamoto H (2001) Effects of irradiance and water temperature on the photosynthesis and growth of the crustose coralline alga *Lithophyllum yessoense* Foslie (Corallinales, Rhodophyceae). Bull Fischeries Sci Hokkaido Univ 52:103–109

Inoue S, Kayanne H, Yamamoto S, Kurihara K (2013) Spatial community shift from hard to soft corals in acidified water. Nat Clim Chang 3:683–687

IPCC (2013) Summary for policymakers. In Climate Change 2013: the physical science basis. Working group I contribution to the fifth assessment report of the intergovernmental panel on climate change. Cambridge University Press (in press)

Jokiel PL, Rodgers KS, Kuffner IB, Andersson AJ, Cox EF, Mackenzie FT (2008) Ocean acidification and calcifying reef organisms: a mesocosm investigation. Coral Reefs 27:473–483

Johnson MD, Carpenter RC (2012) Ocean acidification and warming decrease calcification in the crustose coralline alga *Hydrolithon onkodes* and increase susceptibility to grazing. J Exp Mar Biol Ecol 434–435:94–101

Johnson VR, Russell BD, Fabricius KE, Brownlee C, Hall-Spencer JM (2012) Temperate and tropical brown macroalgae thrive, despite decalcification, along natural CO_2 gradients. Glob Chang Biol 18:2792–2803

Johnson VR, Brownlee C, Rickaby REM, Graziano M, Milazzo M, Hall-Spencer JM (2013) Responses of marine benthic microalgae to elevated CO_2. Mar Biol 160:1813–1824

Kamenos NA, Burdett HL, Aloisio E, Findlay HS, Martin S, Longbone C, Dunn J, Widdicombe S, Calosi P (2013) Coralline algal structure is more sensitive to rate, rather than the magnitude, of ocean acidification. Glob Chang Biol 19:3621–3628

Kamenos NA, Moore PG, Hall-Spencer JM (2004a) Small-scale distribution of juvenile gadoids in shallow inshore waters; what role does maerl play? ICES J Mar Sci 61:422–429

Kamenos NA, Moore PG, Hall-Spencer JM (2004b) Nursery-area function of maerl grounds for juvenile queen scallops *Aequipecten opercularis* and other invertebrates. Mar Ecol Prog Ser 274:183–189

Kato A, Hikami M, Kumagai NH, Suzuki A, Nojiri Y, Saka K (2013) Negative effects of ocean acidification on two crustose coralline species using genetically homogeneous samples. Mar Environ Res. doi:10.1016/j.marenvres.2013.10.010, in press

King RJ, Schramm W (1982) Calcification in the maerl coralline alga *Phymatolithon calcareum*: effects of salinity and temperature. Mar Biol 70:197–204

Koch M, Bowes G, Ross C, Zhang XH (2012) Climate change and ocean acidification effects on seagrasses and marine macroalgae. Glob Chang Biol 19:103–132

Kroeker KJ, Kordas RL, Crim RN, Singh GG (2010) Meta-analysis reveals negative yet variable effects of ocean acidification on marine organisms. Ecol Lett 13(11):1419–1434

Kroeker KJ, Micheli F, Gambi MC (2013) Ocean acidification causes ecosystem shifts via altered competitive interactions. Nat Clim Chang 3:156–159

Kübler JE, Davison IR, Yarish C (1991) Photosynthetic adaptation to temperature in the red algae *Lomentaria baileyana* and *Lomentaria orcadensis*. Br Phycol J 26:9–19

Kuffner IB, Andersson AJ, Jokiel PL, Rodgers KS, Mackenzie FT (2008) Decreased abundance of crustose coralline algae due to ocean acidification. Nat Geosci 1:114–117

Lüning K (1990) Seaweeds. Their environment, biogeography, and ecophysiology. Wiley Interscience, London, 527 pp

McGraw CM, Cornwall CE, Reid MR, Currie KI, Hepburn CD, Boyd P, Hurd CL, Hunter KA (2010) An automated pH-controlled culture system for laboratory-based ocean acidification experiments. Limnol Oceanogr Methods 8:686–694

Martin S, Clavier J, Guarini J-M, Chauvaud L, Hily C, Grall J, Thouzeau G, Jean F, Richard J (2005) Comparison of *Zostera marina* and maerl community metabolism. Aquat Bot 83(3):161–174

Martin S, Castets MD, Clavier J (2006) Primary production, respiration and calcification of the temperate free-living coralline alga *Lithothamnion corallioides*. Aquat Bot 85:121–128

Martin S, Clavier J, Chauvaud L, Thouzeau G (2007) Community metabolism in temperate maerl beds. I. Carbon and carbonate fluxes. Mar Ecol Prog Ser 335:19–29

Martin S, Cohu S, Vignot C, Zimmerman G, Gattuso J-P (2013) One-year experiment on the physiological response of the Mediterranean crustose coralline alga, *Lithophyllum cabiochae*, to elevated pCO_2 and temperature. Ecol Evol 3(3):676–693

Martin S, Gattuso JP (2009) Response of Mediterranean coralline algae to ocean acidification and elevated temperature. Glob Chang Biol 15:2089–2100

Martin S, Rodolfo-Metalpa R, Ransome E, Rowley S, Buia MC, Gattuso JP, Hall-Spencer J (2008) Effects of naturally acidified seawater on seagrass calcareous epibionts. Biol Lett 4:689–692

Milazzo M, Rodolfo-Metalpa R, Chan VBS, Fine F, Alessi C, Thiyagarajan V, Hall-Spencer JM, Chemello R (2014) Ocean acidification impairs vermetid reef recruitment. Sci Rep 4:4189

Nash MC, Opdyke BN, Troitzsch U, Russell BD, Adey WH, Kato A, Diaz-Pulido G, Brent C, Gardner M, Prichard J, Kline DI (2013) Dolomite-rich coralline algae in reefs resist dissolution in acidified conditions. Nat Clim Chang 3:268–272

Nash MC, Troitzsch U, Opdyke BN, Trafford JM, Russell BD, Kline DI (2011) First discovery of dolomite and magnesite in living coralline algae and its geobiological implications. Biogeosciences 8:3331–3340

Noisette F, Egilsdottir H, Davoult D, Martin S (2013a) Physiological responses of three temperate coralline algae from contrasting habitats to near-future ocean acidification. J Exp Mar Biol Ecol 448:179–187

Noisette F, Duong G, Six C, Davoult D, Martin S (2013b) Effects of elevated pCO_2 on the metabolism of a temperate rhodolith *Lithothamnion corallioides* grown under different temperatures. J Phycol 49:746–757

Olabarria C, Arenas F, Viejo RM, Gestoso I, Vaz-Pinto F, Incera M, Rubal M, Cacabelos E, Veiga P, Sobrino C (2013) Response of macroalgal assemblages from rockpools to climate change: effects of persistent increase in temperature and CO_2. Oikos 122:1065–1079

Olischläger M, Bartsch I, Gutow L, Wiencke C (2012) Effects of ocean acidification on different life-cylce stages of the kelp *Laminaria hyperborea* (Phaeophyceae). Bot Mar 55:511–525

Peña V, Barbara I, Grall J, Maggs CA, Hall-Spencer JM (2014) The diversity of seaweeds on maerl in the NE Atlantic. Mar Biodivers. doi:10.1007/s12526-014-0214-7
Porzio L, Buia MC, Hall-Spencer JM (2011) Effects of ocean acidification on macroalgal communities. J Exp Mar Biol Ecol 400:278–287
Porzio L, Garrard SL, Buia MC (2013) The effects of ocean acidification on early algal colonization stages at natural CO_2 vents. Mar Biol 160:2247–2259
Ragazzola F, Foster LC, Form AU, Anderson PSL, Hansteen TH, Fietzke J (2012) Ocean acidification weakens the structural integrity of coralline algae. Glob Chang Biol 18:2804–2812
Ragazzola F, Foster LC, Form AU, Büscher J, Hansteen TH, Fietzke J (2013) Phenotypic plasticity of coralline algae in a high CO_2 world. Ecol Evol 3(10):3436–3446
Raven JA, Beardall J (2003) Carbon acquisition mechanisms of algae: carbon dioxide diffusion and carbon dioxide concentrating mechanisms. In: Larkum AWD, Raven JA, Douglas S (eds) Photosynthesis in algae. Kluwer, Dordrecht, pp 225–244
Raven JA, Johnston AM, Kübler JE, Korb RE, McInroy SG, Handley LL, Scrimgeour CM, Walker DI, Beardall J, Vanderklift MA, Fredriksen S, Dunton KH (2002) Mechanistic interpretation of carbon isotope discrimination by marine macroalgae and seagrasses. Funct Plant Biol 29:355–378
Raven JA, Giordano M, Beardall J, Maberly SC (2012) Agal evolution in relation to atmospheric CO_2: carboxylases, carbon-concentrating mechanisms and carbon oxidation. Phil Trans R Soc B 367:493–507
Ries JB, Cohen AL, McCorkle DC (2009) Marine calcifiers exhibit mixed responses to CO_2-induced ocean acidification. Geology 37:1131–1134
Rodolfo-Metalpa R, Houlbreque F, Tambutte E, Boisson F, Baggini C, Patti FP, Jeffree R, Fine M, Foggo A, Gattuso JP, Hall-Spencer JM (2011) Coral and mollusc resistance to ocean acidification adversely affected by warming. Nat Clim Chang 1:308–312
Roleda MY, Boyd PW, Hurd CL (2012) Before ocean acidification: calcifier chemistry lessons. J Phycol 48:840–843
Russell BD, Thompson JA, Falkenberg LJ, Connell SD (2009) Synergistic effects of climate change and local stressors: CO_2 and nutrient-driven change in subtidal rocky habitats. Glob Chang Biol 15:2153–2162
Russell BD, Passarelli CA, Connell SD (2011) Forecasted CO_2 modifies the influence of light in shaping subtidal habitat. J Phycol 47:744–752
Russell BD, Connell SD, Uthicke S, Muehllehner N, Fabricius KE, Hall-Spencer JM (2013) Future seagrass beds: increased productivity leading to carbon storage? Mar Pollut Bull 73:463–469
Semesi IS, Kangwe J, Bjork M (2009) Alterations in seawater pH and CO_2 affect calcification and photosynthesis in the tropical coralline alga, *Hydrolithon* sp. (Rhodophyta). Estuar Coast Shelf Sci 84:337–341
Smith AD, Roth AA (1979) Effect of carbon dioxide concentration on calcification in the red coralline alga *Bossiella orbigniana*. Mar Biol 52:217–225
Steller DL, Hernandez-Ayon JM, Riosmena-Rodriguez R, Cabello-Pasini A (2007) Effect of temperature on photosynthesis, growth and calcification rates of the free-living coralline alga *Lithophyllum margaritae*. Cienc Mar 33:441–456
Suggett DJ, Hall-Spencer JM, Rodolfo-Metalpa R, Boatman TG, Payton R, Pettay DT, Johnson VR, Warner ME, Lawson T (2012) Sea anemones may thrive in a high CO_2 world. Glob Chang Biol 10:3015–3025
Ware JR, Smith SV, Reaka-Kudla ML (1992) Coral reefs: sources or sinks of atmospheric CO_2? Coral Reefs 11:127–130
Wernberg T, Russell BD, Thomsen MS, Gurgel CFD, Bradshaw CJA, Poloczanska ES, Connell SD (2011) Seaweed communities in retreat from ocean warming. Curr Biol 21:1828–1832
Wiencke C, Bischof K (2012) Seaweed biology: novel insights into ecophysiology, ecology and utilization, vol 219, Ecological Studies. Springer, Berlin/Heidelberg

Wilson S, Blake C, Berges JA, Maggs CA (2004) Environmental tolerances of free-living coralline algae (maerl): implications for European marine conservation. Biol Conserv 120:283–293
Wood R (1999) Reef evolution. Oxford University Press, Inc., New York, 414 pp
Wootton JT, Pfister CA, Forester JD (2008) Dynamic patterns and ecological impacts of declining ocean pH in a high-resolution multi-year dataset. Proc Natl Acad Sci U S A 105:18848–18853

Chapter 4
Economic Importance of Coralline Carbonates

Giovanni Coletti, Daniela Basso, and Alfredo Frixa

Abstract Carbonate materials are important economic resources: limestones are excellent reservoirs and valuable building stones; unconsolidated sediments may be used as a viable source of calcium carbonate for soil conditioning. Since Late Cretaceous coralline algae are one of the most important shallow-water carbonate producers. Sediment production and deposition in carbonate platforms are controlled by physical, chemical and biotic factors. Chemical and biotic factors rule over sediment texture, composition, distribution and early diagenetic processes, and consequently they have a major impact over limestone properties after diagenesis, especially over porosity and permeability. Porosity and permeability in turn control limestone mechanical properties, its durability and its reservoir potential. Thus, understanding the factors controlling formation and fate of coralline-algal carbonate factories is necessary for both sustainable management of the coralline-dominated marine habitats and profitable exploitation of reservoirs and quarries.

4.1 Introduction

Since the Late Cretaceous, and particularly in the Miocene (Bourrouilh Le Jan and Hottinger 1988; Halfar and Mutti 2005; Braga et al. 2010), coralline algae became among the most common and widespread carbonate producers, from polar to tropical oceans. Due to their abundance and widespread distribution, coralline carbonates are a rich source of information to reconstruct and understand the Earth history. In addition, they are also an important economic resource. Coralline-algal limestone

G. Coletti (✉)
Dipartimento di Scienze dell'Ambiente e del Territorio e di Scienze della Terra, Università di Milano Bicocca, Piazza della Scienza 4, 20126 Milano, Italy
e-mail: g.coletti@campus.unimib.it

D. Basso
Department of Earth and Environmental Sciences, Univ. of Milano-Bicocca, Piazza della Scienza 4, 20126 Milan, Italy

A. Frixa
Geologist, Via Bramante 3, 20098 San Giuliano Milanese, Italy

R. Riosmena-Rodríguez et al. (eds.), *Rhodolith/Maërl Beds: A Global Perspective*, Coastal Research Library 15, DOI 10.1007/978-3-319-29315-8_4

is used in the construction sector as building stone and as raw material for cement production (Török et al. 2004; Sassone 2005; Siegesmund et al. 2007; Vrasaljko et al. 2007a, b; Basso et al. 2008; Cultrone et al. 2008; Angrisani et al. 2011; Rajabzadeh et al. 2012). Coralline-algal limestones are important reservoir rocks, with well-known and large oil and gas fields located all over the world (Sattler et al. 2004; Vaziri-Moghaddam et al. 2006; Borromeo et al. 2011). Maërl beds have been largely exploited as a source of calcium carbonate for the conditioning of acid soils (Blunden et al. 1975).

The hydrodynamic conditions, the chemical, physical and biotic factors control the carbonate factory development, the texture, the size and the composition of the sediment produced and also the early diagenesis (i.e.: the processes converting sediment into rock) and consequently limestone porosity and permeability (Flügel 2009). The latter are of the utmost importance since they control limestone mechanical properties, durability and reservoir potential.

With the aim to highlight the economic importance of coralline carbonates, a brief review is presented, with case histories of exploitation in the construction industry, hydrocarbon exploration and agriculture.

4.2 Construction Industry

Sedimentary rocks are some of the most commonly used materials in construction industry (Primavori 1999; Calvo and Regueiro 2010); in particular, limestone and dolomites have the third rank in the amount and value of globally extracted mineral resources (Flügel 2009).

In the past, natural rocks were used extensively for building and decoration purposes (Calvo and Regueiro 2010). More recently they have been mostly replaced by concrete, but they are still widely used for cladding and decoration. Nowadays, in many countries, the largest market segment for the sale of limestone is the construction industry, and the major use of limestone is as crushed rock in construction aggregates (Oates 1998). Cement production is generally the second largest use of limestone (Oates 1998). Coralline-algal limestones, however, are still popular decorative stones for exterior and interior use, because of their pleasing appearance (Flügel 2009).

4.2.1 The Use of Coralline Limestone in Construction Industry

The choice of construction material has always been connected to availability, cost and technical properties.

Especially in Ancient period and Middle Ages, the best material in the nearby areas was picked for the core of the buildings, while fine foreign rocks were saved for the finishing (Borg and Borg 2000). Coralline limestone is easy to work and available in many countries of ancient civilization in Europe (e.g. Southern France, Italy, Iran, Austria, Hungary, Czech Republic, Slovakia, Croatia and Poland). In the

Fig. 4.1 Palais des Papes, Avignon, Southern France

Mediterranean area the outcrops of rhodalgal limestones are abundant, due to a Miocene worldwide spreading of this type of carbonate factory (Halfar and Mutti 2005; Braga et al. 2010). Many of them are easily accessible and workable, and therefore represented an excellent target for exploitation in the past (Calvo and Regueiro 2010). Several monuments of remarkable historic importance are built with coralline limestones (Török et al. 2004; Siegesmund et al. 2007; Vrasaljko et al. 2007a, b; Cultrone et al. 2008; Calvo and Regueiro 2010; Angrisani et al. 2011). Examples include Greek and Roman theaters and temples in Southern Italy (e.g. Greek theater, the Temple of Apollo and the Temple of Athena, Syracuse, Sicily; Roman theater of Minturno, Campania), Roman arenas (e.g. Nîmes, Southern France), medieval fortresses (e.g. Palais des Papes, Avignon, Southern France, Fig. 4.1) and churches (e.g. St. Stephan Church, Vienna; Matthias Church, Budapest; Santa Chiara Monastery, Naples; Zagreb Cathedral, Zagreb), and more recent edifices (e.g. Hungarian Parliament, Budapest).

For building and important masonry purposes dimension stones (i.e.: building stone quarried and prepared in regularly shaped blocks) are employed. Very coarse-grained coralline limestones have generally high porosities often resulting in surfaces that cannot be regularized or smoothed. Since a general condition for exploitation of a rock as dimension stone is the likelihood to extract blocks of regular shape (Primavori 1999), fine-grained and massive coralline-algal limestones are more suitable for building purposes than very coarse or deeply fractured ones. On the other hand, fractured and very coarse limestones, unsuitable as dimension stones, may be used to produce lime or cement.

Magnesium carbonate is the main undesirable impurity in the production of cement; the level of MgO should be less than 3–5 % (Oates 1998). Limestones con-

Table 4.1 General mineralogical composition of some coralline limestones

	Silicate (%)	Carbonate (%)
Pierre de Beaucarie (Gaudon 2010)	1–5	99–95
Pierre de Beaulieu (Gaudon 2010)	2–4	98–96
Leithakalk, Budapes buildings (Török 2002)	2–13.5	86.5–98
Leithakalk Czech Repubblic (Török et al. 2004)	3–5	95–97
Perlato Coreno (Angrisani et al. 2011)	1–2	98–99
Pietra da Cantoni coarse layers	1–8	92–99

taining more than 95 % of $CaCO_3$ are particularly suitable for lime production (Oates 1998). Chemical and mineralogical composition of analyzed coralline limestone is remarkably stable (Table 4.1) and suitable for both uses. Low-Mg calcite account for more than 90 % of the rock material; dolomite and siderite (i.e: iron carbonate) are rare; quartz is common.

In Northwestern Italy, the basal layers of rhodolith rudstones and floatstones (i.e.: coarse-grained limestone according to Dunham 1962) of the "Pietra da Cantoni" limestone Group, were used for cement production since the rock was too irregular and porous (porosity of 25–40 %) for being exploited as dimension stone (AA.VV. 2011). The "Pietra da Cantoni" upper layers, composed of fine-grained limestones (planktonic foraminifers grainstone and marly packstone), were instead largely used as dimension stone in the whole region (Timpanelli 2003; Sassone 2005; Violanti et al. 2008; AA.VV. 2011). The Ternate Formation is a rhodalgal limestone of Central-Northern Italy (Mancin et al. 2001). The rock is not porous and very compact but possesses a dense network of veins. It is quarried as raw material for cement production.

Very coarse and porous coralline limestones are sometimes used for structures which are not in plain sight. For the basement of St. Martin Cathedral and the remedial works of the Castle tower, in Bratislava, coarse-grained limestone was used, while fine-grained limestone was reserved for the facades (Holzer et al. 2004).

4.2.2 *The Properties of Coralline Algae as Dimension Stone*

Among the variety of coralline limestones used as dimension stone, two main groups may be recognized based on different mechanical properties: soft limestone with high porosity and hard limestone with low porosity (Fig. 4.2). Porosity, permeability, geometry and distribution of pores and pore throat diameters greatly influence mechanical properties of limestones. Highly porous limestones are characterized by low values (10–40 Mpa, Fig. 4.2) of compressive strength (i.e the capacity of a material to withstand loads without breaking or receiving permanent deformation) and resistance to abrasion. On the other hand the low-porosity compact limestones show compressive strength in the range of sturdy materials like porphyry and basalts and high abrasion resistance (145–250 Mpa, Fig. 4.2; Primavori 1999).

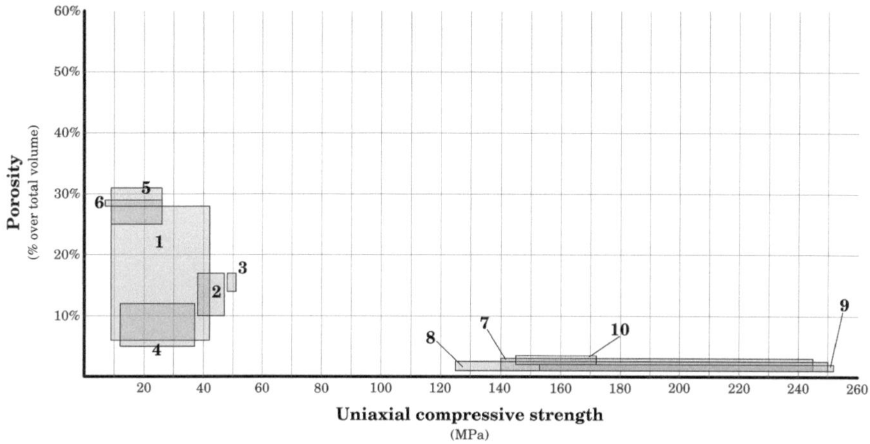

Fig. 4.2 Relation between porosity and uniaxial compressive strength in coralline limestone. *1* Leithakalk of Hungary (Török et al. 2004); *2* Leithakalk of Czech Republic (Török et al. 2004); *3* Leitha Limestone of the Mannersdorf quarry, Austria (Holzer et al. 2004); *4* Carbonate conglomerate of the Hundesheim quarry, Austria (Holzer et al. 2004); *5* Pierre de Beaucarie (Gaudon 2010); *6* Pierre de Beaulieu (Gaudon 2010); *7–10* commercial varieties of Perlato Coreno, *7* Mezza Perla, *8* Perlatino, *9* Botticino, *10* Risatino (Angrisani et al. 2011)

The "Leithakalk" (Austria, Hungary, Czech Republic, Slovakia), also known as Leitha limestone, the "Litavac" (Croatia), the "Pierre de Beaucarie", the "Pierre de Beaulieu", the "Pierre de Pondres" (Southern France) and the "Calcare di Siracusa" (Sicily, Italy) belong to the group of high-porosity coralline limestones (Fig. 4.2). These rocks are composed of coarse fragments of calcareous shells and other remains of organisms that once lived in the warm and shallow waters of the Miocene Tethys, under some degree of water energy (Basso et al. 2008; Reynaud and James 2012). They are principally composed of coralline algae (unattached branches and nodules), bryozoans, large benthic foraminifers, mollusks and rare corals.

The Cusano limestone (commercially known as "Perlato Coreno") is an example of low-porosity coralline limestone especially used as ornamental dimension stone since Roman times, due to its good mechanical properties (Fig. 4.2). It is sold in many different varieties on the basis of their fossiliferous content and color, both in Italy and abroad. This limestone belong to the Miocene "Calcari a Briozoi e Litotamni" Formation of Apennines mountains (Central Italy). The rock is composed of coralline-algae (both fragments and nodules), bryozans and mollusks, embedded in micrite matrix. This benthic association formed in the mesophotic zone, probably in a low-energy setting (Carannante and Simone 1996; Civitelli and Brandano 2005). The Cusano limestone has low porosity and consequently better mechanical properties, despite its skeletal assemblage is similar to those of the high-porosity coralline limestone (Fig. 4.2). The Cusano limestone has an average of five times the compressive strength of high-porosity limestones and very high abrasion resistance (Primavori 1999; Angrisani et al. 2011).

Fig. 4.3 Alteration and decaying features of coralline limestone and similar rocks. (**a**) Budapest limestone surface loss (*sl*), after Siegsmund et al. (2007); (**b**) hollowing and alveolar weathering (*a*) on Budapest limestone, after Siegsmund et al. (2007); (**c**) alteration crust (*c*) on "Pietra bianca di Mellini", after Cultrone et al. (2008); (**d**, **e**) exfoliation on "Pietra da Cantoni" limestone, upper foraminifers-dominated layers. (**f**, **g**) alveolar weathering (*a*) and surface loss (*sl*) on Maltese Globigerina limestone; (**h**) alteration and alveolar watering (*a*) on Avignon coralline limestone, on the left of the panel pristine Avigone limestone (*f*); (**i**) fortified wall of Avignon, altered coralline limestone (alveolar weathering) at the *bottom* of the wall (*a*)

Carbonate rocks used as outdoor building stones are affected by weathering agents, leading to stone decay (Fig. 4.3a, b), a relevant issue on historical buildings. Weathering is a problem, especially where sulfur dioxide and other chemical pollutants produce acid rainwater greatly hastening stone deterioration. Very common decay features of porous limestones are alteration crusts of various morphology and thickness (Vannucci et al. 1994; Rothert et al. 2007; Siegesmund et al. 2007) (Fig. 4.3c). These crusts form through a process of solution and reprecipitation of calcite, which decreases the superficial rock porosity and permeability by deposition of microsparitic and micritic cements (Török 2002). Once formed, the crust protects the rock from further damage. However, crust detachment leads to rapid granular disintegration and extensive surface loss (Siegesmund et al. 2007) (Fig. 4.3). Crusts rich in gypsum, halite, Mg and Na sulfates may crack and disaggregate faster, due to the volume increase connected with salt crystallization and hydration (Rothert et al. 2007).

Weathering resistance depends on many, partly interconnected factors. Porosity, permeability and fabric are primary factors controlling the storage capacity of water and gases (Reijers and Hsü 1986). Pore volume, distribution, degree of pore interconnection, and pore throat diameter influence the flow speed of water inside the stone (Rothert et al. 2007; Cultrone et al. 2008).

Limestones which exhibit very low total porosity values (less than 5 %) are found to be more durable (Miglio et al. 2000). The Cusano limestone, for example, has a greater resistance to weathering compared to high-porosity limestones. In fact, the Cusano limestone had a negligible mass-loss vs a 9 % mass-loss suffered by the Syracuse limestone in the same accelerated aging test (i.e.: a test that artificially speeds up the normal weathering of the rock; for example exposing a rock to a series of fast freeze-thaw cycles; Cultrone et al. 2008; Angrisani et al. 2010).

In the comparison of limestones with similar total porosity, those with larger and more connected pores show a stronger resistance to weathering (Rothert et al. 2007). The "Calcare di Siracusa" and the "Pietra bianca di Mellini" belong to the Oligo-Miocene Monti Climiti Formation of Sicily (Pedley 1981). The Mellini rock is a bioclastic wackestone (Dunham 1962), composed of coralline algae, planktonic and benthic foraminifers, in a micro-sparitic and micritic matrix (Pedley 1981; Cultrone et al. 2008). The "Calcare di Siracusa" is a bioclastic packestone (Dunham 1962) composed of coralline algae (fragments and rhodoliths) and benthic foraminifers in a micro-sparitic and micritic matrix (Pedley 1981; Cultrone et al. 2008). Both rocks show high porosity (>20 % of volume), but the Mellini limestone porosity is 6 % higher than that of the Syracuse limestone (Cultrone et al. 2008). On equally exposed surfaces, the Mellini limestone is always better preserved than the Syracuse limestone (Cultrone et al. 2008). In accelerated aging tests the Mellini limestone showed no evident degradation, while the Syracuse limestone suffered considerable mass loss and erosion (Cultrone et al. 2008). The difference in weathering resistance is caused here by differences in pore interconnection and pore size. The Mellini limestone has numerous pore interconnections, which allow a faster water escape. The Syracuse limestone instead has lower pore interconnections, promoting salt crystallization inside the rock (Cultrone et al. 2008). The high pressure generated inside the pores by the growing salt crystals damages the rock, leading to fast stone decay (Steiger 2005).

4.2.3 Environmental and Diagenetic Controls Over Coralline Limestone Properties

The economic use of limestones depends on their physical and chemical properties, resulting from the sedimentary and diagenetic history of the limestone.

Well sorted and mud-free carbonate sands and gravels, developed in high-energy environments, have high primary porosities that may be preserved during diagenesis, resulting in the formation of high-porosity limestones (Flügel 2009). On the contrary, skeletal carbonates produced in low-energy settings have generally lower primary porosities due to the higher mud content and are likely to become low-porosity limestones. Climate is another important element, since calcium carbonate saturation state of water changes with temperature. In tropical supersaturated water initial processes are generally constructive, while in cold undersaturated water dissolution of carbonate materials generally prevails (Alexandersson 1978; Flügel 2009).

Before diagenesis, coralline-algal remains are composed of high-Mg calcite (Chave 1954), while other associated skeletal grains (mollusk shells, coral fragments and calcareous green-algae) are mostly composed of aragonite. High-Mg calcite and aragonite are unstable in the meteoric environment and frequently also in the shallow pore-waters of platform sediments (Hover et al. 2001; Flügel 2009). High-Mg calcite transforms into low-Mg calcite generally without extensive dissolution and textural alteration, and thus with little effect on the mechanical properties of the limestone (Friedman 1964; Saller and Moore 1989; Canfield and Raiswell 1991; Flügel 2009). On the contrary, skeletal aragonite is replaced by larger crystals of low-Mg calcite. This important textural change may affect limestone porosity and permeability and consequently its mechanical properties (Flügel 2009). Meteoric, marine, and burial diagenesis may either enhance or destroy the original cements, thus increasing or decreasing the original porosity. Mechanical compaction, caused by burial, alters the spatial arrangement of grains and leads to an increase in grain packing paralleled by a decrease in porosity. Following mechanical compaction many rocks undergo chemical compaction, by pressure solution expressed by the formations of stylolites. Pressure solution is believed to be a significant source for the formation of porosity-occluding cements (Tucker 1981).

4.3 Hydrocarbon Exploration

Reservoir rocks are characterized by petrophysical properties favoring the storage of fluids and gases. High porosity is a common feature of coralline limestone and makes this material a potentially good reservoir. Wordwide, many limestones characterized by abundant coralline algae host important hydrocarbon fields.

The Upper Oligocene-Lower Miocene Asmari Formation of Southwestern Iran is the most prolific Iranian reservoir (51 oil reservoir which produce near 90 % of Iran oil; Amirshankarami et al. 2007) and one of the biggest in the world (31 billions of barrel of oil in place; Perry and Choquette 1985). This large carbonate unit originally deposited in a tropical shallow water environment. A complete succession of facies along the paleobathymetric profile has been recognized (Vaziri-Moghaddam et al. 2006; Amirshankarami et al. 2007; Vaziri-Moghaddam et al. 2011; Sahraeyan et al. 2013). Skeletal assemblages are dominated by large benthic foraminifers and coralline algae; bryozoans and corals are present and abundant in some facies. The paleoecological reconstruction suggests the present-day Persian Gulf as the most appropriate model for the Asmari Fm. (Amirshankarami et al. 2007; Sahraeyan et al. 2013). Asmari Fm. is a fractured reservoir (i.e. a reservoir rock where fracture networks control porosity and permeability) (Perry and Choquette 1985; Flügel

2009). Fractures pattern and spacing is controlled by mechanical stratigraphy (i.e. spatial variations in mechanical properties due to changes in lithology) and consequently by facies distribution. Hydrocarbons are best stored within grain-supported, coralline-rich layers, while mud-supported, highly fractured layers represent their preferential transport route (Wennberg et al. 2006).

One of the major oil reservoir in coralline limestone is the Lower-Miocene Zujiang Formation in the South China Sea. These limestones, composed of coralline algae and corals, are the reservoir of Liuhua 11-1 giant oil field (>1 billion barrel of oil in place, Heubeck et al. 2004). The top layers of the succession are mainly composed of rhodoliths and are one of the highest-porosity zones of the reservoir (Sattler et al. 2004). Most of the porosity of this limestone is related to leaching during deep burial diagenesis; corrosive fluids, which developed during late burial, seeped through the formation creating vugs (Sattler et al. 2004). The spatial distribution of porosity was probably influenced by differences in early diagenesis, meteoric diagenesis and compaction, which produced zones with relative low permeability able to divert the corrosive fluids (Sattler et al. 2004). The original mineralogy of skeletal grains may have caused these initial differences in porosity. The final result is that zones rich in aragonitc corals are well cemented and vugs-free, while zones dominated by coralline-algae and other calcitic remains have very high porosity and are the main reservoir (Sattler et al. 2004).

Original porosity may also be enhanced by fresh-water diagenesis. Karaisalı Limestone Formation of the Adana Basin, Southern Turkey, deposited in a tropical environment during Lower Miocene. The formation is the reservoir of the small Bulgurdağ oil field. The skeletal grain association is composed of fragments of coralline-algae, corals and benthic foraminifers (Büyükutku 2009). Among the recognized facies, coralline-algal and benthic foraminifer packstones have the best reservoir potential since they have both high porosity and permeability (Büyükutku 2009). Early dissolution, in the meteoric and phreatic environments, enhanced pre-existing porosity and permeability through pore-size increase and dissolution of skeletal grains (Büyükutku 2009). These processes mainly affected the originally already more porous foraminiferal-algal packstones, and greatly improved their quality as reservoir (Büyükutku 2009).

Other oil fields are located in the "Leitha" Limestone in Hungary and in the Oligocene-Early Miocene Nido Limestone Formation in Philippines (Safti et al. 2003; Fournier and Borgomano 2007).

The Malampaya-Camago gas field (always in the Nido Formation) is another major reservoir present in coralline-limestone (Fournier and Borgomano 2007). The Nido Formation was a reef-bounded isolated platform; the reservoir is composed of reef-related limestones, including red-algal floatstone (Moore 2001). Reservoir quality was strongly influenced by the original position of the carbonate deposits relative to the open ocean. On the ocean-ward margin, porosity-destructive marine cementation was dominant. Reservoir characteristics at Malampaya-Camago field are then predominantly controlled by the platform early diagenetic history, driven by sea level and climate, and are not directly influenced by lithofacies (Moore 2001).

In the "Perla" super-giant gas-field (more than 850 billion cubic meters of natural gas), located offshore the gulf of Venezuela, hydrocarbons are hosted in Lower-Miocene coralline limestones and sandstones (Borromeo et al. 2011). Beside coralline algae fragments and rhodoliths, the limestone skeletal assemblage is composed of large benthic foraminifers, corals, bryozoans, barnacles and mollusks (Borromeo et al. 2011). The primary porosity was enhanced during marine phreatic and shallow burial diagenesis, by dissolution of aragonitic and high-Mg calcite grains (Borromeo et al. 2011). In this case, porosity-destructive marine cementation was less effective in the deep-water, rhodolith-dominated carbonates, resulting in the remarkable reservoir properties of this facies (Borromeo et al. 2011).

4.4 Maërl Exploitation

Extensive biogenic gravel banks, formed by unattached branches and nodules of live and dead coralline algae, called maërl, occur in the Northeastern Atlantic, along British, French and Irish coasts. Maërl is dredged to be used in agriculture and horticulture, mainly as a soil conditioner for acid soils, but also as natural filter, food complement and toothpaste (Blunden et al. 1975).

Acid soils are very common, composing about 40 % of global arable surface (LeNoble et al. 1996). In acid soils, potassium, calcium and magnesium, which are necessary to plant growth, are leached away by rainfall and, due to aluminum solubility, plant growth is limited (aluminum is toxic to plants, it reduces root growth and inhibits nutrients uptake). Calcareous materials increase soil pH and fix calcium and magnesium deficiency.

Contrarily to most limestones, maërl generally have a high magnesium content and is also rich in elements useful for plants, like iron, manganese and boron (Blunden et al. 1997). Ground maërl has a higher porosity than ground non-algal limestone, leading to its faster weathering and speedier effect on soil pH (Blunden et al. 1997). Maërl is then a better soil conditioner than ground limestones, however, these differences would not appear to compensate for the considerably higher cost involved with its exploitation (Blunden et al. 1997).

Due to its porous nature, maërl is also used for filtration of acid waters. This treatment removes important toxic metals by adsorption and precipitation brought by the alkaline pH of maërl (Blunden et al. 1975).

The quantification of the resource is important when a sustainable extraction of maërl is considered. The actual quantity of coralline-algal fragmens, branches and nodules, within any individual deposit, may range between 5 and 90 % (De Grave et al. 2000). Abundance of maërl is controlled by environmental conditions; high concentrations are likely to be found in high-energy environments where currents, mainly tidal, sweep away mud particles. The thickness of the deposit is another important element. This may vary considerably, from some tens of centimeters in exposed areas to several meters in more sheltered regions (De Grave et al. 2000). Environmental conditions not only influence the quality of the maërl deposit but also coralline-algae growth rate and thus the renewability of the resource. On the basis of its low growth-rate maërl is considered a non-renewable resource. Extraction rates range from 5,000 t $year^{-1}$ of Ireland to 500,000 t $year^{-1}$ of France (De Grave et al. 2000). Atlantic maërl banks produce, on average, about 700 g m^{-2} $year^{-1}$ of carbonate sediment (Basso 2012), so, for example, the intensive exploitation of a surface of 7×10^6 m^2 of healthy maërl would be required to keep up with the yearly production of Ireland (that is to say about one tenth of the total surface of Irish maërl banks; De Grave et al. 2000).

In the last years European union greatly restricted maërl exploitation by listing the two main Atlantic maërl-forming species, *Lithothaminon corallioides* and *Phymatolithon calcareum*, in Annex V of the Habitat Directive (i.e. species whose exploitation is subject to management). In 2005 Great Britain further reinforced these limitation ceasing to license maërl extraction (Hall-Spencer 2005). The conservation and protection status are however different across European countries (Salomidi et al. 2012).

4.5 Conclusions

Coralline-algal carbonates are used in various sectors on the basis of their mechanical, physical and chemical properties. They are used in construction industry as dimension stones, crushed rock or raw material for cement production; in agriculture, as soil conditioner. Last but not least, they are also interesting target for hydrocarbon exploration.

Environmental conditions control past and present carbonate factories, their development and fate. Consequently these conditions are responsible for the characteristics of coralline-algae deposits which in turn will be responsible for the heterogeneity of limestone properties.

Hydrodynamic conditions, reflected by grain size, sorting and mud content, are major environmental controls primarily affecting rhodalgal carbonate factories. High-energy environments result in high primary porosity and permeability, possibly preserved or enhance by diagenesis, producing coralline limestones with high porosity and permeability. These rocks are characterized by poor mechanical properties and may have an high reservoir potential. Climatic factors will also control initial porosity and permeability by influencing early diagenesis, with dominant destructive processes in cold climate and dominant constructive processes in tropical waters.

Coralline biology and mineral composition, environmental conditions during life, and after final burial are major controls on the formation and accumulation of modern coralline carbonate sediments, and on the diagenetic processes leading to exploitable limestone quarries and reservoirs. These elements need to be considered not only for the sustainable management of the present-day maërl beds, but also for reservoir characterization, economically-efficient quarrying, and for the restoration of historic buildings.

Acknowledgements The authors acknowledge the COCARDE – European Research Network (ESF) for the opportunity to discuss and share the knowledge and the concepts that inspired this manuscript. We would like to thank an anonymous referee for his constructive revision of an early draft of the manuscript, and Alessandro Cavallo, Milano-Bicocca University, for his useful suggestions and support. The first author has been funded through a PhD fellowship from University of Milano-Bicocca.

References

AA.VV. (2011) L'industria della calce e del Cemento in Ozzano. La Terra Promessa, Novara

Alexandersson T (1978) Destructive diagenesis of carbonates sediments in the eastern Skagerrak, North sea. Geology 6:324–327

Amirshankarami M, Vaziri-Moghaddam H, Taheri A (2007) Sedimentary facies and sequence stratigraphy of the Asmari Formation at Chaman-Bolbol, Zagros Basin, Iran. Asian J Earth Sci 29:947–959

Angrisani AC, Calcaterra D, Colella A, De Gennaro M (2010) Stone properties and weathering phenomena of the Miocene Cusano Limestone (a.k.a. Perlato Royal Coreno): the case of the basament of Santa Chiara Monastery bell tower (Naples-Italy). Period Mineral Spec Issue 79:1–10

Angrisani AC, Calcaterra D, Cappelletti P, Colella A, Parente M, Přikryl R, De Gennaro M (2011) Geological features, technological characterization and weathering phenomena of the Miocene Bryozoan and *Lithothamnion* limestones (central-southern Italy). Ital J Geosci 130(1):75–92

Basso D (2012) Carbonate production by calcareous red algae and global change. Geodiversitas 34(1):13–33

Basso D, Vrsaljko D, Grgasović T (2008) The coralline flora of a Miocene maërl: the Croatian "Litavac". Geol Croat 61(2–3):333–340

Blunden G, Binns WW, Perks F (1975) Commercial collection and utilization of Maërl. Econ Bot 29:140–145

Blunden G, Sheelagh A, Campbell A, Smith JR, Guiry MD, Heission CC, Griffin RL (1997) Chemical and physical characterization of calcified red algal deposits known as maërl. J Appl Phycol 9:11–17

Borg G, Borg BE (2000) From small quarries to large temples, the enigmatic source of limestone for the Apollo Temple at Didyma, W-Anatolia. In: Lazzarini L (ed) Interdisciplinary studies on ancient stone. ASMOSIA, VI international conference, S: 427–436

Borromeo O, Miraglia S, Sartorio D, Bolla EM, Ortenzi A, Sandro R, Castellanos C, Villalobos R (2011) The Perla world-class giant gas field, Gulf of Venezuela; depositional and diagenetic controls on reservoir quality in early Miocene carbonates. AAPG International Conference and Exhibition, Milan

Bourrouilh Le Jan FG, Hottinger LC (1988) Occurrence of rhodoliths in the tropical Pacific – a consequence of Mid-Miocene paleo-oceanographic change. Sediment Geol 60:355–367

Braga JC, Bassi D, Piller WE (2010) Paleoenvironmental significance of Oligocene-Miocene coralline red algae – a review. In: Mutti M, Piller WE, Betzler C (eds) Carbonate system during the Oligocene-Miocene Climatic Transition. Int Assoc Sedimentol Spec Publ 42

Büyükutku AG (2009) Reservoir properties of Karaisalı formation in the Adana Basin, Southern Turkey. J Pet Sci Eng 65:33–44

Calvo JP, Regueiro M (2010) Carbonate rocks in the Mediterranean region – from classical to innovative use of building stone. In: Smith BJ, Gomez-Heras M, Viles HA, Cassar J (eds) Limestone in the built environment: present-day challenges for the preservation of the past. Geol Soc Spec Publ 331: 27–35

Canfield DE, Raiswell (1991) Carbonate precipitation and dissolution, its relevance to fossil preservation. In: Allison PA, Briggs DEG (eds) Taphonomy: releasing the data locked in the fossil record. Topics in Geobiology, vol 9

Carannante G, Simone L (1996) Rhodolith facies in the central-southern Apennines mountains, Italy. In: Franseen EK, Esteban M, Ward W, Rouchy J (eds) Models for carbonate stratigraphy from Miocene reef complexes of Mediterranean regions. SEPM Concepts Sedimentology and Paleontology 5

Chave KE (1954) Aspects of biogeochemistry of magnesium. 1. Calcareous marine organism. J Geol 62:266–599

Civitelli G, Brandano M (2005) Atlante delle litofcies e modello deposizionale dei Calcari a Briozoi e Litotamni nella Piattaforma carbonatica laziale-abruzzese. Boll Soc Geol Ital 124:611–643

Cultrone G, Russo LG, Calabrò C, Uroševič M, Pezzino A (2008) Influence of pore system characteristic on limestone vulnerability: a laboratory study. Environ Geol 54:1271–1281

De Grave S, Fazakerley H, Kelly L, Guiry MD, Ryan M, Walshe J (2000) A study of selected Maërl banks in Irish waters and their potential for sustainable extraction. Marine Resource Series 10, Marine Institute

Dunham RJ (1962) Classification of carbonate rocks according to depositional texture. In: Ham WE (ed) Classification of carbonate rocks. American Association of Petroleum Geologist Memoir. 1:108–121

Flügel E (2009) Microfacies of carbonate rocks: analysis interpretation and application. Springer, 1007 pp

Fournier F, Borgomano J (2007) Geological significance of seismic reflections and imaging of the reservoir architecture in the Malampaya gas field (Philippines). AAPG Bull 91:235–258

Friedman GM (1964) Early diagenesis and lithification in carbonate sediments. J Sediment Petrol 34(4):777–813

Gaudon P (2010) Guide Durabilité & Compatibilité de pierres régionales. École des Mines d'Alès, http://pierresud.brgm.fr. Accessed 10 Oct 2013

Halfar J, Mutti M (2005) Global dominance of coralline red-algal facies: a response to Miocene oceanographic events. Geology 33(6):481–484

Hall-Spencer J (2005) Ban on maërl extraction. Mar Pollut Bull 50:121–124

Heubeck C, Story K, Peng P, Sullivan C, Duff S (2004) An integrated reservoir study of the Liuhua 11-1 field using high-resolution three-dimensional seismic data set. AAPG Mem 81:149–168

Holzer R, Laho M, Durmenková T (2004) Ancient building stone sources of Bratislava's monuments. In: Přikryl R (ed) Dimension stone-new perspective for a traditional building material. Balkema Publishers

Hover VC, Walter LM, Peacor DR (2001) Early marine diagenesis of biogenic aragonite and Mg-calcite new constraints from high-resolutions STEM and AEM analyses of modern platform carbonates. Chem Geol 175:221–248

LeNoble ME, Blevins DG, Sharp RE, Cumbie BG (1996) Prevention of aluminum toxicity with supplemental boron, maintenance of root elongation and cellular structure. Plant Cell Environ 19:1132–1142

Mancin N, Ceriani A, Tagni F, Brambilla G (2001) The Ternate formation (northern Italy): micropaleontological content and sedimentary petrography. Atti Ticinesi Sci Terra 42:37–46

Miglio BF, Richardson DM, Yates TS, West D (2000) Assessment of the durability of porous limestones: specification and interpretation of test data in UK practice. In: Hoigard KR (ed) Dimension stone cladding: design, construction, evaluation and repair. American Soc for Testing and Material, West Conshohocken

Moore CH (2001) Carbonate reservoirs: porosity evolution and diagenesis in a sequence stratigraphic framework. Dev Sedimentol 55

Oates JAH (1998) Lime and limestone. Chemistry and technology. Production and uses. Wiley, 455 pp

Pedley HM (1981) Sedimentology and paleoenvironment of the southeast Sicilian Tertiary platform carbonates. Sediment Geol 28:273–291
Perry OR, Choquette PW (1985) Carbonate petroleum reservoirs. Springer, New York
Primavori P (1999) Pianeta Pietra. Giorgio Zusi, Verona
Rajabzadeh MA, Moosavinasab Z, Rakhshandehroo G (2012) Effects of rock classes and porosity on the relation between uniaxial compressive strength and some rock properties for carbonate rocks. Rock Mech Rock Eng 45:113–122
Reijers TJA, Hsü KJ (1986) Manual of carbonate sedimentology: a lexicographical approach. Academic Press, London
Reynaud J, James NP (2012) The Miocene Sommières basin, SE France; Bioclastic carbonates in a tide dominated depositional system. Sediment Geol 282:360–373
Rothert E, Eggers T, Cassar J, Ruedrich J, Fitzner B, Siegsmund S (2007) Stone properties and weathering induced by salt crystallization of Maltese Globigerina Limestone. In: Přikryl R, Smith BJ (eds) Building stone decay, from diagnosis to conservation. Geol Soc London Spec Publ, London, 271
Safti B, Veli J, Sztanó O, Juhász G, Ivković Z (2003) Tertitary subsurface facies, source rocks and hydrocarbon reservoirs in the SW part of the Pannonian Basin (Northern Croatia and South-Western Hungary). Geol Croat 56(1):101–122
Sahraeyan M, Bahrami M, Arzaghi S (2013) Facies analysis and depositional environment of the Oligo-Miocene Asmari Formation, Zagros Basin, Iran. Geosci Front 5:103–112
Saller AH, Moore CH (1989) Meteoric diagenesis, marine diagenesis and microporosity in Pleistocene and Oligocene limestones, Enewetak Atoll, Marshall Islands. Sediment Geol 63:253–272
Salomidi M, Katsanevakis S, Borja A, Braeckman U, Damalas D, Galparsoro I, Mifsud R, Mirto S, Pascual M, Pipitone C, Rabaut M, Todorova V, Vassilopoulou V, Vega Fernàndez T (2012) Assessement of goods and services, vulnerability and conservation status of European seabed biotopes: a stepping stone towards ecosystem-based marine spatial management. Mediterr Mar Sci 13(1):49–88
Sassone P (2005) La “Pietra da Cantoni” del Monferrato Casalese (AL): ipotesi di ripresa produttiva per la conservazione della tradizione edilizia locale. GEAM, June–Sept 2005: 15–24
Sattler U, Zampetti V, Schlager W, Immenhauser A (2004) Late leaching under deep burial conditions: a case of study from the Miocene Zhujiang Carbonate Reservoir, South China Sea. Mar Pet Geol 21:977–992
Siegesmund S, Török A, Hüpers A, Müller C, Klemm W (2007) Mineralogical, geochemical and microfabric evidences of gypsum crusts: a case study from Budapest. Environ Geol 52:385–397
Steiger M (2005) Crystal growth in porous materials II: influence of crystal size on the crystallization pressure. J Cryst Growth 282:470–481
Timpanelli M (2003) La Pietra da Cantoni del Monferrato Casalese: salvaguardia e valorizzazione di un patrimonio scientifico, storico e culturale. GEAM, Dec 2003: 17–26
Török A (2002) Oolitic limestone in polluted atmospheric environment in Budapest: weathering phenomena and alteration in physical properties. In: Siegsmund S, Weiss T, Vollbrecht A (eds) Natural stones, weathering phenomena, conservation strategies and case studies. Geol Soc London Spec Publ London, 205
Török A, Rozgonyi N, Přikryl R, Přikrylová J (2004) Leithakalk: the ornamental and building stone of Central Europe, an overview. In: R Přikryl (ed) Dimension stone-new perspective for a traditional building material. Balkema Publishers
Tucker ME (1981) Sedimentary petrology: an introduction. Blackwell Scientific Publications
Vannucci S, Alessandrini G, Cassar J, Tampone G, Vannucci ML (1994) The prehistoric megalithic temples of the Maltese Islands: causes and processes of deterioration of Globigerina limestone. In: V Fassina, H Ott, F Zezza (eds) Conservation of monuments in the Mediterranean Basin, Proceedings of the 3rd international symposium, Soprintendenza di Beni Artistici e Storici di Venezia, Venice, Italy

Vaziri-Moghaddam H, Kimiagari M, Taheri A (2006) Depositional environment and sequence stratigraphy of the Oligo-Miocene Asmari Formation in SW Iran. Facies 52:41–51

Vaziri-Moghaddam H, Kalanat B, Taheri A (2011) Sequence stratigraphy and depositional environment of the Oligocene deposits at Firozabad section, southwest of Iran based on microfacies analysis. Geopersia 1(1):71–82

Violanti D, Frixa A, Natta C, Sassone P, Trenkwalder S (2008) Geologia e micropaleontologia della "Pietra da Cantoni" nel Monferrato Casalese (Piemonte, Italia NW): Esperienze di divulgazione, turismo culturale e tutela del territorio: Le giornate di Paleontologia, Siena, 9–12 Sept 2008

Vrsaljko D, Hećimović I, Avanić R (2007a) Miocene deposits of Northern Croatia. In: Grgasović T, Vlahović I (eds) Ninth international symposium on fossil algae, field trip guidebook and abstracts. Croat Geol Survey, pp 143–153

Vrsaljko D, Marković S, Grgasović T (2007b) Cathedral of the assumption of the Blessed Virgin Mary. In: Grgasović T, Vlahović I (eds) Ninth international symposium on fossil algae, field trip guidebook and abstracts. Croat Geol Survey, p 155

Wennberg OP, Svånå T, Azizzadeh A, Aqrawi AMM, Brockbank P, Lyslo KB, Ogilvie S (2006) Fracture intensity vs mechanical stratigraphy in platform top carbonates the Aquitanian of the Asmari Formation, Khaviz Anticline, Zagros, SW Iran. Pet Geosci 12:235–246

Part II
The Role of Rhodolith/Maërl Beds in Historical Oceans

Chapter 5
Rhodoliths and Rhodolith Beds in the Rock Record

Julio Aguirre, Juan C. Braga, and Davide Bassi

Abstract Calcareous coralline algae (Rhodophyta; Corallinales, Hapalidiales, and Sporolithales; corallines hereafter) constitute one of the most widespread and successful groups of marine macrophytes. They occur as crusts partially coating hard or soft substrates, as laminar thalli growing directly on the seabed, or forming structures rolling freely on the substrate with an inner nucleus or without it. These latter structures are called rhodoliths. They can be one of the most abundant components in carbonate platform deposits, forming the so-called rhodalgal facies. In assessments of the rhodoliths, internal and external algal growth morphology, rhodolith external form, rhodolith inner arrangement, and assemblages of organisms forming the rhodoliths can provide valuable information for reconstructing palaeoenvironmental and palaeoclimatic conditions. Rhodoliths can occur massively concentrated in beds several meters thick. These concentrations are referred as rhodolith beds. These rhodolith beds may be the result of biotic (autochthonous rhodolith beds), abiotic (allochthonous rhodolith beds) concentrations or due to a mixture of processes (paraautochthonous rhodolith beds). Taphonomic and facies analyses, as well as faunal assemblages, can provide the information needed to confidently differentiate among them. The rock record offers unique information to envisage the founding conditions and the long-term maintenance of the rhodolith beds. In this chapter, we review and update the information on fossil rhodoliths and rhodolith beds, and discuss their value for palaeoenvironmental and palaeoclimatic reconstructions. Also, we discuss the sedimentary and the sequence stratigraphy contexts in which rhodolith beds are preferentially formed and developed.

J. Aguirre (✉) • J.C. Braga
Departamento de Estratigrafía y Paleontología, Universidad de Granada,
Campus Fuentenueva, 18002 Granada, Spain
e-mail: jaguirre@ugr.es

D. Bassi
Dipartimento di Fisica e Scienze della Terra, Università di Ferrara,
via Saragat 1, 44122 Ferrara, Italy

R. Riosmena-Rodríguez et al. (eds.), *Rhodolith/Maërl Beds: A Global Perspective*, Coastal Research Library 15, DOI 10.1007/978-3-319-29315-8_5

5.1 Introduction

Calcareous coralline algae show a high quality and a long fossil record starting about 140 Ma ago (mid-Valanginian, Early Cretaceous) (Chatalov et al. 2015). According to literature, rarefied data of coralline algal diversity show that the group diversified up to a peak in the Early Miocene, some 20 Ma ago, after which diversification slightly decreased with very minor fluctuations (Aguirre et al. 2000). Throughout their evolutionary history, calcareous corallines have been one of the most common and most widely distributed groups of fossil marine benthic algae. They are one of the major algal components both in tropical and in cool-water carbonates.

The oldest known coralline algae are small fragments from the Valanginian (Chatalov et al. 2015) or relatively thin crusts partially coating corals from lower Hauterivian coral reefs from the central Tethys (Arias et al. 1995; Tomás et al. 2007). By the end of the Early Cretaceous (Aptian-Albian) coralline algae formed multispecies rhodoliths in many marine settings (i.e. Beckmann and Beckmann 1966; Lemoine 1970; Simone et al. 2012). Since then, coralline algae have been an evolutionarily successful group of autotrophic marine organisms. However, little is known about the long-term production of rhodolith beds throughout the geological record. Halfar and Mutti (2005) compiled information on rhodolith-dominated deposits during the Miocene. However, no suitable data on the abundance of rhodolith beds for older time intervals are available. Long-term macroevolutionary history of coralline algae, including data of phylogeny, molecular clocks, and the fossil record, indicates that the group underwent radiation events during the Late Cretaceous and then during the Eocene (Aguirre et al. 2010). Whether these diversification events are correlated with widespread rhodolith bed formation remains to be satisfactorily assessed.

During their long geological history, coralline algae colonized almost any marine setting within the photic zone worldwide, as their present-day representatives. Precipitation of high-magnesium calcite in the cell walls, together with specific anatomical traits, makes coralline algae particularly resistant to extremely rough environmental conditions (Adey and Macintyre 1973; Round 1981; Wilson et al. 2004), as well as to high rates of herbivory (Steneck 1983, 1985, 1986; Maneveldt and Keats 2008; Burkepile and Hay 2010). In summary, these physiological attributes, together with their capability to thrive in poorly illuminating settings, account for their adaptability to inhabit a wide variety of marine settings and seem to have contributed to the ecological and evolutionary success of coralline algae (Aguirre et al. 2000).

In the rock record, coralline algae are present in almost all carbonate platform deposits, from coastal to outer shelf, as well as in siliciclastic and mixed carbonate-siliciclastic sediments. They can be major components of fossil assemblages, in the so-called rhodalgal lithofacies (Carannante et al. 1988), occurring as crusts coating both hard and soft substrates, as laminar growths directly on the seabed, or forming rhodoliths. The study of rhodoliths and rhodolith beds (their stratigraphic distribution, the taxonomy of the fossil coralline algal components, the growth morphology of the coralline algae, associated fauna, etc.) can provide key information on the past environmental conditions. In this respect, the fossil record is a unique historical archive that offers the possibility of understanding the foundation conditions for the formation of rhodolith beds and assessing the long-term processes involved in the development and maintenance of these ecosystems. In this chapter, we present a review of fossil rhodoliths and rhodolith beds (see a terminological discussion below), and their use to reconstruct ancient environments. The aims are: (1) to provide an overview of the palaeoecological and palaeoclimatic significance of rhodoliths; (2) to review the environmental factors that catalyse the formation of rhodolith beds, as well as their development through time up to their end; and (3) to discern the sequence stratigraphic contexts in which the formation and development of rhodolith beds are favoured.

5.2 Nomenclatural Background

5.2.1 Rhodolith

Non-geniculate coralline algae can occur attached to both hard and soft substrates, either organic or inorganic in origin. While growing, the algal thalli may eventually overgrow the whole settlement area completely covering the substrate in all directions and forming nodular structures freely rolling on the seabed. Coralline algae may also occur as free-living, isolated branches or forming nodular structures without an apparent inner nucleus. In their pioneering paper on the coralline algal nodules from Bermuda Islands, Bosellini and Ginsburg (1971, p. 670) proposed the name rhodolites for those "nodules and detached branched growths with a nodule form composed principally of coralline algae". Previously, similar structures were identified as *pralinés* in the French literature (Molinier 1956; Pérès and Picard 1958). These coralline algal nodules were also identified as oncolites (McMaster and Conover 1966; Blanc 1968), a term used mostly for coated grains generated by bacterial activity. Bosellini and Ginsburg (1971) emphasized that the term rhodolite was suitable to differentiate this sort of algal nodule from other similar structures but constructed by different algae or bacteria. Later, Binda (1973) questioned the use of the term rhodolite since it was commonly and traditionally used for a variety of garnet. Consequently, Ginsburg and Bosellini (1973) proposed the name rhodolith, which Barnes et al. (1970, p. 268) had already introduced but had not defined, to substitute rhodolite. Since then, rhodolith has been the most accepted name in the geological literature (Adey and Macintyre 1973; Bosence 1983a, 1991), although the term rhodolite has been sporadically used (i.e., Orszag-Sperber et al. 1977; Montaggioni 1979). Flügel (1978), however, maintained the general name oncolite for any nodular structure biologically constructed, whether algae, animals or cyanobacteria (see also Richter and Sedat 1983).

Despite the general and widespread usage of the term rhodolith, some authors have used different names to refer to the same coralline algal nodular structures.

Peryt (1983), citing the rhodolites of Bosellini and Ginsburg (1971), introduced the name rhodoid to refer to unattached nodular structures made up mostly of coralline algae (i.e. Burgess and Anderson 1983). Other authors have used this name to refer to nodular structures formed by other encrusting calcareous rhodophytes, such as peyssonneliaceans (Buchbinder and Halley 1985; Rasser and Piller 2004). Bosence (1983a) discarded the use of rhodoid based on its etymological unsuitability and the general acceptance of rhodolith. Even so, Flood (1983) and Flügel (2004) preferred to maintain the general term rhodoid instead of rhodolith.

Since rhodoliths can be built up by diverse encrusting organisms, further particular names have been proposed for those nodular structures in which other encrusting organisms represent the same proportion, or even greater than coralline algae. Richter and Sedat (1983) described nodules formed by a combination of the cyanobacteria *Rivularia haematites* and the coralline *Lithoporella* sp. from Pleistocene terraces of Greece as oncolites. Reid and Macintyre (1988) introduced the name foraminiferal-algal nodules to refer to nodular structures from the eastern Caribbean in which the encrusting benthic foraminifer *Gypsina* sp. equals to, or is more abundant than coralline algae as a major builder (Fig. 5.1a). Later, Prager and Ginsburg (1989), studying similar *Gypsina*-coralline algal nodular structures from the Florida shelf called them for-algaliths, a term that has been subsequently used by other authors (i.e. Martín and Braga 1993; Braga and Aguirre 2004; McNeil and Pisera 2010). Aguirre et al. (1993) adopted the term serpulid nodules for those nodular structures in which serpulids are significantly more abundant than corallines (Fig. 5.1b).

Hottinger (1983) used the term macroid to define any free-moving coated grain made up by encrusting organisms (see also Baarli et al. 2012). He applied this term to nodules dominated by either the benthic foraminifer *Acervulina* or coralline algae. Thus, a rhodolith would be a macroid built up exclusively or mostly by coralline algae. Nonetheless, Hottinger (1983) also used the term rhodoids for coralline algal-dominated nodules. Peryt (1983) restricted the term macroid to those coated grains (either chemically or biogenically produced) larger than 10 mm in diameter. Bassi et al. (2011, 2012), following Hottinger, used the term macroid for nodules formed mostly by *Acervulina* (up to 80 % of the nodule).

Steneck (1986) suggested a different nomenclatorial framework to distinguish the wide morphological variation of coralline algal appearances. This author identified three types, depending upon the degree of adherence of the algal thalli (Fig. 5.2): (1) adherent, when the thalli are completely attached to the substrate, having three distinguishable morphological end-points (thick, thin and branched); (2) leafy, when the margin of the algal thallus forms expansions (lateral branches) that are not in contact with the substrate; and (3) free-living or totally nonadherent, when corallines are unattached. The two first categories refer to coralline algal crusts while the latter includes coatings, rhodoliths, nodules, and maërls (= marls or gravels) (Fig. 5.2). This fourfold division of free-living coralline algae was in turn established depending upon the presence/absence of an evident nucleus (coatings and rhodoliths versus nodules and maërls, respectively) or the taxonomic composition (nodules and maerls are monospecific). In coatings, coralline algae completely cover a nucleus but the algae represent less than 50 % of the total structure (including the

Fig. 5.1 (**a**) Macroid of *Acervulina*, (**b**) serpulid nodule according to Aguirre et al. (1993)

alga and the nucleus). By contrast, in rhodoliths, the algal cover constitutes more than 50 % of the total structure (Fig. 5.2). For algal structures without a nucleus, maërl refers to monospecific-isolated branches, and nodules are monospecific "more densely branched, spherical-to-ellipsoid plants" (Steneck 1986, p. 280) (Fig. 5.2). The criteria for separating these four categories of free-living corallines might be difficult to apply and consequently should be discarded. On this point, assessing the presence/absence of a nucleus to distinguish nodules from rhodoliths needs sectioning of the algal structures. The same applies to estimate the percentage of algal cover. Finally, the four types of structures might be formed by either a single or several species, and thus the monospecific composition is not exclusive of nodules and maërl despite the definition of Steneck (1986). To avoid terminological confusions and the difficulty of applying nomenclatorial frameworks, we suggest keeping the original definition of Bosellini and Ginsburg (1971) and thereby maintaining the historical stability of the term.

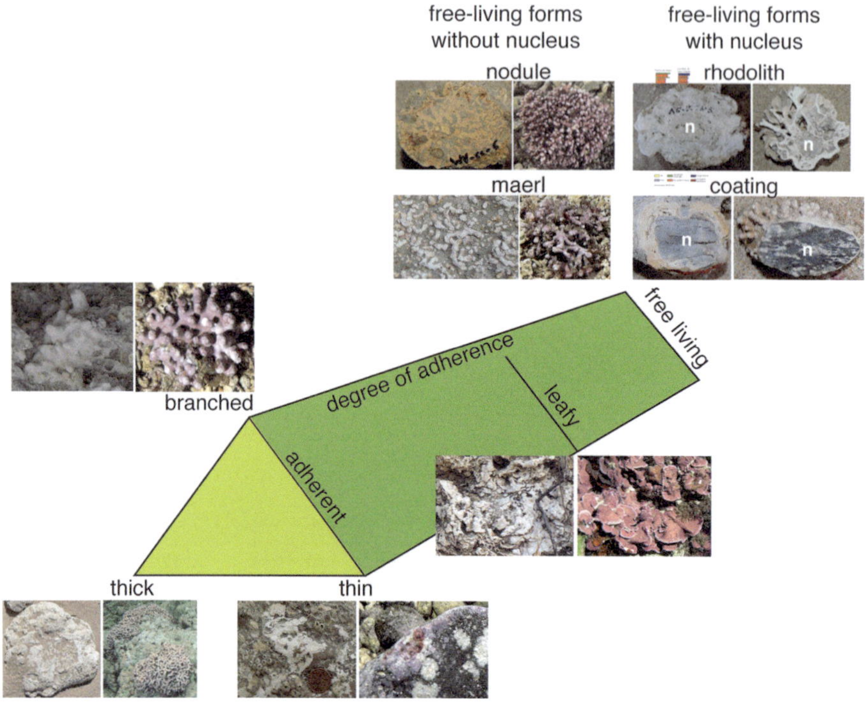

Fig. 5.2 Classification of coralline algal appearances depending upon the degree of adherence of the algal thalli (Modified from Steneck 1986). In all pairs of photographs, picture on the right is a recent exemplar and the image on the left is a fossil counterpart

5.2.2 *Rhodolith Beds*

Rhodoliths can form extensive concentrations, giving rise to the so-called rhodolith beds. In present-day oceans, rhodolith beds occur worldwide, from the Equator to circumpolar latitudes and from the intertidal down to more than 120 m in water depth (Molinier 1956; Pérès and Picard 1958, 1964; McMaster and Conover 1966; Blanc 1968; Adey 1979, 1986; Steneck 1986; Prager and Ginsburg 1989; Littler et al. 1991; Foster 2001; Basso 1998; Riosmena-Rodríguez et al. 2010; Lund et al. 2000; Steller et al. 2003, 2009; Ballesteros 2006; Amado-Filho et al. 2010, 2012a; Matsuda and Iryu 2011; Foster et al. 2013). They constitute hotspots of marine biodiversity in present-day oceans since many marine organisms use rhodolith beds as nurseries, growth habitats or refuges (see a review in Nelson 2009). In addition, some of these rhodolith-bed inhabitants are of particular interest as they are resources for humans (Steller and Cáceres 2009; Steller et al. 2009; Riosmena-Rodríguez et al. 2010).

Lemoine (1910) used the Breton name maërl (also maerl in the literature) to refer to unattached coralline algal concentrations along the Brittany coast. According to

Fig. 5.3 Examples of fossil rhodolith beds. (**a**) Tortonian deposits of the Almanzora Corridor Basin (Almería, SE Spain), (**b**) Serravallian rhodolith bed of Alessandria (Piedmont Basin, NW Italy), (**c**) upper Pliocene deposits of the Cabo de Roche area (Cádiz, SW Spain)

OSPAR Commission (2010), "maërl is a collective term for various species of non-jointed coralline red algae (Corallinaceae) that live unattached. These species can form extensive beds, mostly in coarse clean sediments of gravels and clean sands or muddy mixed sediments…. Maërl beds may be composed of living or dead maërl or varying proportions of both". It, therefore, bears mentioning that in recent literature "rhodolith beds" and "maërl beds" are used indistinctly to refer to extensive concentrations of free-living structures built up mostly by non-geniculate coralline algae. According with the French literature, the term *fons à praliné* was also used (Molinier 1956).

Coralline algae can also densely occur in coastal to deep-subtidal settings but originating bioconstructions: algal ridges, algal frameworks, coralligènes, trottoirs, and corniches (Pérès and Picard 1958, 1964; Blanc 1968; Adey 1978; Bosence 1985a; Freiwald and Henrich 1994; Freiwald 1998; Macintyre et al. 2001; Nalin et al. 2006; Ballesteros 2006; Georgiadis et al. 2009; Aguirre et al. 2014). In all these cases, encrusting coralline algae form a rigid structure.

In the rock record, rhodoliths can form extensive and thick concentrations (Fig. 5.3). Authors have referred to these accumulations variously as beds, pavements, maërls, banks, algal gravels, mounds, biostromes, rhodolith facies, or rhodolith rudstones-floatstones (Nebelsick et al. 2005; Basso et al. 2012). For simplicity, and as an analogue of the present-day rhodolith beds, we will use the same name.

5.3 Classification of Rhodoliths

Bosence (1976, 1983a) summarized a general classification scheme for rhodoliths based on the taxonomic composition (monospecific versus multispecific), the external morphology (spheroidal, ellipsoidal, and discoidal), branch density (I: single branch, II: open branches, III: frequent branches; IV: densely branched), and algal growth forms (laminar, either concentric or boxwork, branching, and columnar). This author further established the methodological procedure to describe the external morphology of the rhodoliths by using the triangular diagram for the classification of pebble shape of Sneed and Folk (1958) (Bosence 1976, 1983a). The three extreme morphologies considered in the diagram are spheroidal, ellipsoidal, and discoidal (Fig. 5.4). These plots are still currently used. However regarding the coralline algal growth form on the surface of the rhodoliths, palaeontologists have been increasingly using the terminological classification of Woelkerling et al. (1993) (Fig. 5.5).

Branch density has been considered in a different way depending upon the authors. Steller and Foster (1995, p. 203) quantified the branch density as "the mean number of apical tips counted in five haphazardly placed 1 cm^2 quadrats". Basso et al. (2009) preferred the use of protuberance degree in order to avoid the word density since it can have a different meaning. In fossil examples, both branch density and protuberance degree have been used (i.e. Bosence and Pedley 1982;

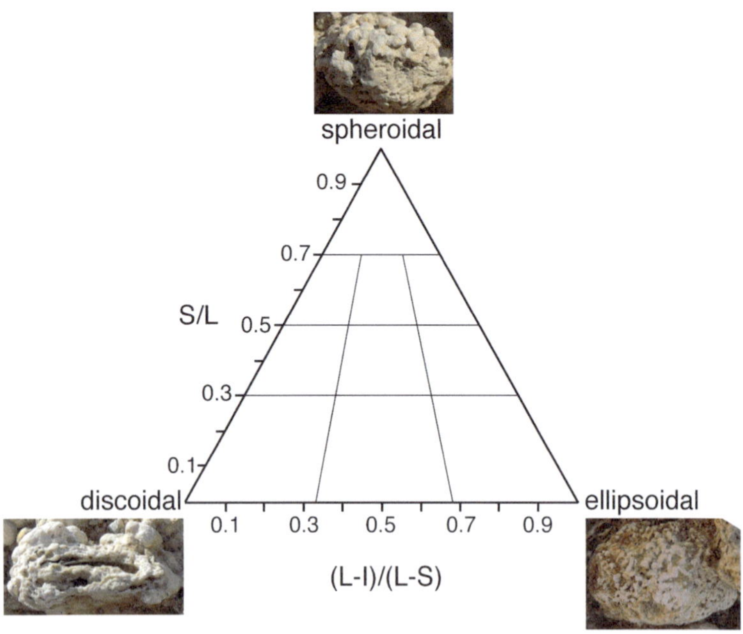

Fig. 5.4 Triangular diagram used for the representation of the rhodolith shape. L, I and S indicate the large, intermediate and short axes, respectively, of the rhodoliths

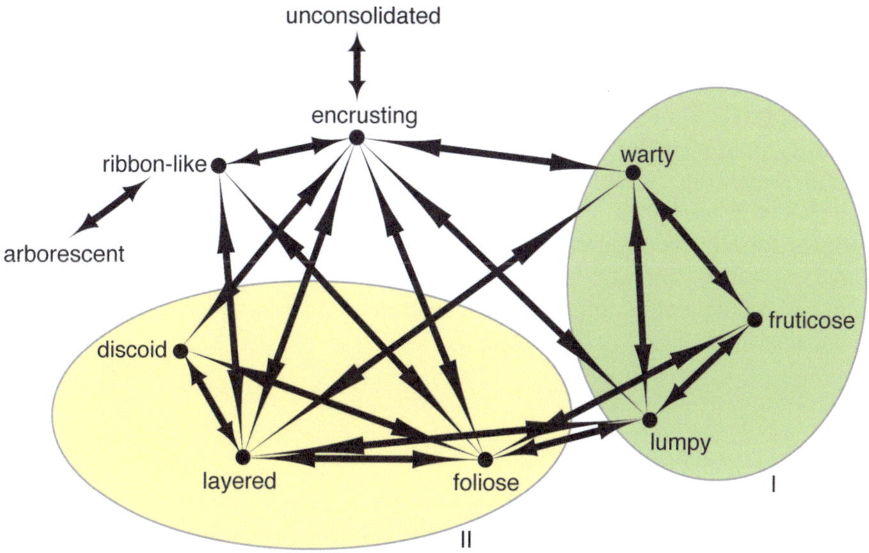

Fig. 5.5 Names of the different types of coralline algal thallus growth forms. The thalli with some sorts of protuberances are included in the green ellipse, while the thalli encompassing lamellar growths are clustered in the yellow ellipse (Modified from Woelkerling et al. 1993)

Bosence 1985b; Basso 1998; Basso et al. 2009; Quaranta et al. 2012, Brandano this volume).

Basso (1998) classified rhodoliths within three morphological groups depending on the size, the inner structure, the external shape, the algal growth forms, and the taxonomic composition: (1) Boxwork: multispecific rhodoliths larger than 4 cm across, irregular to elliptical in shape, formed by laminar to columnar algal growth forms, and with a conspicuous nucleus and numerous internal voids filled up by sediment. (2) Prâline: small (2–4 cm in diameter), massive, monospecific rhodoliths irregular to spheroidal in shape built up by laminar to branching or columnar algal thalli, and with an evident nucleus. (3) Unattached branches: small (1–5 cm in length or diameter), non-nucleated (or with a very small nucleus), monospecific rhodoliths of any possible shape made up by branched algal thalli without voids. According to Basso (1998), the unattached branched rhodoliths include the four types of branch density of Bosence (1976, 1983a), being the denser branching (types III–IV) produced in shallow settings.

This scheme proposed by Basso (1998) is difficult to apply for the rapid classification of rhodoliths in the field. Identifying the algal components building up the rhodoliths very often requires observation under the microscope. Thus, the taxonomic composition (whether monospecific or multispecific) should not be a classification criterion. In addition, as commented above, sectioning is needed to ascertain whether rhodoliths are nucleated. In terms of algal growth form, there is a clear overlap between boxwork and prâline since both can be made up of laminar to columnar algal thalli (Basso 1998).

Taking into consideration that the presence of a differentiated nucleus and the taxonomic composition are two criteria that are very difficult to use in the field, we propose to adopt a classification scheme based on morphological characters at three different levels: (1) Rhodolith morphology. This can be determined using the triangular diagram mentioned above with the three end points ellipsoidal, spheroidal, and discoidal (Fig. 5.4). (2) Growth morphology of the algal thalli on the surface of the rhodolith, using the terminology proposed by Woelkerling et al. (1993). (3) Growth form of the algal thalli in the interior of the rhodolith. In latter case, the three categories proposed by Bosence (1983a) would clearly include the whole range of possible algal forms: laminar (concentric or non-concentric [boxwork]), branching (types I to IV), and columnar.

5.4 Estimation of Rhodolith Abundance

An important issue is to determine rhodolith abundance and density within rhodolith beds. Several approaches have been used. Density, expressed as number of rhodoliths by surface, can be quantitatively estimated directly by counting rhodoliths in a specified area (Ballantine et al. 2000; Amado-Filho et al. 2010). Basso et al. (2012) proposed to quantify rhodolith abundance in rhodolith beds with image analysis, as done by Amado-Filho et al. (2012a). Abundance can be also semi-quantitatively determined by applying charts for visual comparison of the proportion of fossil components (rhodoliths in this case) per rock volume (Kidwell and Holland 1991). Abundance can be further qualitatively approached by estimating rhodolith packing using the three categories of the comparison tables of Kidwell and Holland (1991) (Fig. 5.6): (1) dispersed: rhodoliths are separated from each other by a distance of more than the largest diameter of the rhodoliths; (2) loosely packed: rhodoliths are close each other but not in contact; and (3) densely packed: rhodoliths are in direct contact with each other throughout the rhodolith bed.

Within individual rhodoliths, the relative abundance of coralline algal taxa can be estimated by point counting the cross-sectional areas that they occupy in thin sections following the method proposed by Perrin et al. (1995).

5.5 Rhodoliths as Palaeoenvironmental Indicators

Fossil coralline algae have been known since the nineteenth century and have long been studied intensively from a taxonomic perspective (Aguirre and Braga 2005) but without considering their potential use to make palaeoecological interpretations (Adey and Macintyre 1973; Bosence 1983b). Bosellini and Ginsburg (1971) came to the conclusion, for the first time, that rhodolith morphology, as well as inner algal arrangement and algal growth forms, record valuable information of the environmental conditions in which these nodular structures grow. Therefore, these authors stressed that the study of fossil rhodoliths would be useful to make palaeoenvironmental interpretations.

Fig. 5.6 (**a**, **b**) dispersed rhodolith beds, (**c**, **d**) loosely packed rhodolith beds, (**e**, **f**) densely packed rhodolith beds

Adey and Macintyre (1973) updated the general knowledge concerning living rhodoliths and discussed their potential use in geological studies. They further emphasized that taxonomic problems and the limited knowledge of environmental distribution of recent rhodoliths are major hindrances to apply a strict uniformitarian approach to interpret fossil rhodoliths. Bosence (1991) highlighted the necessity of establishing an uniform and consistent taxonomic framework for fossil coralline algae comparable to the taxonomic standards of present-day corallines. Braga et al. (1993) subsequently demonstrated that many of the taxonomic criteria used to identify present-day corallines were preserved in the fossil counterparts, influencing the latter advance of fossil coralline taxonomy.

Adey and Macintyre (1973) also argued that despite the widespread geographic and bathymetric distribution of coralline algae as a whole, particular taxonomic groups occur in characteristic latitudinal and shelf habitats. This was tested in different regions from the tropics to cold-temperate latitudes (Adey 1979, 1986; Adey et al. 1982; Bosence 1983b, 1991; Minnery et al. 1985; Minnery 1990). The realization that groups of coralline algae preferentially inhabit particular environments is critical for palaeoecological and palaeoclimatology-palaeolatitudinal inferences. More recently, the use of geochemical analyses has increased the value of rhodoliths for palaeoenvironmental reconstructions (Kamenos et al. this volume).

5.5.1 Water Energy

It has often been stated that the morphology of rhodoliths is closely related to water movement. Bosellini and Ginsburg (1971) established five types of rhodolith morphologies that can be found along a water-energy gradient. Spheroidal and ellipsoidal rhodoliths would be characteristic of moderate- to high-energy settings. Under these conditions, water movement promotes a frequent overturning of rhodoliths, and coralline algae grow in all possible directions. Flat, discoidal, and ameboidal rhodoliths would be more frequent in calm waters, as certain stability is required to produce these morphologies. In contrast, Adey and Macintyre (1973, p. 900) concluded that "rhodolith morphology should not be related to energy conditions", questioning the rhodolith shape-water energy relationship proposed by Bosellini and Ginsburg (1971).

Bosence (1976) experimentally demonstrated using a wave tank that ellipsoidal rhodoliths are more easily transported than spheroidal ones, and that discoidal forms are the most stable morphologies. According with this experiment, discoidal rhodoliths would be expected to be more abundant in quiet settings while ellipsoidal and spheroidal rhodoliths should be dominant in more exposed environments. However, in rhodolith beds from the Mannin Bay (Connemara, Ireland), ellipsoidal and spheroidal rhodoliths did not show any characteristic energy-dependent environmental distribution (Bosence 1976). More recently, observations made in the Gulf of California (Steller and Foster 1995; Steller et al. 2009) and off Fraser Island in the Great Barrier Reef (eastern Australia) (Lund et al. 2000) have also demonstrated that there is no clear relationship between rhodolith morphology and water energy.

Regardless of the lack of correspondence between experimental tests and field observations, morphology of fossil rhodoliths has been usually interpreted according to the original laboratory results of Bosence (1976). Hence, discoidal rhodoliths are often interpreted as formed in calm water settings and ellipsoidal and spheroidal ones are usually considered characteristic of high-energy conditions (e.g. Bassi 1995, 1998; Bassi and Nebelsick 2010; Checconi et al. 2010). Nonetheless, rhodolith shape is not directly correlated with water energy (e.g. Brandano et al. 2005; Bassi et al. 2006). A combined analysis of facies and sedimentary structures, together with rhodolith shapes, is needed to correctly infer the palaeoenvironmental turbulence.

This lack of relationship may be due to different factors. The movement of rhodoliths could be due directly to the activity of vagrant organisms moving around algal nodules and not to currents or waves (Marrack 1999). This author carried out experimental and observational work on rhodolith morphology in rhodolith beds at different sites and water depths from the Gulf of California. Using video camera and scuba surveys, Marrack (1999) showed that bioturbation was the key factor causing displacements of rhodoliths deeper than 12 m, where water energy is low enough or negligible to overturn rhodoliths. At this water depth, rhodoliths were only sporadically moved by currents during severe storms.

Rhodolith morphology can also largely depend upon the inherited shape of the nucleus if the algal cover is poorly developed (Braga and Martín 1988; Aguirre et al. 1993). Large flat nuclei encrusted by corallines producing a thin coating due to a limited time of growth would produce discoidal rhodoliths in shallow, high-energy settings. On the other hand, long-term exposure of rhodoliths on the seafloor with enough time to develop thick algal covers with respect to the nucleus would generate ellipsoidal and spheroidal rhodoliths. This has been shown in recent rhodoliths off Fraser Island (NE Australia) (Lund et al. 2000), as well as in upper Miocene rhodolith beds of the Almanzora Corridor Basin (Almería, SE Spain) (Braga and Martín 1988) and in upper Pliocene rhodolith beds of Cabo de Roche (Cádiz, SW Spain) (Aguirre et al. 1993).

Other features of rhodoliths have been related to water energy and considered reliable tools in palaeoenvironmental analyses. The internal coralline algal growth forms do relate to the degree of water energy (Bosellini and Ginsburg 1971; Adey and Macintyre 1973; Bosence 1983a, b, 1991; Steneck 1986). According to these authors, rhodoliths formed by thin, laminar algal thalli occur in calm waters. Fruticose, delicate branching algal thalli are related to low or intermediate water movement. Under higher turbulence, branch density increases (Bosence 1976; Basso et al. 2009). Additionally, the branch tips in fruticose thalli break and are eroded, producing shorter columns with broadening of tips that can fuse together to better withstand moderate water movements (Steneck 1986). Thus, under moderate energy, algal thalli acquire lumpy-warty growth forms. These variations in thallus morphology related to turbulence have been recorded in different fossil examples (Bosence and Pedley 1982; Braga and Martín 1988; Aguirre et al. 1993; Bassi 1995, 2005; Bassi and Nebelsick 2010).

Growth forms, however, can be genetically controlled. Foliose and fruticose thalli of the rhodolith-forming species *Lithophyllum margaritae* Riosmena-Rodríguez et al. are actually genetic variants (Schaefer et al. 2002). These authors suggested that these two genetic entities would correspond to different species.

The percentage of constructional voids in rhodoliths is also a good approach to infer water energy (Minnery et al. 1985). Open inner arrangements, with a great proportion of voids among algal thalli, are generally indicative of quiet waters. An increase in turbulence implies a progressive massiveness of the internal rhodolith structure. The relationship between massiveness and water energy has been shown in different fossil examples (Bassi 1995, 2005; Bassi and Nebelsick 2010).

5.5.2 Water Depth

Coralline algae, like all photosynthetic organisms, depend on the sunlight to survive. Light is progressively absorbed as it penetrates the water column. Coralline algae have phycobiliproteins as accessory pigments (Lobban and Harrison 1994; van der Hoeck et al. 1995; Lee 2008; Graham et al. 2009) that provide them the ability to colonize the deepest settings among all algal groups (Littler et al. 1985).

Light irradiance might be considerably reduced with increasing suspended material, either inorganic (particles of sediment) or organic (food particles). Therefore, turbid waters limit the bathymetric distribution of coralline algae (Adey and Macintyre 1973; Bosence 1983b; Martindale 1992; Koop et al. 2001; Steller et al. 2009). In shallow settings, turbidity might be favoured by resuspension of sediment due to waves and bottom currents, siliciclastic input, and refractory organic matter supply from rivers. Offshore, light penetration might decline in areas of high productivity linked to upwelling centers.

As mentioned above, corallines occupy benthic habitats at any depth within the photic zone. Nonetheless, some taxonomic groups of coralline algae show particular depth distributions (Adey and Macintyre 1973; Adey 1979, 1986; Bosence 1983b, 1991; Minnery et al. 1985; Minnery 1990) (Fig. 5.7). The depth ranges of the taxa, however, vary depending on the latitude (see next Sect. 5.5.3). In low latitudes, shallow-water coralline algal assemblages are mostly dominated by 'mastophoroids' (sensu Harvey et al. 2003) followed by lithophylloids (Gordon et al. 1976; Adey 1979; Adey et al. 1982; Bosence 1984; Minnery et al. 1985; Minnery 1990; Verheij and Erftemeijer 1993; Iryu et al. 1995; Payri et al. 2000; Ringeltaube and Harvey 2000; Littler and Littler 2003). Lithophylloids, with subordinate 'mastophoroids', dominate in temperate seas, such as in the Mediterranean (Hamel and Lemoine 1953; Comarci et al. 1985; Adey 1986; Di Geronimo et al. 1993; Braga and Aguirre 2009). Members of the family Sporolithaceae are also present in shallow, warm waters but they are scarce and limited to cryptic habitats (i.e. Braga and Bassi 2007). Melobesioids, as well as sporolithaceans, increase with waters depth (Adey 1979, 1986; Adey et al. 1982; Minnery et al. 1985; Minnery 1990; Verheij and Erftemeijer 1993; Iryu et al. 1995; Rasser and Piller 1997; Lund et al. 2000; Payri et al. 2000; Littler and Littler 2003) (Fig. 5.7). In high latitudes, melobesioids is the overwhelmingly dominant group (Adey 1979, 1986; Adey et al. 1982; Steneck 1986).

Taking into consideration the latitudinal variation in relative abundance of taxa with respect to their bathymetric ranges, the depth/coralline-taxa distribution relationship has been used to infer palaeobathymetry. The basic idea is to combine coralline algae that inhabit particular water-depth ranges into ecologically homogeneous groups and then analyse relative abundances of these taxa in geological sections to infer palaeobathymetry, as well as related sea-level changes through time. Qualitative estimations of bathymetrically controlled taxa were used to infer palaeodepth (Buchbinder 1977; Braga and Martín 1988; Aguirre et al. 1993). Braga and Aguirre (2001) used quantitative data for the first time to infer palaeodepth

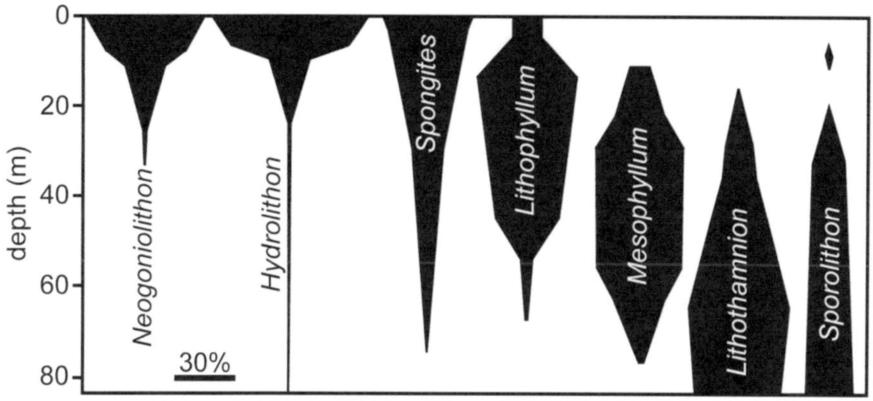

Fig. 5.7 Abundance of typical coralline genera-forming rhodoliths in relation with water depth in low latitudes

changes in upper Miocene-upper Pliocene temperate and warm-water carbonate deposits in different basins in southern Spain. Subsequent studies have applied this method in different deposits from disparate regions (Braga and Aguirre 2004; Brandano et al. 2005; Kroeger et al. 2006; Braga et al. 2009; Benisek et al. 2009; Aguirre et al. 2012; Nebelsick et al. 2013).

5.5.3 *Palaeolatitude and Palaeoclimate*

Corallines were long erroneously considered as warm, low-latitude organisms in the fossil record (Adey and Macintyre 1973). However, as commented above, they are found in warm and cool-temperate carbonates. Despite their cosmopolitan occurrence, particular taxonomic groups of corallines (subfamilies) from shallow-water settings show preferential latitudinal distributions (Fig. 5.8) (Adey and Macintyre 1973; Adey et al. 1982; Adey 1986; Minnery et al. 1985; Bosence 1991; Perrin et al. 1995). Among coralline algae, members of the subfamily 'Mastophoroideae' show a preferential tropical distribution (Steneck 1986; Woelkerling 1996a). Representatives of the order Sporolithales show a similar latitudinal dispersal (Verheij 1993; Woelkerling 1996b). Within Sporolithales, the genus *Sporolithon* has maintained this latitudinal preference through time (Johnson 1963; Fravega et al. 1989; Braga and Bassi 2007). Lithophylloids are abundant mainly in intermediate, warm, and temperate latitudes (Woelkerling 1996c). Finally, the subfamily Melobesoideae is dominant mostly at high latitudes (Woelkerling 1996d).

The upper Neogene deposits in different basins from southeastern Spain are characterized by an alternation of reefal and temperate carbonates (Martín and Braga 1994; Brachert et al. 1996). The temperature control of this alternation was firstly demonstrated using $\delta^{18}O$ isotope values measured in planktonic and benthic foraminiferal shells collected from the time-equivalent marly sediments (Sánchez-

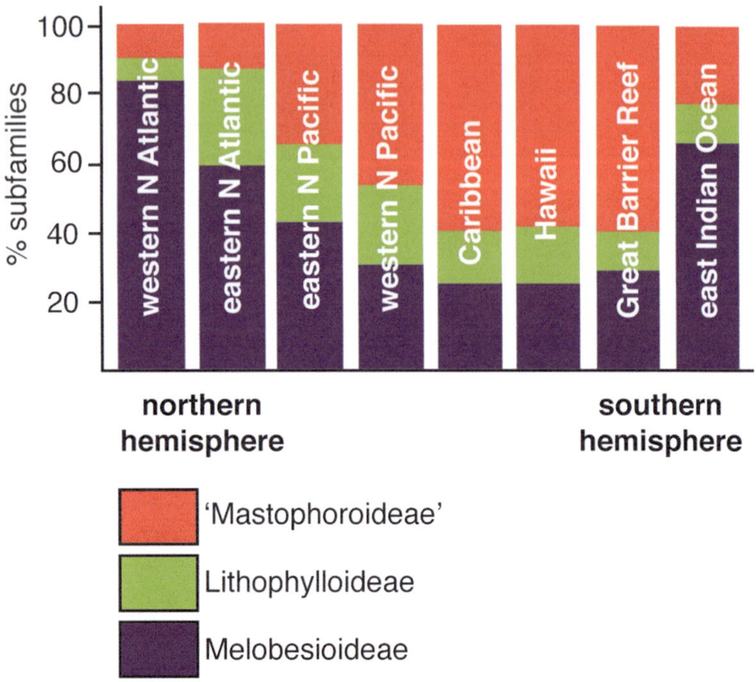

Fig. 5.8 Latitudinal distribution of algal subfamilies in shallow-water assemblages (Modified from Steneck 1986). The subfamily 'Mastophoroideae' is here considered according with Harvey et al. (2003)

Almazo et al. 2001). 'Mastophoroid'-rich assemblages (accompanied by Sporolithales) dominated in reefal carbonate deposits while lithophylloid-dominated assemblages did so in the shallow-water temperate carbonates (Braga and Aguirre 2001). In deeper settings, both temperate and reefal carbonates are characterized by abundant melobesioids. Later studies have followed this approach to infer palaeoclimatic changes (Brandano et al. 2005; Kroeger 2007; Nalin et al. 2008).

Geochemical studies have added new insights on the use of corallines as palaeothermometers. Kamenos et al. (this volume) review the applications of these approaches, including elemental analyses as well as oxygen isotopes. Nonetheless, any of these geochemical analyses as palaeoclimatic proxies in deep time (millions of years) have, so far, not been applied. Diagenetic alterations, such as early cementation, early transformation of the high-Mg calcite into low-Mg calcite (Alexandersson 1974, 1977, 1978; Freiwald and Henrich 1994), most likely bias the original geochemical signature, impeding the use of these proxies in palaeontological samples. Stable oxygen isotopes are also difficult to apply as palaeoclimatic/palaeotemperature indicators in fossil coralline algal skeletons due to a strong vital effect, that is, the preference uptake of a specific oxygen isotope due to metabolic (photosynthesis and respiration) and kinetic (isotope diffusion) effects (Wefer and

Berger 1991; Halfar et al. 2000; Lee and Carpenter 2001; but see Rahimpour-Bonab et al. 1997).

5.6 Rhodoliths as Palaeocommunities

Rhodoliths are multispecific structures formed by the successive growth of encrusting organisms, with the accompanying non-sessile ones, that can therefore be considered communities by themselves. Space is a key ecological factor structuring communities of encrusting organisms (Begon et al. 1990; Buss 1990; Hochberg and Lawton 1990; Leigh 1990). Competition for space might be critical for those organisms growing in a limited space during formation of rhodoliths. For rhodoliths with a nucleus, the first step is the colonization of an empty space. Eventually, these colonizers might come into contact during their growth, thus beginning intra- and interspecific competition for the space (Adey and Macintyre 1973; Bosence 1983c, 1985b; Steneck 1985; Prager 1987). Coralline algae grow very slowly in relation with other potential competitors. In this regard, thin thalli of coralline algae that might grow quickly would behave as primary colonizers or generalist epiphytic species interacting with other fast-growing encrusting organisms, such as serpulids, bryozoans or benthic foraminifers (Fig. 5.9) (Adey and Macintyre 1973; Adey and Vassar 1975; Steneck 1985; Woelkerling 1988; Keats et al. 1993; Figueiredo et al. 1997).

However, inferring space competition among organisms forming rhodoliths in fossil examples is difficult, since many factors should be taken into consideration: (1) longevity of the different organisms; (2) the importance of other possible ecological interactions, such as mutualism or parasitism; (3) the role played by non-preserved soft-body epiphytes or epizoos; (4) effect of herbivorous; (5) chemical defenses among competitors; (6) subtle preferences for abiotic factors (light intensity, nutrients, currents, turbulence, etc.); (7) growth rate; and (8) temporal interruption of growth. All these factors might have played a key role structuring rhodolith communities, influencing the distribution and success of all encrusting organisms involved in rhodolith formation.

Analyses of the changes in the growth forms of the coralline algae, as well as in the taxonomic composition from the nucleus to the surface of the rhodoliths can provide valuable information of palaeoenvironmental shifts during their development. Some authors have interpreted these taxonomic transitions as ecological successions by facilitation (Adey and Macintyre 1973; Bosence 1983b, c). Prager (1987), in contrast, highlighted that the taxonomic changes from the nucleus to the outer surface of rhodoliths might be due to physical or biotic changes through time not related to ecological successions (see also Basso et al. 2009). Additionally, rhodoliths do not grow continuously, precluding any inference of ecological succession during rhodolith growth. Bioerosion, erosion, early lithification, or abrasion of rhodolith builders can reflect complex taphonomic histories that record important palaeoenvironmental changes. This is the case of rhodoliths from different regions that

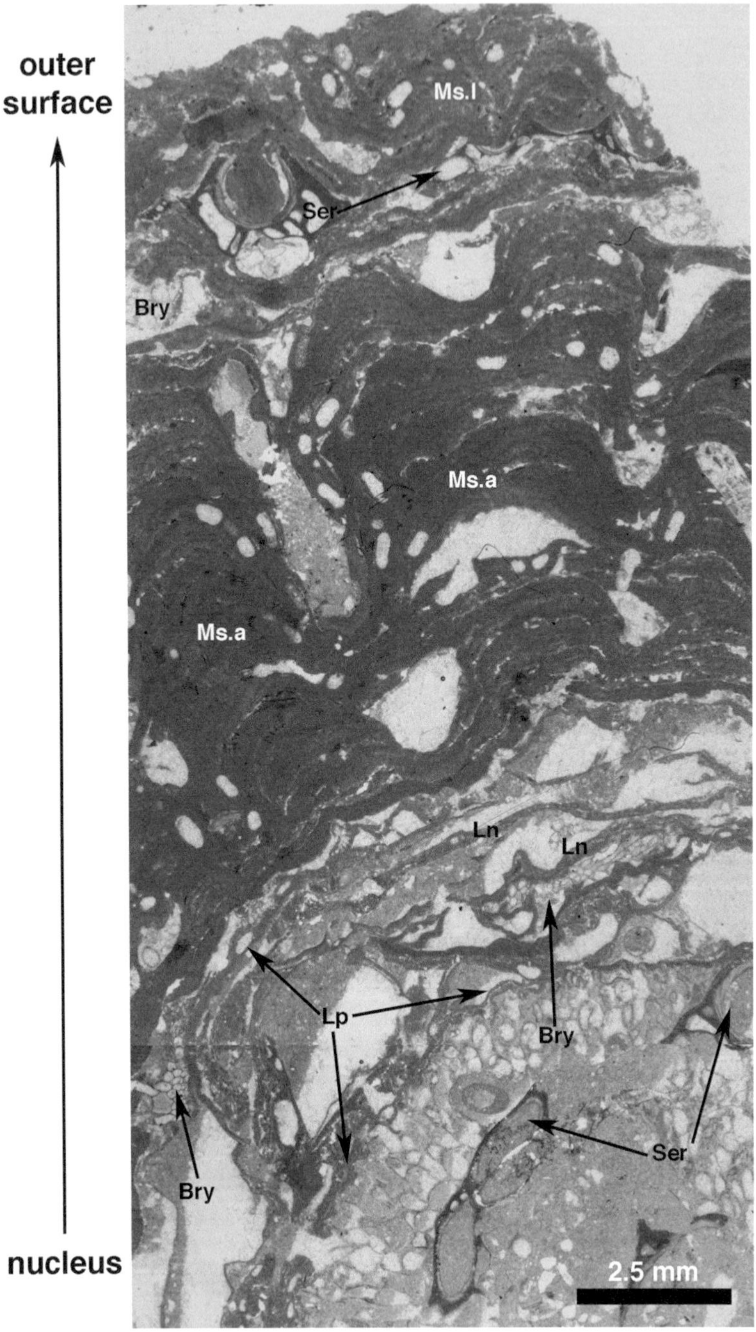

Fig. 5.9 Thin section through a fossil rhodolith from the nucleus to the outer surface showing interactions among the major builders of the structure. The nucleus is a nodular bryozoan colony. In the outer surface of the nucleus, serpulids (*Ser*) intergrew with the bryozoan up to an eventual outcompetition of the serpulids. The serpulids were then colonized by thin laminar, fast-growing thalli of *Lithophyllum pustulatum* (Lamouroux) Foslie (Lp) and *L. nitorum* Adey and Adey (Ln) that interacted with bryozoans (*Bry*) and serpulids. Finally, serpulids became progressively less abundant and warty/fruticose thalli of *Mesophyllum alternans* (Foslie) Mendoza and Cabioch (Ms.a), which are, in turn covered by serpulids, bryozoans and encrusting thalli of *Mesophyllum lichenoides* (Ellis) Lemoine (Ms.l)

consist of two well-differentiated parts, a highly altered, lithified, and bored inner part and an outer pristine one (McMaster and Conover 1966; Focke and Gebelein 1978; Reid and Macintyre 1988; Minnery 1990; Littler et al. 1991). According to radiocarbon dating, the separation between the two parts is characterized by an interruption in the rhodolith growth averaging about 500 years.

Going back into the deep time, when radiometric dating methods are not applicable to rhodoliths, identifying erosive surfaces in the interior of the rhodoliths affecting the whole nodule is the best evidence to detect prolonged interruptions during their formation. Examples of multistory rhodoliths have been described in the fossil record (Checconi et al. 2010; Aguirre and Braga 2012). In rhodolith assemblages from the middle Miocene hemipelagic *Orbulina* marls from the Southern Apennines (Italy), Checconi et al. (2010) distinguished two rhodolith growth stages separated by an erosive and bored surface: (1) nucleation and growth of the rhodoliths, and (2) a final growth stage before burial. Nucleation is characterized by melobesioids and subordinate 'mastophoroids', with rare sporolithaceans and lithophylloids. The rhodolith growth (main increase in size) is represented by abundant melobesioids and rare to common 'mastophoroids'; very rare sporolithaceans are also present. The final growth stage is dominated by melobesioids with rare 'mastophoroids' and very rare sporolithaceans.

The Pliocene deposits in Cabo de Roche area (Cádiz, SW Spain) consist of three unconformably superimposed units (Fig. 5.10a, b). The upper unit is made up of densely packed concentrations of rhodoliths formed in a relatively deep and sheltered palaeobay (Aguirre 1992; Aguirre et al. 1993; Aguirre and Braga 2012) (Figs. 5.3c, 5.6e, and 5.10b). More than 20 % of these rhodoliths (n = 140 rhodoliths) show two distinct phases separated by an erosion surface (Fig. 5.10b). The inner phase is eroded and bored and consists of thin laminar growths of *Lithophyllum pustulatum* intergrowing with serpulid worm tubes, which are the dominant components. The outer phase is characterized by warty-lumpy to fruticose growths of *Phymatolithon calcareum* (Pallas) Adey and McKibbin (Fig. 5.10b). Occasionally, laminar thalli of *Mesophyllum lichenoides* and/or, to a lesser extent, *Lithophyllum incrustans* Philippi overgrow *P. calcareum* branches. The differentiation of the two stages points to a complex history of rhodolith development. These rhodolith-dominated deposits unconformably overlie the middle Pliocene unit (Fig. 5.10b), consisting of bioclastic calcirrudites with nodular structures dominated by serpulids encrusted by thin thalli of *L. pustulatum*, the so-called serpulid nodules by Aguirre et al. (1993) (Figs. 5.1b, 5.10b). This suggests that the inner parts of the complex rhodoliths of the upper unit are serpulid nodules reworked from these older deposits constituting the nuclei for *P. calcareum* settlement and growth (Aguirre and Braga 2012). Thus, in this case, an intricate sedimentological and taphonomic history of growth, burial, reworking, and regrowth can be inferred during long-term rhodolith development.

In addition to encrusters contributing to the construction of rhodoliths, endolithic organisms use them to live inside bores, thus promoting the destruction of these hard living substrates. Bassi et al. (2011, 2012) studied this destructive guild in recent nodules of *Acervulina*-coralline algae from deep fore-reef settings (60–100 m water depth) off Kikai-jima (S Japan). They described an ichnocoenosis dominated

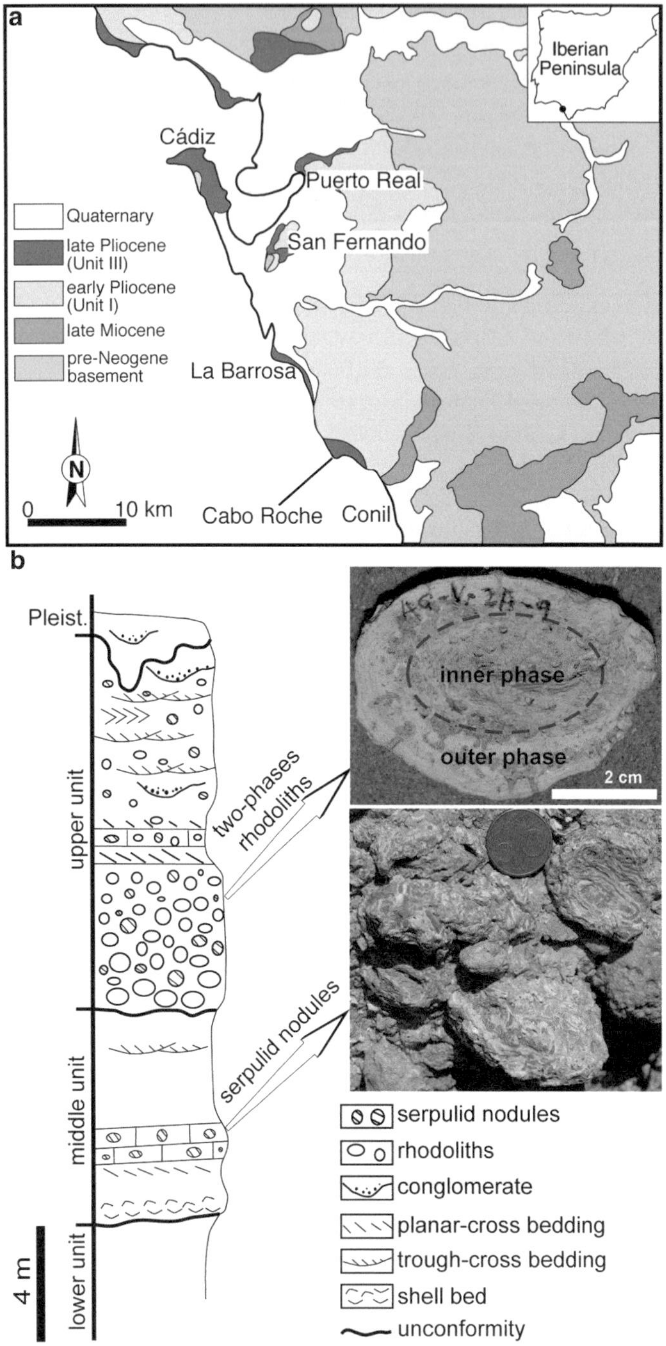

Fig. 5.10 (**a**) location of the Pliocene deposits close to the Cabo de Roche area (Cádiz, SW Spain), (**b**) synthetic stratigraphic log of the Pliocene deposits of the Cabo de Roche area, consisting of three unconformably overlying units. A bed formed by serpulid nodules intercalates in the middle unit. The upper unit starts with a dense rhodolith bed, up to 4 m thick, in which more than 20 % of the rhodoliths consist of two phases separated by and eroded and bored surface. The inner phase represents serpulid nodules eroded from the middle unit that constituted the nucleus for the algal growths of the outer phase. (Pleist. = Pleistocene)

by *Entobia* and *Gastrochaenolites*, with accessory *Trypanites* and *Meandropolydora* as well as unidentified microborings. These ichnogenera preferentially inhabit in settings characterized by low sedimentation rates and high turbulence. This ichnoassemblage has been traditionally considered diagnostic of very shallow waters (less than 10 m) and, therefore, an excellent sea level indicator. This general assumption is patently challenged by the findings of Bassi et al. (2011, 2012). Regarding ecological controls on borers, Bassi et al. (2011, 2012) concluded that while small *Entobia*, made by the sponge *Cliona*, is present in nodules of any size, large *Gastrochaenolites*, produced by bivalves, occurs in the late growth of larger nodules (6–7 cm in diameter).

A decrease in nodule size and thickness of coralline thalli with depth is observed in present-day coralline algal nodules off Fraser Island (eastern Australia) from the inner to the uppermost outer shelf (28–117 m water depth) (Bassi et al. 2013). Such trends most likely account for the decrease in diversity of ichnogenera with depth, as the largest bioeroders are excluded from the ichnoassemblages. Lower coralline growth rates favour higher density of bioerosion in deeper algal nodules.

5.7 Sedimentological and Sequence Stratigraphic Significance of Rhodolith Beds

Fossil rhodolith beds can be formed by either sedimentological (allochthonous) or biological (autochthonous) processes. Allochthonous concentrations are the result of either offshore or onshore transport from the production areas of rhodoliths. Bassi et al. (this volume) discuss examples of basinward-reworked rhodolith beds in several Neogene basins of Italy (central Mediterranean) associated with submarine lobes and channels. Johnson et al. (2012, 2013, this volume) present study cases of offshore reworked rhodolith-dominated deposits, and onshore export of rhodoliths from deeper settings linked to extremely high-energy events, such as hurricanes or tsunamis, in Macaronesian archipelagos.

Here, we focus on autochthonous accumulations of rhodoliths formed by in situ production. Pure autochthonous rhodolith beds are very unusual because long-term time-averaged processes prevail in the rock record (Kidwell and Bosence 1991). As a consequence, some reworking might be expected in the areas of rhodolith production. As commented above, autochthonous rhodolith beds can potentially form throughout the platform along a depth gradient, from the coast to the outer-shelf. However, some environmental conditions, as discussed in this section, inhibit the development of rhodolith beds.

Growth rates of coralline algae are very low, being greater in the tropics than in the warm and cold-temperate realms (Adey and Macintyre 1973; Kamenos et al. 2008; Halfar et al. 2011). Consequently, they are very sensitive to sediment influx, and long-term rhodolith bed formation requires sediment starvation. Off Brazil, rhodoliths form one of the largest rhodolith bed known so far (Amado-Filho et al. 2012b). However, terrigenous discharges linked to Amazon and Parana rivers inhibit rhodolith bed formation (Milliman 1977, p. 236). Experimental work has demonstrated the deleterious effect of sediment burying rhodoliths, even under a millimetre-thick layer of fine-grained sediment (Riul et al. 2008; Villas-Bôas et al. 2014). Rhodoliths made up of *Mesophyllum engelhartii* bleached after being buried 75 days and those formed by *Lithothamnion* sp. died even earlier, after 41 days (Villas-Bôas et al. 2014). Burial of rhodoliths by fine-grained sediment with high content of organic matter might also lead to anoxic conditions that decimate rhodolith growth (Wilson et al. 2004; Hall-Spencer et al. 2006). Sediment supply also muddies the water column thus limiting light penetration and inhibiting the healthy growth of rhodoliths (Adey and Macintyre 1973; Bosence 1983b; Steller and Foster 1995; Foster et al. 1997; Steller et al. 2009). The harmful influence of terrigenous sedimentation on rhodolith bed development has been proven in Pliocene deposits in SE Spain (Aguirre et al. 2012).

As stated above, rhodoliths are able to cope with a wide spectrum of environmental conditions, from shallow high-energy to deep low-energy settings. Nonetheless, successful and continuous development of dense rhodolith beds takes place under moderate water energy (Bosence 1983b; Hottinger 1983; Foster et al. 1997; Steller et al. 2009; Aguirre et al. 2012). In different Neogene basins of SE Spain, rhodolith beds formed in mid to outer ramp settings, in a facies belt nearly parallel to palaeocoast and basinward of a shoal facies belt (Braga et al. 2006; Puga-Bernabéu et al. 2007; Martín et al. 2004; Aguirre et al. 2012). In Miocene warm-temperate to tropical carbonate platforms in central Mediterranean (Italy), rhodolith beds occur from inner to mid platform, both shoreward and offshore of *Porites* patch reefs (Benisek et al. 2010; Brandano et al. 2010). Under very low water energy, very fine-grained sediment can be deposited in these settings, suffocating coralline algal development. Steller et al. (2009) experimentally removed rhodoliths from a rhodolith bed to deeper areas in the Gulf of California and found that they became buried rapidly and, consequently, were killed off. In contrast, moderate water energy allows to: (1) wash out fine particles and excess of organic matter, reducing fouling; (2) rock the rhodoliths, allowing coralline algae to grow in all directions; (3) permit free movement of vagrant organisms, which slowly displace rhodoliths; and (4) preclude the formation of algal bridging among rhodoliths, preventing the growth of a more rigid structure (Bosence 1983c; Steller and Foster 1995; Foster et al. 1997; Freiwald and Henrich 1994; Harvey and Bird 2008; Peña and Bárbara 2008, 2009; Steller et al. 2009).

The best conditions for rhodolith bed formation and persistence are relatively low-moderate water hydrodynamic, good oxygenation, and low sedimentation rate. The long-term maintenance of these environmental conditions is essential for the formation of thick and dense rhodolith beds like those found in the geological record. In a sequence stratigraphic context, the appropriate conditions are usually reached during transgressive periods (Friebe 1993; Nalin et al. 2008; Leszczynski et al. 2012). When the sea level raises the coastal settings, where high-energy conditions prevail and siliciclastic sedimentation is favoured, move onland widening the areas of carbonate production in the platforms where rhodolith beds can be established.

Although it is largely accepted that thick rhodolith beds form preferentially during the transgressions, rhodolith beds also develop during highstand stages (Aguirre et al. 2012). The required condition for rhodolith bed formation during highstand stages is the absence of terrigenous sedimentation. During lowstand stages, large terrestrial areas are exposed and the base line of erosion in fluvial systems moves downwards. Consequently, erosion and fluvial sediment discharges generally increase, precluding the healthy long-term formation of rhodolith beds. In any case, as stated above, when terrigenous starvation and moderate hydrodynamics concur in a particular area, a sustained rhodolith bed formation is possible.

In the geological record, densely packed rhodolith beds (rhodolith-supported) can be up to several meters thick (Fig. 5.3). Since coralline algal growth is very low (although highly variable from the tropics to high latitudes), the formation of rhodolith beds should imply long time spans. Steller et al. (2009) qualitatively estimated that a rhodolith 3–8 cm in diameter at the Gulf of California would remain on the seabed from years to decades. According to Reid and Macintyre (1988), the 1–3 cm external cover of rhodoliths in different localities, including northern Gulf of Mexico, the Bermuda platform, the eastern Caribbean, and insular shelves of the Canary Islands, has been continuously growing on average for 500 years (see also Littler et al. 1991 for examples in the Florida Keys). A simple calculation implies that to produce a 1-m-thick rhodolith bed due to the continuous accumulation of constantly growing rhodoliths 3 cm across would require about 16,500 year to form. Schäfer et al. (2011, table 6) tabulated growth rates of corallines from different localities and water depths. They range from 0.05 to 5.2 mm year^{-1} (excluding one value of 0.01 mm year^{-1}). Using these end values, respectively, a rhodolith 3 cm in diameter grow from 600 to 5.8 years. Thus, a 1-m-thick rhodolith bed would be formed between 20,000 and 193 years. Bosence and Wilson (2003) compiled information of accumulation rates of maërl beds from different temperate and tropical areas. Their results show accumulation rates ranging from 0.08 to 1.4 m year^{-1} (Table 5.1). According to these minimum and maximum values, a 1-m-thick rhodolith bed would form between ~10,000 and 714 years, in agreement with the results estimated above using algal growth rates. A 20-cm thick rhodolith bed on the sea floor on the Alborán Ridge in the western Mediterranean that has been developing for the last 800–1,000 years (Betzler et al. 2011) lies within this range. All calculations imply continuous rhodolith growth and a continuous rhodolith accretion, so that these figures show the minimum time lapse involved in the formation of 1-m-thick rhodolith bed. Obviously, the time lapse involved in the formation of such

Table 5.1 Accumulation rates of rhodolith beds from different places

Site	Thickness	Accumulation rate	Reference
St Mawes Bank (Ireland)	2 m	0.5 m kyr^{-1}	Bosence (1976)
Orkney sounds (N Scotland)	0.5 m	0.08 m kyr^{-1}	Farrow et al. (1984)
Kvaløy (Norway)		1.4 m kyr^{-1}	Freiwald (1998)
Tromsøn Island (Norway)		0.8–1 m kyr^{-1}	Freiwald (1998)
Tavernier Keys (Florida)	2 m	0.5–1.5 m kyr^{-1}	Bosence (1985b)

a rhodolith bed would be longer since neither rhodolith growth nor rhodolith accumulations are continuous. In conclusion, thick fossil rhodolith beds most likely form in hundreds to tens of thousands of years on average.

In contrast to the geological record, where rhodolith beds of several meters in thickness are found, living rhodolith beds appear to be only superficial features, except for those examples shown by Bosence and Wilson (2003) (Table 5.1). Several hypotheses might be envisaged to account for this "paradox of the present" regarding the long-term rhodolith production in continental shelves: (1) The present-day sea level corresponds to the highstand following the pre-Holocene transgression. As commented above, these are not the best environmental conditions to produce the long-term rhodolith beds and they occur only in areas with low terrigenous input. (2) Thick rhodolith beds developed during the last sea-level rise (Bosence and Wilson 2003) and they are now relic deposits in starved platforms. (3) Present-day rhodolith beds can be thicker than observed (see examples by Bosence and Wilson 2003) but no reliable information on subsurface data is available. Core boring through living rhodolith beds has to be performed.

5.8 Concluding Remarks

Rhodoliths can provide valuable information to reconstruct past environments, as well as palaeoclimatic conditions. Taxonomic composition of the coralline algal assemblages, growth morphology of algal thalli (both in the interior and surface of the rhodoliths), rhodolith morphology, internal arrangement of rhodolith-builders, taphonomic signatures, and organism interactions (including ichnoassemblages) are key data for the correct palaeoecological interpretations. A uniformitarian approach is important as a source for these inferences. Nonetheless, taphonomic and facies analyses, as derived exclusively from the geological record itself, are required to complement uniformitarian palaeoenvironmental inferences. In this respect, it would be of considerable help to improve taphonomic assessments, both experimental and field-observational analyses, to learn more concerning the preservation behaviour of particular coralline algal taxa, coralline algal growth forms and rhodolith forms. The fossil record also offers the unique possibility of understanding the long-term processes involved in the rhodolith bed dynamics (origination and development). Geological studies demonstrate that the local foundation conditions for the sustained development of rhodolith beds are low sedimentation rates and relatively sheltered settings with intermediate water energy. The maintenance of these stable conditions over time would guarantee the growth of healthy and thick rhodolith beds.

Finally, both short- and long-term studies on coralline algae and rhodolith beds require a precise taxonomic analysis. Bosence (1991), as previously done by Adey and Macintyre (1973), highlighted the necessity of establishing a uniform and consistent taxonomic framework for fossil coralline algae comparable to the taxonomic standards of present-day corallines. Braga et al. (1993) subsequently demonstrated the fact that many of the taxonomic criteria used to identify present-day corallines have been preserved in the fossil counterparts, thus critically influencing the later advance of fossil coralline taxonomy (Braga and Aguirre 1995; Braga 2003). Specific data regarding the fossil coralline taxa forming rhodoliths in Palaeogene and Neogene rhodolith beds are provided by Braga, Brandano and Bassi and coworkers in other chapters of the book.

Acknowledgements This work was funded by the research projects CGL2013-47236-P, of the Ministerio de Ciencia e Innovación of Spain, and RNM-190 of the Junta de Andalucía. We thank David Nesbitt for the correction of the English text.

References

Adey WH (1978) Algal ridges of the Caribbean Sea and West Indies. Phycologia 17:361–367

Adey WH (1979) Crustose coralline algae as microenvironmental indicators in the tertiary. In: Gray J, Boucot AJ (eds) Historical biogeography, plate tectonics and the changing environment. Oregon State University Press, Corvallis, pp 459–464

Adey WH (1986) Coralline algae as indicators of sea-level. In: van de Plassche (ed) Sea-level research: a manual for the collection and evaluation of data. Free University, Amsterdam, pp 229–280

Adey WH, Macintyre IG (1973) Crustose coralline algae: a re-evaluation in the geological sciences. Geol Soc Am Bull 84:883–904

Adey WH, Vassar JM (1975) Colonization, succession and growth rates of tropical crustose coralline algae (Rhodophyta, Cryptonemiales). Phycologia 14:5–69

Adey WH, Townsend R, Boykins W (1982) The crustose coralline algae of the Hawaiian Islands. Smithson Contrib Mar Sci 15:1–74

Aguirre J (1992) Evolución de las asociaciones fósiles del Plioceno marino de Cabo Roche (Cádiz). Rev. Española Paleontol. (Extra) 3–10

Aguirre J, Braga JC (2005) The citation of nongeniculate fossil coralline red algal species in the twentieth century literature: an analysis with implications. Rev Esp Micropaleontol 37:57–62

Aguirre J, Braga JC (2012) Upper Pliocene multistory rhodoliths from Cádiz (Atlantic S Spain). In: Aguirre J, Rösler A, Braga JC (eds) IV international rhodolith workshop, abstract volume and field trip guide. Granada. Sept 2012, p 5

Aguirre J, Braga JC, Martín JM (1993) Algal nodules in the upper Pliocene deposits at the coast of Cadiz (S Spain). In: Barattolo F, De Castro P, Parente M (eds) Studies on fossil benthic algae. Bull Soc Paleontol Ital Spec 1:1–7

Aguirre J, Riding R, Braga JC (2000) Diversity of coralline red algae: origination and extinction patterns from the Early Cretaceous to the Pleistocene. Paleobiology 26:651–667

Aguirre J, Perfectti F, Braga JC (2010) Integrating phylogeny, molecular clocks and the fossil record in the evolution of coralline algae (Corallinales, Rhodophyta). Paleobiology 36:519–533

Aguirre J, Braga JC, Martín JM, Betzler C (2012) Palaeoenvironmental and stratigraphic significance of Pliocene rhodolith beds and coralline algal bioconstructions from the Carboneras Basin (SE Spain). Geodiversitas 34:115–136

Aguirre J, Beláustegui Z, Domènech R, de Gibert JM, Martinell J (2014) Snapshot of a lower Pliocene *Dendropoma* reef from Sant Onofre (Baix Ebre Basin, Tarragona, NE Spain). Palaeogeogr Palaeoclimatol Palaeoecol 395:9–20

Alexandersson T (1974) Carbonate cementation in coralline algal nodules in the Skagerrak, north Sea; biochemical precipitation in undersaturated waters. J Sediment Petrol 44:7–26

Alexandersson T (1977) Carbonates cementation in recent coralline algal constructions. In: Flügel E (ed) Fossil algae. Recent results and developments. Springer, Berlin, pp 261–269

Alexandersson T (1978) Destructive diagenesis of carbonate sediments in the eastern Skagerrak, north Sea. Geology 6:324–327

Amado-Filho GM, Maneveldt GW, Pereira-Filho GH, Manso RCC, Bahia RG, Barros-Barreto MB, Guimaraes SMPB (2010) Seaweed diversity associated with a Brazilian tropical rhodolith bed. Cien Mar 36:371–391

Amado-Filho GM, Pereira-Filho GH, Bahia RG, Abrantes DP, Veras PC, Matheus Z (2012a) Occurrence and distribution of rhodolith beds on the Fernando de Noronha archipelago of Brazil. Aquat Bot 101:41–45

Amado-Filho GM, Moura RL, Bastos AC, Salgado LT, Sumida PY, Guth AZ, Francini-Filho RB, Pereira-Filho GH, Abrantes DP, Brasileiro PS, Bahia RG, Leal RN, Kaufman L, Kleypas JA, Farina M, Thompson FL (2012b) Rhodolith beds are major $CaCO_3$ bio-factories in the tropical south west Atlantic. PLoS One 7:e35171. doi:10.1371/journal.pone.0035171

Arias C, Masse JP, Vilas L (1995) Hauterivian shallow marine calcareous biogenic mounds: SE Spain. Palaeogeogr Palaeoclimatol Palaeoecol 119:3–17

Baarli BG, Santos A, da Silva CM, Ledesma-Vázquez J, Mayoral E, Cachao M, Johnson ME (2012) Diverse macroids and rhodoliths from the upper Pleistocene of Baja California Sur, Mexico. J Coast Res 28:296–305

Ballantine DL, Bowden-Kerby A, Aponte NE (2000) *Cruoriella* rhodoliths from shallow-water back reef environments in La Parguera, Puerto Rico (Caribbean Sea). Coral Reefs 19:75–81

Ballesteros E (2006) Mediterranean coralligenous assemblages: a synthesis of present knowledge. Oceanogr Mar Biol 44:123–195

Barnes J, Bellamy DJ, Jones DJ, Whitton BA (1970) Sublittoral reef phenomena of Aldabra. Nature 225:268–269

Bassi D (1995) Crustose coralline algal pavements from late Eocene – Colli Berici of northern Italy. Riv Ital Paleontol Stratigr 101:81–92

Bassi D (1998) Coralline algal facies and their palaeoenvironments in the late Eocene of northern Italy (Calcare di Nago). Facies 39:179–202

Bassi D (2005) Larger foraminiferal and coralline algal facies in an upper Eocene storm influenced, shallow water carbonate platform (Colli Berici, north-eastern Italy). Palaeogeogr Palaeoclimatol Palaeoecol 226:17–35

Bassi D, Nebelsick JH (2010) Components, facies and ramps: redefining upper Oligocene shallow water carbonates using coralline red algae and larger foraminifera (Venetian area, northeast Italy). Palaeogeogr Palaeoclimatol Palaeoecol 295:258–280

Bassi D, Carannante G, Murru M, Simone L, Toscano F (2006) Rhodalgal/bryomol assemblages in temperate type carbonate, channelised depositional systems: the Early Miocene of the Sarcidano area (Sardinia, Italy). In: Pedley HM, Carannante G (eds) Cool-water carbonates: depositional systems and palaeoenvironmental control. Geol Soc Lond Spec Publ 255:35–52

Bassi D, Humblet M, Iryu Y (2011) Recent ichnocoenoesis in deep water macroids, Ryukyu islands, Japan. Palaios 26:232–238

Bassi D, Iryu Y, Humblet M, Matsuda H, Machiyama H, Sasaki K, Matsuda S, Arai K, Inoue T (2012) Recent macroids on the Kikai-jima shelf, central Ryukyu islands, Japan. Sedimentology 59:2024–2041

Bassi D, Iryu Y, Braga JC, Takayanagi H, Tsuji T (2013) Bathymetric distribution of ichnocoenoses from recent subtropical algal nodules off Fraser Island, eastern Australia. Palaeogeogr Palaeoclimatol Palaeoecol 369:58–66

Bassi D, Simone L, Nebelsick, JH (this volume) Re-sedimented rhodoliths in channelized depositional systems: synopsis of examples from middle Eocene and early-middle Miocene. In: Riosmena-Rodríguez R, Nelson W, Aguirre J (eds) Rhodolith/maerl beds: a global perspective. Springer-Verlag, Berlin

Basso D (1998) Deep rhodolith distribution in the Pontian Islands, Italy: a model for the palaeoecology of a temperate sea. Palaeogeogr Palaeoclimatol Palaeoecol 137:173–187

Basso D, Nalin R, Nelson CS (2009) Shallow-water *Sporolithon* rhodoliths from north island (New Zealand). Palaios 24:92–103

Basso D, Quaranta F, Vannucci G, Piazza M (2012) Quantification of the coralline carbonate from a Serravallian rhodolith bed of the tertiary Piedmont Basin (Stazzano, Alessandria, NW Italy). Geodiversitas 34:137–149

Beckmann JP, Beckmann R (1966) Calcareous algae from the Cretaceous and Tertiary of Cuba. Schweiz Paläont Abh 85:1–45

Begon M, Harper JL, Tonwsend CR (1990) Ecology: individuals, populations and communities. Blackwell, Oxford, p 1068

Benisek M-F, Betzler C, Marcano G, Mutti M (2009) Coralline-algal assemblages of a Burdigalian platform slope: implications for carbonate platform reconstruction (northern Sardinia, western Mediterranean Sea). Facies 55:375–386

Benisek M-F, Marcano G, Betzler C, Mutti M (2010) Facies and stratigraphic architecture of a Miocene warm-temperate to tropical fault-block carbonate platform, Sardinia (central Mediterranean Sea). In: Mutti M, Piller WE, Betzler C (eds) Carbonate systems during the Oligocene-Miocene climatic transition, vol 42. Spec Publ Int Assoc Sediment, Blackwell Publishing, pp 107–128

Betzler C, Braga JC, Jaramillo-Vogel D, Römers M, Hübscher C, Schmiedl G, Lindhorst S (2011) Late Pleistocene and Holocene cool-water carbonates of the western Mediterranean Sea. Sedimentology 58:643–669

Binda PL (1973) Form and internal structure of recent algal nodules (rhodolites) from Bermuda: a discussion. J Geol 81:283

Blanc JJ (1968) Sedimentary geology of the Mediterranean Sea. Oceanogr Mar Biol Ann Rev 6:377–454

Bosellini A, Ginsburg RN (1971) Form and internal structure of recent algal nodules (rhodolites) from Bermuda. J Geol 79:669–682

Bosence DWJ (1976) Ecological studies on two unattached coralline algae from western Ireland. Palaeontology 19:365–395

Bosence DWJ (1983a) Description and classification of rhodoliths (rhodoids, rhodolites). In: Peryt TM (ed) Coated grains. Springer, Berlin, pp 217–224

Bosence DWJ (1983b) The occurrence and ecology of recent rhodoliths – a review. In: Peryt TM (ed) Coated grains. Springer, Berlin, pp 225–242

Bosence DWJ (1983c) Coralline algae from the Miocene of Malta. Palaeontology 26:147–173

Bosence DWJ (1984) Construction and preservation of two recent coralline algal reefs, St. Croix, Caribbean. Palaeontology 27:549–574

Bosence DWJ (1985a) The "Coralligéne" of the Mediterranean – a recent analogue for tertiary coralline algal limestones. In: Toomey DF, Nitecki MH (eds) Paleoalgology: contemporary research and applications. Springer, Berlin, pp 216–225

Bosence DWJ (1985b) Preservation of coralline algal reef frameworks. 5th Int Symp Coral Reefs Tahiti 6:623–628

Bosence DWJ (1991) Coralline algae: mineralization, taxonomy, and palaeoecology. In: Riding R (ed) Calcareous algae and stromatolites. Springer, Berlin, pp 98–113

Bosence DWJ, Pedley HM (1982) Sedimentology and palaeoecology of a Miocene coralline algal biostrome from the Maltese Islands. Palaeogeogr Palaeoclimatol Palaeoecol 38:9–43

Bosence DWJ, Wilson J (2003) Maerl growth, carbonate production rates and accumulation rates in the northeast Atlantic. Aquat Conserv 13:S21–S31

Brachert TC, Betzler C, Braga JC, Martín JM (1996) Record of climatic change in neritic carbonates: turnover in biogenic associations and depositional modes (upper Miocene, southern Spain). Int J Earth Sci (Geol Rundsch) 85:327–337

Braga JC (2003) Application of botanical taxonomy to fossil coralline algae (Corallinales, Rhodophyta). Acta Micropaleontol Sin 20:47–56

Braga JC, Aguirre J (1995) Taxonomy of fossil coralline algal species: Neogene Lithophylloideae (Rhodophyta, Corallinaceae) from southern Spain. Rev Paleobot Palynol 86:265–285

Braga JC, Aguirre J (2001) Coralline algal assemblages in upper Neogene reef and temperate carbonates in southern Spain. Palaeogeogr Palaeoclimatol Palaeoecol 175:27–41

Braga JC, Aguirre J (2004) Coralline algae indicate Pleistocene evolution from deep, open platform to outer barrier reef environments in the northern great barrier reef margin. Coral Reefs 23:547–558
Braga JC, Aguirre J (2009) Algas calcáreas del Parque Natural de Cabo de Gata-Níjar. Guía de campo. In: Villalobos M, Pérez-Muñoz AB (eds) ACUMED y Consejería de Medio Ambiente (Junta de Andalucía). Aguas de la Cuenca Mediterránea, Madrid, p 206
Braga JC, Bassi D (2007) Neogene history of *Sporolithon* Heydrich (Corallinales, Rhodophyta) in the Mediterranean region. Palaeogeogr Palaeoclimatol Palaeoecol 243:189–203
Braga JC, Martín JM (1988) Neogene coralline-algal growth-forms and their palaeoenvironments in the Almanzora River Valley (Almeria, S.E. Spain). Palaeogeogr Palaeoclimatol Palaeoecol 67:285–303
Braga JC, Bosence DW, Steneck RS (1993) New anatomical characters in fossil coralline algae and their taxonomic implications. Palaeontology 36:535–547
Braga JC, Martín JM, Betzler C, Aguirre J (2006) Models of temperate carbonate deposition in neogene basins in SE Spain: a synthesis. In: Pedley HM, Carannante G (eds) Cool-water carbonates: depositional systems and palaeoenvironmental controls. Geol Soc Lond Spec Publ 255:121–135
Braga JC, Vescogni A, Bosellini FR, Aguirre J (2009) Coralline algae (Corallinales, Rhodophyta) in western and central Mediterranean Messinian reefs. Palaeogeogr Palaeoclimatol Palaeoecol 275:113–128
Brandano M (this volume) Oligocene rhodolith beds in the central Mediterranean area. In: Riosmena-Rodríguez R, Nelson W, Aguirre J (eds) Rhodolith/maerl beds: a global perspective. Springer, Berlin
Brandano M, Vannucci G, Pomar L, Obrador A (2005) Rhodolith assemblages from the lower Tortonian carbonate ramp of Menorca (Spain): environmental and paleoclimatic implications. Palaeogeogr Palaeoclimatol Palaeoecol 226:307–323
Brandano M, Corda L, Castorina F (2010) Facies and sequence architecture of a tropical foramol-rhodalgal carbonate ramp: Miocene of the central Apennines (Italy). In: Mutti M, Piller WE, Betzler C (eds) Carbonate systems during the Oligocene-Miocene climatic transition, vol 42. Spec Publ Int Assoc Sediment, Blackwell Publishing, pp 107–128
Buchbinder B (1977) Systematic and palaeoenvironment of the calcareous algae from the Miocene (Tortonian) Tziqlag Formation, Israel. Micropaleontology 23:415–435
Buchbinder B, Halley RB (1985) Occurrence and preservation of Eocene squamariacean and coralline rhodoliths: Eua, Tonga. In: Toomey DF, Nitecki MH (eds) Paleoalgology: contemporary research and applications. Springer, Berlin, pp 248–256
Burgess CJ, Anderson JM (1983) Rhodoids in temperate carbonates from the Cenozoic of New Zealand. In: Peryt TM (ed) Coated grains. Springer, Berlin, pp 243–258
Burkepile DE, Hay ME (2010) Impact of herbivore identity on algal succession and coral growth on a Caribbean reef. PLoS One 5(1):e8963. doi:10.1371/journal.pone.0008963
Buss LW (1990) Competition within and between encrusting clonal invertebrates. Trends Ecol Evol 5:352–356
Carannante G, Esteban M, Milliman JD, Simone L (1988) Carbonate lithofacies as paleolatitude indicators: problems and limitations. Sediment Geol 60:333–346
Chatalov A, Bonev N, Ivanova D (2015) Depositional characteristics and constraints on the mid-Valanginian demise of a carbonate platform in the intra-Tethyan domain, Circum- Rhodope Belt, northern Greece. Cretac Res 55:84–115
Checconi A, Bassi D, Monaco P, Carannante G (2010) Re-deposited rhodoliths in the middle Miocene hemipelagic deposits of Vitulano (southern Apennines, Italy): coralline assemblage characterization and related trace fossils. Sediment Geol 225:50–66
Comarci M, Furnari G, Giaccone G, Colonna P, Mannino AM (1985) Metodo sinecologico per la valutazione degli apporti inquinanti nella rada di Augusta (Siracusa). Bull Acad Gioenia Sci Nat 18:829–850

Di Geronimo R, Alongi G, Giaccone G (1993) Formacione organogene a *Lithophyllum lichenoides* Philippi (Rhodophyta, Corallinales) nel Mesolitorale di Capo S. Alessio (Sicilia orientale). Bull Acad Gioenia Sci Nat 26:145–172
Farrow GE, Allen NH, Akpan EB (1984) Bioclastic carbonate sedimentation on a high-latitude, tide-dominated shelf: northeast Orkney Islands, Scotland. J Sediment Res 54:373–393
Figueiredo MAO, Norton TA, Kain JM (1997) Settlement and survival of epiphytes on two intertidal crustose coralline algae. J Expl Mar Biol Ecol 213:247–260
Flood PG (1983) Coated grains from the great barrier reef. In: Peryt TM (ed) Coated grains. Springer, Berlin, pp 561–565
Flügel E (1978) Mikrofazielle untersuchungsmethoden von Kalken. Springer, Berlin
Flügel E (2004) Microfacies of carbonate rocks, analysis, interpretation and application. Springer, Berlin
Focke TW, Gebelein CD (1978) Marine lithification of reef rock and rhodolites at the fore-reef slope locality (50 m) off Bermuda. Geol Mijnb 57:163–171
Foster MS (2001) Rhodoliths: between rocks and soft places. J Phycol 87:659–667
Foster MS, Riosmena-Rodríguez R, Steller DS, Woelkerling WJ (1997) Living rhodolith beds in the Gulf of California and their implications for palaeoenvironmental interpretation. In: Johnson ME, Ledesma-Vázquez J (eds) Pliocene carbonates and related facies flaking the Gulf of California, Baja California. Geol Soc Am Spec Pap 318:27–139
Foster MS, Amado-Filho GM, Kamenos NA, Riosmena-Rodríguez R, Steller DL (2013) Rhodoliths and rhodolith beds. In: Lange M (ed) Smithsonian contributions to the marine sciences no. 39. Smithsonian Institution, pp 143–155
Fravega P, Piazza M, Vannucci G (1989) *Archaeolithothamnium* Rothpletz indicatore ecologico-stratigrafico? In: Di Geronimo I (ed) Atti del 3° Simposio di Ecologia e Paleoecologia delle Comunità Bentoniche. Catania, pp 729–743
Freiwald A (1998) Modern nearshore cold-temperate calcareous sediment in the Troms District, northern Norway. J Sed Res A 68:763–776
Freiwald A, Henrich R (1994) Reefal coralline algal build-ups within the arctic circle: morphology and sedimentary dynamics under extreme environmental seasonality. Sedimentology 41:963–984
Friebe JG (1993) Sequence stratigraphy in a mixed carbonate-siliciclastic depositional system (middle Miocene, Styrian Basin, Austria). Geol Rundsch 82:281–294
Georgiadis M, Papatheodorou G, Tzanatos E, Geraga M, Ramfos A, Koutsikopoulos C, Ferentinos G (2009) Coralligene formations in the eastern Mediterranean Sea: morphology, distribution, mapping and relation to fisheries in the southern Aegean Sea (Greece) based on high-resolution acoustics. J Exp Mar Biol Ecol 368:44–58
Ginsburg RN, Bosellini A (1973) Form and internal structure of recent algal nodules (rhodolites) from Bermuda: a reply. J Geol 81:239
Gordon DC, Masaki T, Akioka H (1976) Floristic and distributional account of the common crustose coralline algae on Guam. Micronesica 12:247–277
Graham LE, Graham JM, Wilcox LW (2009) Algae. Pearson Benjamin Cummings, San Francisco
Halfar J, Mutti M (2005) Global dominance of coralline red-algal facies: a response to Miocene oceanographic events. Geology 33:481–484
Halfar J, Zack T, Kronz A, Zachos JC (2000) Geochemical signals of rhodoliths (coralline red algae) – a new biogenic archive. J Geophys Res 105:22107–22116
Halfar J, Hetzinger S, Adey W, Zack T, Gamboa G, Kunz B, Williams B, Jacob DE (2011) Coralline algal growth-increment widths archive north Atlantic climate variability. Palaeogeogr Palaeoclimatol Palaeoecol 302:71–80
Hall-Spencer JM, White N, Gillespi E, Gillham K, Foggo A (2006) Impact of fish farms on maerl beds in strongly tidal areas. Mar Ecol Progr Ser 326:1–9
Hamel G, Lemoine MP (1953) Corallinacées de France et d'Afrique du Nord. Arch Mus Hist Nat Paris Sér 7(1):15–136

Harvey AS, Bird FL (2008) Community structure of a rhodolith bed from cold-temperate waters (southern Australia). Aust J Bot 56:437–450

Harvey AS, Broadwater ST, Woelkerling WJ, Mitrovski PJ (2003) *Choreonema* (Corallinales, Rhodophyta): 18S rDNA phylogeny and resurrection of the Hapalidiaceae for the subfamilies Choreonematoideae, Austrolithoideae, and Melobesioideae. J Phycol 39:988–998

Hochberg ME, Lawton JH (1990) Competition between kingdoms. Trends Ecol Evol 5:367–371

Hottinger L (1983) Neritic macroid genesis, an ecological approach. In: Peryt TM (ed) Coated grains. Springer, Berlin, pp 38–55

Iryu Y, Nakimori T, Matsuda S, Abe O (1995) Distribution of marine organisms and its geological significance in the modern reef complex of the Ryukyu Islands. Sed Geol 99:243–258

Johnson JH (1963) The genus *Archaeolithothamnium* and its fossil representatives. J Paleontol 37:175–211

Johnson ME, Baarli GB, Cachão M, da Silva CM, Ledesma-Vázquez J, Mayoral EJ, Ramalho RS, Santos A (2012) Rhodoliths, uniformitarianism, and Darwin: Pleistocene and recent carbonate deposits in the Cape Verde and Canary archipelagos. Palaeogeogr Palaeoclimatol Palaeoecol 329–330:83–100

Johnson ME, Baarli G, da Silva CM, Cachão M, Ramalho RS, Ledesma-Vázquez J, Mayoral EJ, Santos A (2013) Coastal dunes with high content of rhodolith (coralline red algae) bioclasts: Pleistocene formations on Maio and São Nicolau in the Cape Verde archipelago. Aeolian Res 8:1–9

Johnson ME, Ledesma-Vázquez J, Ramalho RS, da Silva CM, Santos A, Baarli G, Mayoral EJ, Cachão M (this volume) Taphonomic range and sedimentary dynamics of modern and fossil rhodolith beds: Macaronesian realm (North Atlantic Ocean). In: Riosmena-Rodríguez R, Nelson W, Aguirre J (eds) Rhodolith/maerl beds: a global perspective. Springer, Berlin

Kamenos NA, Cusack M, Moore PG (2008) Red coralline algae are global paleothermometers with bi-weekly resolution. Geochim Cosmochim Acta 72:771–779

Kamenos NA, Burdett HL, Darrenougue N (this volume) Coralline algae as recorders of past climatic and environmental conditions. In: Riosmena-Rodríguez R, Nelson W, Aguirre J (eds) Rhodolith/maerl beds: a global perspective. Springer, Berlin

Keats DW, Groener A, Chamberlain YM (1993) Cell sloughing in the littoral zone coralline alga, *Spongites yendoi* (Foslie) Chamberlain (Corallinales, Rhodophyta). Phycologia 32:143–150

Kidwell SM, Bosence DWJ (1991) Taphonomy and time-averaging of marine shelly faunas. In: Allison PA, Briggs DEG (eds) Taphonomy: releasing the data locked in the fossil record, vol 9, Topics in Geobiology. Plenum Press, New York, pp 115–209

Kidwell SM, Holland SM (1991) Field description of coarse bioclastic fabrics. Palaios 6:426–434

Koop K, Booth D, Broadbent A, Brodie J, Bucher D, Capone D, Coll J, Dennison W, Erdmann M, Harrison P, Hoegh-Guildberg O, Hutchings P, Jones GB, Larkum AWD, O'Neil J, Steven A, Tentori E, Ward S, Williamson J, Yellowlees D (2001) Encore: the effect of nutrient enrichment on coral reefs. Synthesis of results and conclusions. Mar Pollut Bull 42:91–120

Kroeger KF (2007) Upper Miocene coralline red algal associations of central Crete (Greece): taxonomy and palaeoenvironmental implications. N Jb Geol Paläont Abh 244:143–171

Kroeger KF, Reuter M, Brachert TC (2006) Palaeoenvironmental reconstruction based on non-geniculate coralline red algal assemblages in Miocene limestone of central Crete. Facies 52:381–409

Lee RE (2008) Phycology. Cambridge University Press, Cambridge

Lee D, Carpenter SJ (2001) Isotopic disequilibrium in marine calcareous algae. Chem Geol 172:307–329

Leigh EG (1990) Community diversity and environmental stability: a re-examination. Trends Ecol Evol 5:340–344

Lemoine PM (1910) Répartition et mode de vie du maërl coralline alga (*Lithothamnium calcareum*) aux environs de Concameau (Finistère). Ann Inst Océanogr Paris 1:1–29

Lemoine PM (1970) Les algues floridées calcaires du Crétacé du Sud de la France. Arch Mus Nat Hist Nat Paris Sér 7(10):129–240

Leszczyński S, Kołodziej B, Bassi D, Malata E, Gasiński MA (2012) Depositional history of mixed siliciclastic-carbonate flysch deposits: Paleocene–Eocene transition, Silesian Nappe, Polish Outer Carpathians. Facies 58:367–387
Littler DS, Littler MM (2003) South Pacific reef plants. A divers' guide to the plant life of South Pacific coral reefs. Offshore Graphics, Washington
Littler MM, Littler DS, Blair SM, Norris JN (1985) Deepest known plant life discovered on an uncharted seamount. Science 227:57–59
Littler MM, Littler DS, Hanisak MD (1991) Deep-water rhodolith distribution, productivity, and growth history at sites of formation and subsequent degradation. J Mar Biol Ecol 150:163–182
Lobban CS, Harrison PJ (1994) Seaweed ecology and physiology. Cambridge University Press, Cambridge
Lund M, Davies PJ, Braga JC (2000) Coralline algal nodules off Fraser Island, eastern Australia. Facies 42:25–34
Macintyre IG, Glynn PW, Steneck RS (2001) A classic Caribbean algal ridge, Holandes Cays, Panama: an algal coated storm deposit. Coral Reefs 20:95–105
Maneveldt GW, Keats DW (2008) Effects of herbivore grazing on the physiognomy of the coralline alga *Spongites yendoi* and on associated competitive interactions. Afr J Mar Sci 30:581–593
Marrack EC (1999) The relationship between water motion and living rhodolith beds in the southwestern gulf of California, Mexico. Palaios 14:159–171
Martín JM, Braga JC (1993) Eocene to Pliocene Coralline Algae in the Queensland Plateau (Northeastern Australia). In: McKenzie JA, Davies PJ, Palmer-Julson A, et al (eds) Proceedings ocean drilling program: scientific results. College Station, TX, 133:67–74
Martín JM, Braga JC (1994) Messinian events in the Sorbas basin in southeastern Spain and their implications in the recent history of the Mediterranean. Sed Geol 90:257–268
Martín JM, Braga JC, Aguirre J, Betzler C (2004) Contrasting models of temperate carbonate sedimentation in a small Mediterranean embayment: the Pliocene Carboneras Basin, SE Spain. J Geol Soc Lond 161:387–399
Martindale W (1992) Calcified epibionts as palaeoecological tools: examples from the recent and Pleistocene reefs of Barbados. Coral Reefs 11:167–177
Matsuda S, Iryu Y (2011) Rhodoliths from deep fore-reef to shelf areas around Okinawa-jima, Ryukyu Islands, Japan. Mar Geol 282:215–230
McMaster RL, Conover JT (1966) Recent algal stromatolites from the Canary Islands. J Geol 74:647–652
McNeil DF, Pisera A (2010) Neogene lithofacies evolution on a small carbonate platform in the Loyalty Basin, Maré, New Caledonia. In: Morgan WA, George AD, Harris PM, Kupecz JA, Sarg JA (eds) Cenozoic carbonate systems of Australasia, vol 95. SEPM Spec Publ, pp 243–255
Milliman JD (1977) Role of calcareous algae in Atlantic continental margin sedimentation. In: Flügel E (ed) Fossil algae. Recent results and developments. Springer, Berlin, pp 232–247
Minnery GA (1990) Crustose coralline algae from the Flower Garden Banks, northwestern gulf of Mexico: controls on distribution and growth morphology. J Sediment Petrol 60:992–1007
Minnery GA, Rezak R, Bright TJ (1985) Depth zonation and growth form of crustose coralline algae: Flower Garden Banks, northwestern gulf of Mexico. In: Toomey DF, Nitecki MH (eds) Paleoalgology: contemporary research and applications. Springer, Berlin, pp 237–247
Molinier R (1956) Les fonds à laminaires du Grand Banc de Centuri (Cap Corse). Com Rendus Acad Sci 342:939–941
Montaggioni LF (1979) Environmental significance of rhodolites from the Mascarene Reef Province, western Indian Ocean. Bull Centres Rech Explor Prod Elf-Aquitaine 3:713–723
Nalin R, Basso D, Massari F (2006) Pleistocene coralline algal build-ups (coralligéne de plateau) and associated bioclastic deposits in the sedimentary cover of Cutro marine terrace (Calabria, southern Italy). In: Pedley HM, Carannante G (eds) Cool-water carbonates: depositional systems and palaeoenvironmental controls. Geol Soc Lond Spec Publ 255:11–22

Nalin R, Nelson CS, Basso D, Massari F (2008) Rhodolith-bearing limestones as transgressive marker beds: fossil and modern examples from north Island, New Zealand. Sedimentology 55:249–274
Nebelsick JH, Rasser M, Bassi D (2005) Facies dynamics in Eocene to Oligocene circumalpine carbonates. Facies 51:197–216
Nebelsick JH, Bassi D, Lempp L (2013) Tracking palaeoenvironmental changes in coralline algal dominated carbonates of the Lower Oligocene Calcareniti di Castelgomberto formation (Monti Berici, Italy). Facies 59:133–148
Nelson W (2009) Calcified macroalgae – critical to coastal ecosystems and vulnerable to change: a review. Mar Freshw Res 60:787–801
Orszag-Sperber F, Poignant AF, Poisson A (1977) Paleogeographic significance of rhodolites: some examples from the Miocene of France and Turkey. In: Flügel E (ed) Fossil algae. Recent results and developments. Springer, Berlin, pp 286–294
OSPAR Commission (2010) Background document for maërl beds. Biodivers Ser, p 34
Payri C, N'Yeurt AR, Orempuller J (2000) Algues de Polynésie française. Au Vent des Îles, Singapour
Peña V, Bárbara I (2008) Biological importance of an Atlantic maerl bed off Benencia Island (northwest Iberian Peninsula). Bot Mar 51:493–505
Peña V, Bárbara I (2009) Distribution of the Galician maerl beds and their shape classes (Atlantic Iberian Peninsula): proposal of areas in future conservation actions. Cah Biol Mar 50:353–368
Pérès JM, Picard J (1958) Recherches sur les peuplements benthiques de la Méditerranée nord-orientale. Ann Inst Océanogr Monaco 34:213–291
Pérès JM, Picard J (1964) Nouveau Manuel de Bionomie benthique de la mer Méditerranée. Rec Trav Stat Mar Endoume 31(47):1–137
Perrin C, Bosence DWJ, Rosen B (1995) Quantitative approaches to palaeozonation and palaeobathymetry of corals and coralline algae in Cenozoic reefs. In: Bosence DWJ, Allison PA (eds.) Marine palaeoenvironmental analysis from fossils. Geol Soc Lond Geol Soc Spec Publ 83:181–229
Peryt TM (1983) Classification of coated grains. In: Peryt TM (ed) Coated grains. Springer, Berlin, pp 3–6
Prager EJ (1987) The growth and structure of calcareous nodules (for-algaliths) on Florida's outer shelf. Thesis, Univ Miami, p 65
Prager EJ, Ginsburg RN (1989) Carbonate nodule growth on Florida's outer shelf and its implications for fossil interpretations. Palaios 4:310–317
Puga-Bernabéu A, Braga JC, Martín JM (2007) High-frequency cycles in upper-Miocene ramp-temperate carbonates (Sorbas Basin, SE Spain). Facies 53:329–345
Quaranta F, Tomassetti L, Vannucci G, Brandano M (2012) Coralline algae as environmental indicators: a case study from the Attard Member (Chattian, Malta). Geodiversitas 34:151–166
Rahimpour-Bonab H, Bone Y, Moussavi-Harami R, Turnbull K (1997) Geochemical comparisons of modern cool-water calcareous biota, Lacepede Shelf, south Australia. In: James NP, Clarke JAD (eds) Cool-water carbonates. Spec Publ SEPM Soc Sediment Geol 56:77–92
Rasser MW, Piller WE (1997) Depth distribution of calcareous encrusting associations in the northern Red Sea (Safaga, Egypt) and their geological implications. Proc 8th Int Coral Reef Symp 1:743–748
Rasser MW, Piller WE (2004) Crustose algal frameworks from the Eocene Alpine Foreland. Palaeogeogr Palaeoclimatol Palaeoecol 206:21–39
Reid RP, MacIntyre IG (1988) Foraminiferal–algal nodules from the Eastern Caribbean: growth history and implications on the value of nodules as paleoenvironmental indicators. Palaios 3:424–435
Richter DK, Sedat R (1983) Brackish-water oncoids composed of blue-green and red algae from a Pleistocene terrace near Corinth, Greece. In: Peryt TM (ed) Coated grains. Springer, Berlin, pp 299–307

Ringeltaube P, Harvey A (2000) Non-geniculate coralline algae (Corallinales, Rhodophyta) on Heron Reef. Great Barrier Reef (Australia) Bot Mar 43:431–454
Riosmena-Rodríguez R, Steller DL, Hinojosa-Arango G, Foster MS (2010) Reefs that rock and roll: biology and conservation of rhodolith beds in the gulf of California. In: Brusca RC (ed) The gulf of California biodiversity and conservation. The University of Arizona Press and The Arizona-Sonora Desert Museum, Tucson, pp 49–71
Riul P, Targino CH, Farias JDN, Visscher PT, Horta PA (2008) Decrease in *Lithothamnion* sp. (Rhodophyta) primary production due to the deposition of a thin sediment layer. J Mar Biol Assoc UK 88:17–19
Round EF (1981) The ecology of algae. Cambridge University Press, Cambridge
Sánchez-Almazo IM, Spiro B, Braga JC, Martín JM (2001) Constraints of stable isotope signatures on the depositional palaeoenvironments of upper Miocene reef and temperate carbonates in the Sorbas basin, SE Spain. Palaeogeogr Palaeoclimatol Palaeoecol 175:153–172
Schaefer TN, Smith J, Foster MS, De Tomaso A (2002) Genetic differences between two growth-forms of *Lithophyllum margaritae* (Rhodophyta) in Baja California Sur, Mexico. J Phycol 38:1090–1098
Schäfer P, Fortunato H, Bader B, Liebetrau V, Bauch T, Reijmer JJG (2011) Growth rates and carbonate production by coralline red algae in upwelling and non-upwelling settings along the Pacific coast of Panama. Palaios 26:420–432
Simone L, Bassi D, Carannante G, Cherchi A (2012) Rudist-bearing rhodalgal facies in the post-Turonian recovery of the periTethyan carbonate systems: the case history from the Nurra Region (northwestern Sardinia, Italy). Geodiversitas 34:167–187
Sneed ED, Folk RL (1958) Pebbles in the lower Colorado River, Texas: a study in particle morphogenesis. J Geol 66:114–150
Steller DL, Cáceres C (2009) Coralline algal rhodoliths enhance larval settlement and early growth of the Pacific calico scallop *Argopecten ventricosus*. Mar Ecol Progr Ser 396:49–60
Steller DL, Foster MS (1995) Environmental factors influencing distribution and morphology of rhodoliths in Bahía Concepción, B.C.S., México. J Exp Mar Biol Ecol 194:201–212
Steller DL, Riosmena-Rodríguez R, Foster MS, Roberts CA (2003) Rhodolith bed diversity in the gulf of California: the importance of rhodolith structure and consequences of disturbance. Aquat Conserv 13:S5–S20
Steller DL, Foster MS, Riosmena-Rodríguez R (2009) Living rhodolith bed ecosystems in the gulf of California. In: Johnson JM, Ledesma-Vázquez J (eds) Atlas of coastal ecosystems in the gulf of California: past and present. University of Arizona Press, Tucson, pp 72–82
Steneck RS (1983) Escalating herbivory and resulting adaptive trends in calcareous algal crusts. Paleobiology 9:44–61
Steneck RS (1985) Adaptations of crustose coralline algae to herbivory: patterns in space and time. In: Toomey D, Nitecki M (eds) Paleoalgology: contemporary research and applications. Springer, Berlin, pp 352–366
Steneck RS (1986) The ecology of coralline algal crusts: convergent patterns and adaptive strategies. Ann Rev Ecol Syst 17:273–303
Tomás S, Aguirre J, Braga JC, Martín-Closas C (2007) Late Hauterivian coralline algae (Rhodophyta, Corallinales) from the Iberian Chain (E Spain). Taxonomy and the evolution of multisporangial reproductive structures. Facies 53:79–95
van der Hoeck C, Mann DG, Jahns HM (1995) Algae. An introduction to phycology. Cambridge University Press, Cambridge
Verheij E (1993) The genus *Sporolithon* (Sporolithaceae fam. nov., Corallinales, Rhodophyta) from the Spermonde Archipelago, Indonesia. Phycologia 32:184–196
Verheij E, Erftemeijer PLA (1993) Distribution of seagrasses and associated macroalgae in south Sulawesi, Indonesia. Blumea 38:45–64
Villas-Boas AB, Tâmega FTS, Coutinho MAR, Figueiredo MAO (2014) Experimental effects of sediment burial and light attenuation on two coralline algae of a deep water rhodolith bed in Rio de Janeiro, Brazil. Cryptogam Algol 35:67–76

Wefer G, Berger WH (1991) Isotope paleontology: growth and composition of extant calcareous species. Mar Geol 100:207–248
Wilson S, Blake C, Berges JA, Maggs CA (2004) Environmental tolerances of free-living coralline algae (maerl): implications for European marine conservation. Biol Conserv 12:283–293
Woelkerling WJ (1988) The coralline red algae: an analysis of the genera and subfamilies of non-geniculate corallinaceae. Oxford University Press, Oxford, 268 pp
Woelkerling WJ (1996a) Subfamily Mastophoroideae Setchell 1943. In: Womersley HBS (ed) The marine benthic flora of southern Australia. Rhodophyta. Part IIIB, Gracilariales, Rhodymeniales, Corallinales and Bonnemaisoniales. Aust Biol Resources Study, Canberra, pp 237–283
Woelkerling WJ (1996b) Family Sporolithaceae. In: Womersley HBS (ed) The marine benthic flora of southern Australia. Rhodophyta. Part IIIB, Gracilariales, Rhodymeniales, Corallinales and Bonnemaisoniales. Aust Biol Resources Study, Canberra, pp 153–158
Woelkerling WJ (1996c) Subfamily Lithophylloideae Setchell 1943. In: Womersley HBS (ed) The marine benthic flora of southern Australia. Rhodophyta. Part IIIB, Gracilariales, Rhodymeniales, Corallinales and Bonnemaisoniales. Aust Biol Resources Study, Canberra, pp 214–237
Woelkerling WJ (1996d) Subfamily Melobesioideae Bizzozero 1885. In: Womersley HBS (ed) The marine benthic flora of southern Australia. Rhodophyta. Part IIIB, Gracilariales, Rhodymeniales, Corallinales and Bonnemaisoniales. Aust Biol Resources Study, Canberra, pp 164–210
Woelkerling WJ, Irvine LM, Harvey AS (1993) Growth-forms in non-geniculate coralline red algae (Corallinales, Rhodophyta). Aust Syst Bot 6:277–293

Chapter 6
Re-sedimented Rhodoliths in Channelized Depositional Systems

Davide Bassi, Lucia Simone, and James H. Nebelsick

Abstract The knowledge of re-sedimented rhodolith deposits has always lagged behind that of in situ deposits, which can be formed in shallow and deeper water carbonate and mixed siliciclastic-carbonate depositional settings. A combination of detailed outcrop analyses from three published case studies reveals a series of palaeobiological and taphonomic signals that are used to identify fossil re-sedimented rhodoliths. The re-sedimented rhodolith deposits of the middle Eocene carbonates in the Venetian area (northeast Italy), the lower Miocene carbonates from southern Sardinia (Italy), and the lower–middle Miocene carbonates from Southern Apennines (southern Italy) are described in terms of rhodolith morphology, coralline algal assemblages, inner arrangement, outer growth-forms, and taphonomic signatures. In all the cases, shallow water rhodolith beds were redeposited to feed offshore deposits through submarine channel systems. The sedimentological features, rhodolith characteristics and taphonomic signatures of the rhodolith deposits are compared from the carbonate factory, through the shelf-margin to the proximal and distal parts of the tributary belt. Within submarine channelized carbonate settings, complex relationship patterns of autochthonous/parautochthonous and allochthonous rhodolith deposits were governed by the interplay of changes in environmental factors such as water energy, light irradiance, substrate characteristics, and residence time on the sediment-water interface.

D. Bassi (✉)
Dipartimento di Fisica e Scienze della Terra, Università di Ferrara,
via Saragat 1, 44122 Ferrara, Italy
e-mail: bsd@unife.it

L. Simone
Dipartimento di Scienze della Terra, University of Napoli Federico II,
Largo S. Marcellino 10, 80138 Naples, Italy

J.H. Nebelsick
Department of Geosciences, University of Tübingen,
Sigwartstraße 10, D-72076 Tübingen, Germany

R. Riosmena-Rodríguez et al. (eds.), *Rhodolith/Maërl Beds: A Global Perspective*, Coastal Research Library 15, DOI 10.1007/978-3-319-29315-8_6

6.1 Introduction

Re-sedimented rhodolith deposits are less well documented than their *in situ* counterparts, originating either from shallow or deeper water carbonate and mixed siliciclastic-carbonate depositional settings. As they are in most cases developed on carbonate shelves, they can provide information with respect to the original depositional environment and its development within these shallow-water settings (compare Bosence 1983a; Bassi et al. 2012). Facies models have been proposed for inner and marginal shelf and slope settings where rhodolith beds thrived (e.g., Adey and MacIntyre 1973; Bosence 1983a; Carannante et al. 1988; Basso and Tomaselli 1994; Nebelsick et al. 2005; Braga et al. 2006, 2010a, b; Aguirre et al. 2012), as well as for point-sourced and/or line-sourced calciclastic submarine fans and/or channels (e.g., Reijmer and Everaars 1991; Cherchi et al. 2000; Wonham et al. 2000; Braga et al. 2001; Carannante and Vigorito 2001; Payros and Pujalte 2008). On the whole, in the rhodolith bed-bearing systems, a significant part of the fossil benthic communities represents re-mobilized skeletal debris and only a minor part of the produced skeletal grains is normally preserved as *in situ* deposits (e.g., Davaud and Septfontaine 1995; Yordanova and Hohenegger 2002; Renema et al. 2013). Most of the studies on fossil rhodolith accumulations have focused on autochthonous and parautochthonous deposits, described in terms of coralline algal assemblages, facies distribution and succession, and depositional geometries (e.g., Nebelsick et al. 2005; Braga et al. 2010a, b). Herein re-sedimented rhodolith deposits are defined as those that encompass shallow-water rhodoliths, which were re-worked offshore (i.e., in middle- to proximal outer ramp, shelf margin, slope, and toe of the slope deposits).

In this paper, we present a synopses of three case studies from Italy (Fig. 6.1): (1) the middle Eocene carbonates in the Venetian area, (2) the lower Miocene carbonates from southern Sardinia and (3) the lower–middle Miocene carbonates from the Southern Apennines. These carbonates have been interpreted as re-sedimented rhodolith deposits in submarine channelized complexes. These interpretations are based on facies analysis and mapping of the three-dimensional facies architecture. Several unusual features, however, have been found in all three case studies that make them of special interest. Firstly, they crop out in extensive areas providing a three dimensional depiction of the stacked channelized units of each analysed submarine setting. The quality of the outcrops makes it possible to obtain quantitative data on specific characteristics of particular facies and sedimentary bodies. Therefore, facies and architectural data from individual rhodolith deposits can be integrated into a more general facies model. Secondly, coralline algal and rhodolith assemblages have been all analysed by applying the same methodology in order to provide a consistent database of palaeontological information.

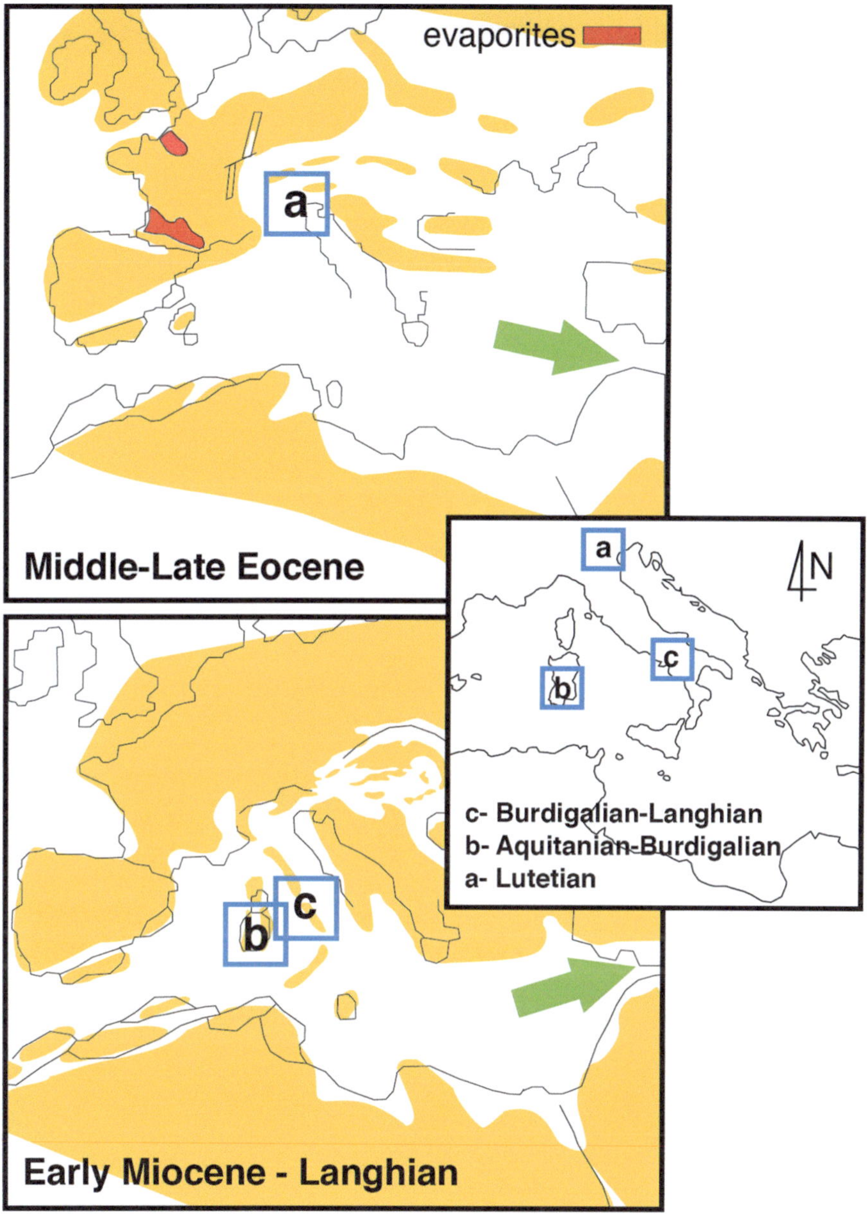

Fig. 6.1 Location of the Italian study areas within the changing Paleogene/Neogene palaeogeography of the western Tethys (Mediterranean area). (**a**) Middle Eocene, Monti Berici; (**b**) early Miocene, southern Sardinia; (**c**) early–middle Miocene, Southern Apennines. *Arrows point* to the Indo-Pacific seaway. Palaeomaps after Rögl (1998)

6.2 Examples of Re-sedimented Rhodolith Successions

Re-sedimented rhodolith successions occur since the Late Cretaceous (e.g., Simone et al. 2012), throughout the Paleogene and Neogene sedimentary record. These recurrent deposits are often inferred to relate to periods of regional tectonic activity and/or sea level oscillation, although the forcing mechanisms and frequency of such re-sedimentations are difficult to determine. Re-sedimented rhodolith successions from proximal to distal shelf settings represent allochthonous deposits that, after deposition, modify the background substrate and change habitats for different benthic communities.

In the following sections, three examples of re-sedimented rhodolith-dominated deposits in channelized systems are reviewed: (1) off-shore re-sedimented rhodolith and larger *Nummulites* deposits in a middle Eocene carbonate ramp depositional system, (2) erosive tributary channels and deeper main channel in a lower Miocene carbonate system, and (3) channel networks on tectonically active open-shelf depositional system (early–middle Miocene). The discussion of the three case studies includes: (1) coralline taxonomic composition, (2) rhodolith size and shape, (3) inner arrangement of rhodoliths, (4) outer growth forms, and (5) taphonomic signatures.

Detailed descriptions of geographic locations, materials and methods carried out in the case studies are reported in Bassi et al. (2006, 2010, 2013). Coralline algal growth-form terminology follows Woelkerling et al. (1993). Family and sub-family circumscription follows Braga (2003), Harvey et al. (2003), Iryu et al. (2012) and Nelson et al. (2015). Figure 6.2 illustrates the Sneed and Folk's (1958) ternary diagram showing the three categories of shapes forming the apices of the triangle, namely, spheroidal, discoidal, and ellipsoidal. Accordingly, the mathematical relationships s/L, I/L, and (L-I)/ (L-s) is used to calculate deviations in sphericity between apices discoidal and spheroidal, spheroidal and ellipsoidal and discoidal and ellipsoidal, respectively (e.g., Graham and Midglay 2000) (Fig. 6.2).

6.2.1 Offshore Re-deposition in a Middle Eocene Ramp (Lutetian, Monti Berici, Venetian Area)

The Monti Berici, encompassing the Monte Baldo and Monti Lessini, the Colli Euganei as well as the Vincentin Pre-Alps, are located in the western Paleogene basin of the Venetian area, northeast Italy. This basin is characterised by widespread Paleocene, lower-middle Eocene and Oligocene volcanic rocks and by the presence of shallow-water Paleogene platform (the Lessini Shelf) and related basinal units (e.g., Bassi et al. 2008). The Monti Berici occur in the eastern sector of the Alpone–Chiampo Graben extended along a NNW-SSE axis in the central-eastern Lessini

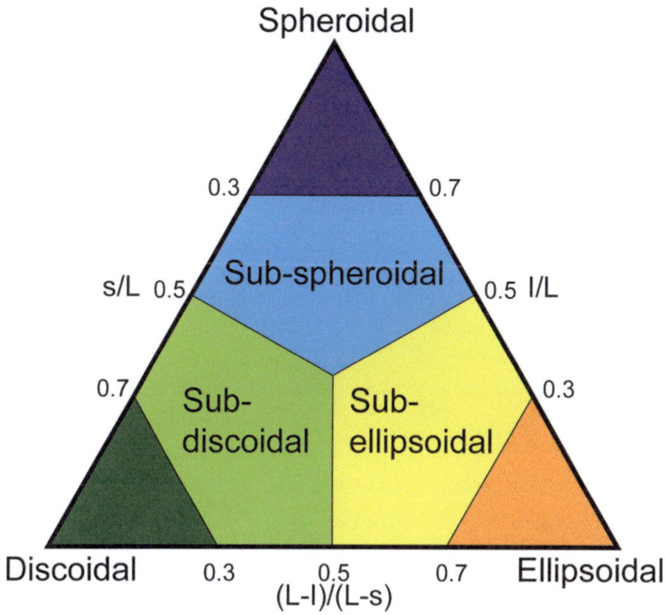

Fig. 6.2 Sneed and Folk's (1958) triangular diagram for rhodolith sphericity illustrating the areas encompassing the sub-spheroidal, sub-ellipsoidal, and sub-discoidal shapes. Rhodolith axes: *L* longest, *I* intermediate, *s* shortest

Shelf (Figs. 6.3 and 6.4). From middle Eocene times, in the southeast Monti Berici an important tectonic threshold separated the Berici–Lessini area, which comprises mainly shallow-marine deposits, from the Euganean Basin characterised by deeper water sedimentation (Bassi et al. 2000).

In the Monti Berici, lower Eocene turbiditic tuffs, tuffaceous marls and marls a few metres thick are overlain by marly calcarenites (Pietra di Nanto formation; Fabiani 1915). These calcarenites are early Lutetian in age and grade upwards into larger *Nummulites* calcarenites, up to 50–90 m thick, traditionally named as the Calcari nummulitici formation (e.g., Bassi et al. 2000). In the studied eastern Monti Berici (Mossano area), the *ca.* 90 m-thick Calcari nummulitici formation extends up to the Bartonian–Priabonian boundary (Schweighauser 1953; Ungaro 1969; Papazzoni and Sirotti 1995; Bassi et al. 2000). Towards the Colli Euganei, these shallow-water deposits pass offshore into the hemipelagic marly Marne Euganee formation. The studied carbonate succession, cropping out in the eastern Monti Berici, in the well-known Mossano area, belongs to the lowermost part of the Calcari nummulitici formation (Lutetian–Bartonian). The stratigraphy of the studied succession has been described in Schweighauser (1953), Ungaro (1969), and Bassi et al. (2000).

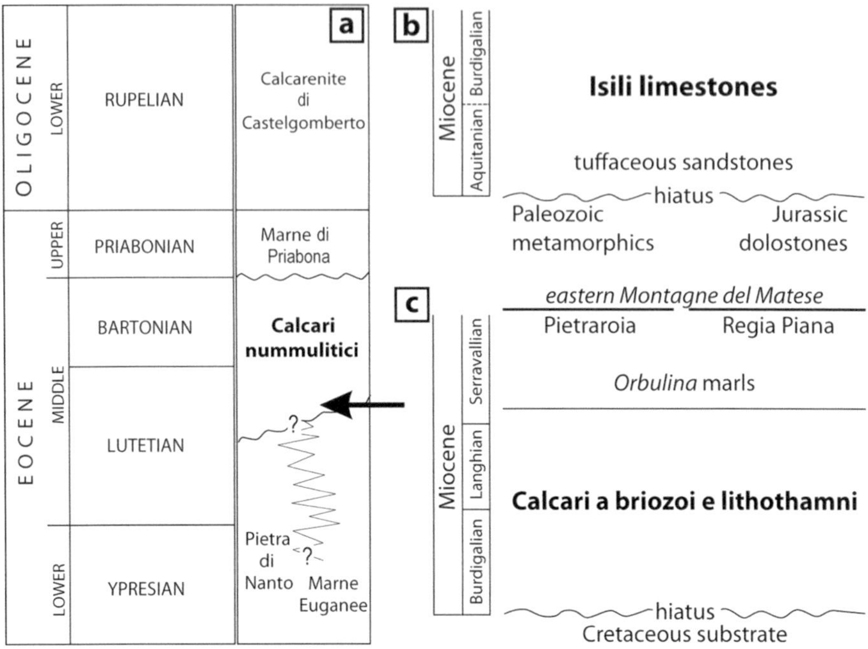

Fig. 6.3 Stratigraphical settings of the three studied sedimentary successions characterized by re-sedimented rhodolith deposits. (**a**) In the Lutetian-Bartonian Calcari nummulitici formation, the re-sedimented rhodoliths occur toward the top of the formation (*arrow*; Bassi et al. 2000). (**b**) In the Isili area, patchy localized carbonate factories (i.e., Isili limestones) developed locally at the top of faulted-blocks along syn-rift sub-basin margins, which span from Palaeozoic metamorphics to Jurassic dolostones (Cherchi et al. 2000). (**c**) In the Montagne del Matese area, the Burdigalian–Langhian shallow-water carbonate succession (Calcari a briozoi e lithothamni) is overlain by hemipelagic marls (Marne ad *Orbulina;* Bassi et al. 2010)

6.2.1.1 Coralline Algal and Rhodolith Assemblages

In the studied area, the Calcari nummulitici formation crops out with four different carbonate facies: (1) bioclastic calcarenite, (2) rhodolith and *Nummulites* rudstone, (3) larger foraminiferal packstone, and (4) coralline debris and larger *Nummulites* rudstone (Bassi and Nebelsick 2005; Bassi et al. 2013). The facies association represents middle- and proximal outer ramp environments. The middle-ramp is characterized by coralline debris and larger lenticular and lenticular-inflated *Nummulites* rudstone on shallow submerged palaeohighs. These palaeohighs created presumably a complex topography with intra-palaeohighs pathways where currents were strong enough to generate sandy mobile substrates with free-living rhodoliths (i.e., rhodolith and *Nummulites* rudstone facies, larger foraminiferal packstone facies). The high amount of micrite in the larger foraminiferal packstone reflects a relatively less water energy than that interpreted for the rhodolith and *Nummulites* rudstone facies. The components of this latter facies were removed from middle-ramp intra-palaeohigh settings and re-deposited on proximal outer-ramp areas as bioclastic

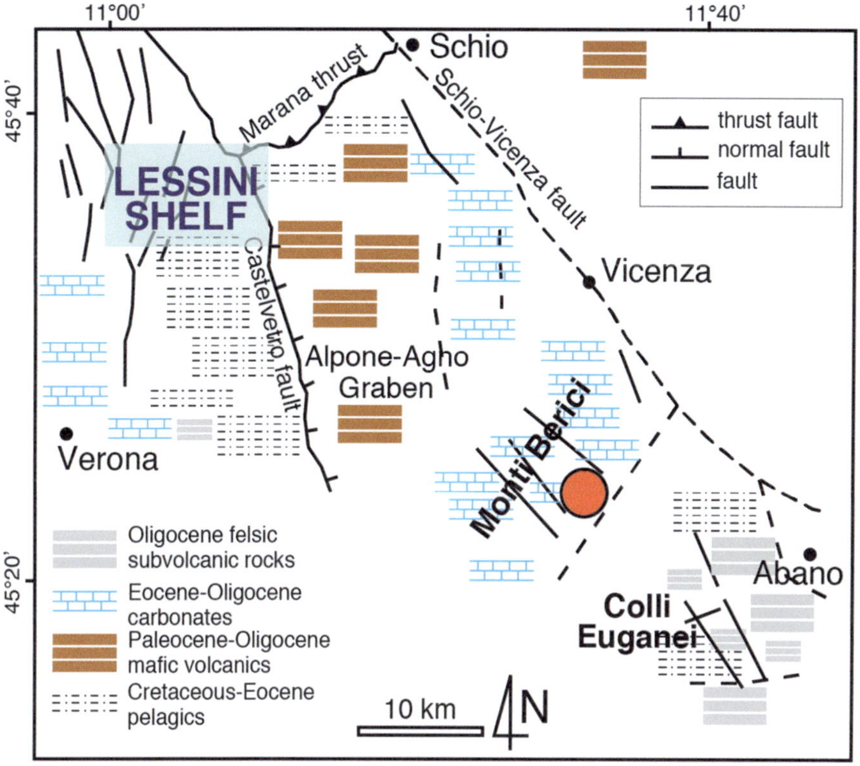

Fig. 6.4 Location of the study area (*orange circle*) within the shallow-water Eocene–Oligocene sedimentary and volcanic successions in the southern Lessini Shelf (Veneto, northeast Italy). The middle Eocene Monti Berici represented the southernmost part of the wider Lessini Shelf area bordering on the Colli Euganei pelagic basin (modified after Márton et al. 2011). The study area is located near the south-easternmost margin of a carbonate shelf, which grew up to the early Oligocene (further details in Bassi et al. 2008)

grainstone calcarenite and marly silty packstone. The presence of channelized deposits in the deepest part of the carbonate ramp (cropping out in the San Pancrazio area) indicates that part of the offshore sediment transport from the *Nummulites* palaeohighs and their intra-areas took place across a submarine channel excavated into the ramp. The channel-fill deposits were formed by high-density turbidity currents that could have been generated from near-bottom return currents during storm conditions (Bassi et al. 2013).

Coralline red algae are dominant components in the coralline debris and larger *Nummulites* rudstone facies, in the rhodolith and *Nummulites* rudstone facies, and in the larger foraminiferal packstone facies. Corallines are present as rhodoliths, encrusting thallial growth forms on bioclasts as well as coralline algal debris (Fig. 6.5a–c).

The coralline debris and larger *Nummulites* rudstone facies with packstone matrix ranges in thickness from *ca.* 20 to 150 cm. Abundant coralline red algae (up

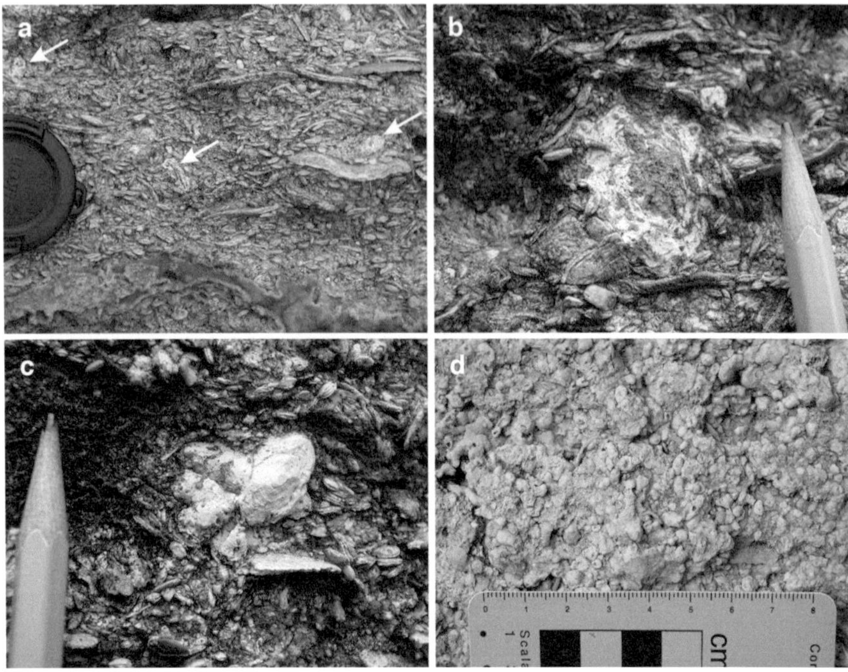

Fig. 6.5 Coarse carbonate skeletal debris. (**a**–**c**) Middle Eocene rhodolith and *Nummulites* rudstone facies (**a**) formed in the shallower middle ramp and re-sedimented offshore on hemipelagic planktonic foraminiferal marls, Monti Berici, northeast Italy. (**b**) Rare sub-spheroidal warty rhodoliths (*arrows*) with dense laminar inner arrangement. (**c**) Warty outer growth forms are dispersed in the larger foraminiferal rudstone; camera cap is 4.5 cm in diameter and pencil is 0.8 cm in diameter. (**d**) Lower Miocene coralline red algal debris constituted of fragmented fruticose corallines represents the shallow-water parautochthonous carbonate factory sediment formed in the Isili basin, southern Sardinia

to total 55 %) are present as coralline debris of mainly fruticose thalli and subordinate warty and encrusting forms. Small sub-spheroidal rhodoliths, up to 4 cm in mean diameter, consisting of *Sporolithon* with lumpy to fruticose outer growth-forms are scattered in the packstone matrix. The fruticose thalli are fragmented and abraded. Bioerosion is absent.

The rhodolith and *Nummulites* rudstone facies, from 10 to 60 cm in thickness, contains abundant larger foraminifera and coralline red algae (rhodoliths, coralline debris), common pectinids, echinoids, thin-branched solitary corals and large ostreids (*Pycnodonta*). Rare crustaceans (*Ranina marestiana*) and bryozoans are also present. Corallines are fragments of encrusting and fruticose plants of *Mesophyllum* and *Lithoporella*, while thin encrusting thalli of *Polystrata alba* can locally be present. The rhodoliths, approximately 1–4 cm in mean diameter, are dispersed within the packstone matrix. Most common rhodolith morphologies are laminar sub-spheroidal, with a delicate, laminar symmetrical inner growth stage and a warty or lumpy phase extending out up to the outer rhodolith surface. Nuclei can consist of dense coralline thalli or bryozoans. The few constructional voids are infilled by sediment matrix. The thin algal crusts either float in the packstone matrix or encrust larger foraminifera and other bioclasts. Bioerosion is present only in the larger rhodoliths as *Trypanites*, rarely small-sized *Gastrochaenolites* and micro-galleries.

In the larger foraminiferal packstone facies, comprising up to 1 m-thick massive to poorly bedded calcarenites, corallines show encrusting and fragmented warty and fruticose growth-forms. Two mastophoroids (*Lithoporella*, *Spongites*), one melobesioid (*Mesophyllum*) and one sporolithacean (*Sporolithon*) were identified. Rare sub-spheroidal warty rhodoliths with low percentage of constructional voids reach sizes up to 4 cm in diameter and show symmetrical inner arrangement. They are dispersed throughout the matrix. This facies overlies and pass laterally into the rhodolith and *Nummulites* rudstone facies.

In the rhodolith and *Nummulites* rudstone facies and in the larger foraminiferal packstone facies coralline thalli are often abraded and fragmented. Encrustation can also be present on large-sized bivalve shells.

6.2.2 Lower Miocene Erosive Tributary Channels and Deeper Main Channel Deposits (Aquitanian–Burdigalian, Southern Sardinia)

The Sardinia Rift basin (central Mediterranean Sea, Italy) formed in the western Mediterranean region during the Oligocene–Miocene extensional phases ("Fossa sarda" *Auctorum*, Vardabasso 1963). This south-to-north Cenozoic basin develops from Cagliari to Sassari areas (Fig. 6.1), extending northward into the Asinara Gulf and offshore into the western Corsica (Funedda et al. 2000). Oligo–Miocene mainly siliciclastic and subordinately carbonate deposits fill in the tectonic-driven basin. The Sardinia Rift basin comprises several minor sub-basins, which formed on grabens and opposing-polarity half-grabens (Cherchi and Montadert 1982; Casula et al. 2001). Active tectonics, relative sea level oscillations and localized ecological factors combined to control the rate and type of sediment input to these minor sub-basins as well as the location, development and demise of patchy localized carbonate factories (Cherchi et al. 2000; Vigorito et al. 2005, 2006, 2010). These latters developed locally at the top of faulted-blocks along the submerged margins of the syn-rift sub-basins where the environmental and topographic settings matched the ecological behaviours of the benthic communities as sediment producers. Tectonics largely controlled the physiography of the sub-basins and in turn the location and trend of sediment pathways that locally included large submarine channel-systems (Murru et al. 2001, 2015; Vigorito et al. 2005, 2006, 2010).

The studied sedimentary successions are located in one of the resulting small sub-basins (the Isili Basin), in which fault-related structures confined a narrow NNW/SSE-trending, south-plunging trough (the Isili Trough). The margins of the Isili Trough were punctuated by coarse-grained fan-deltas, which passed through

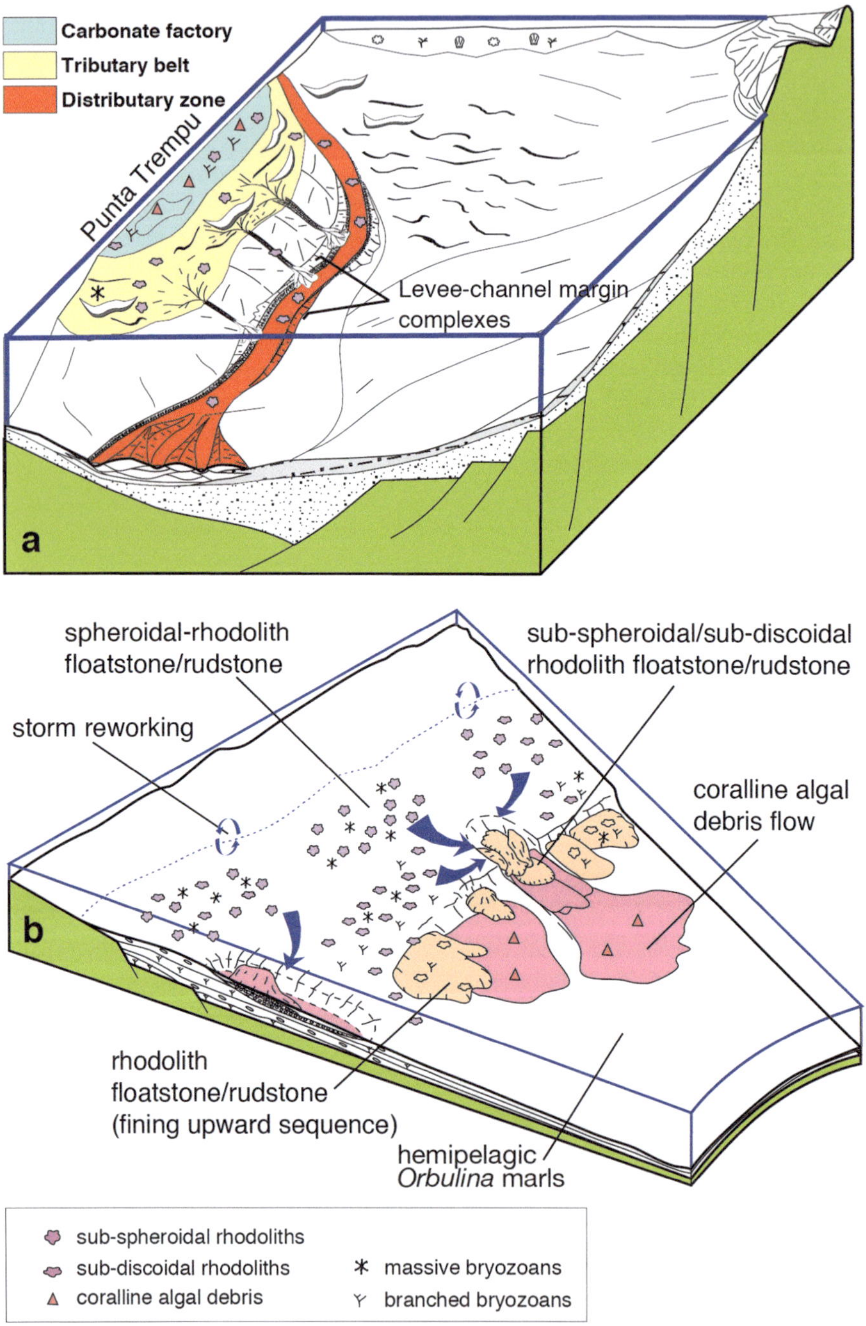

Fig. 6.6 Palaeoenvironmental models of depositional systems with the location of the Miocene re-sedimented rhodolith deposits. (**a**) Lower Miocene temperate-type carbonate-siliciclastic successions were deposited in basins located on the graben and half-graben along the Oligocene–Miocene Sardinia Rift Basin (Isili, southern Sardinia). The Isili basin's depositional system was

and overlain onto shallow-marine tuffaceous sandstones mixed with terrigenous/carbonate sandy/pebbly deposits (Fig. 6.3). These deposits are overlain by rhodolith rudstone that formed large trough-ward prograding sedimentary wedges (Cherchi et al. 2000). Since the Aquitanian in the Isili area active tectonics and sea-level variation drove the development of patchily distributed small shallow-water carbonate production areas and the offshore remobilization of carbonate sediments. These carbonate factories, and in particular the Punta Trempu structural high, were periodically eroded mainly during sea-level drops and the removed sediments were funnelled into the basin through a complex submarine channel network that included a tributary belt, one main channel (the Isili Channel) and the related fan (Cherchi et al. 2000; Vigorito et al. 2005; Fig. 6.6).

The studied rhodolith deposits belong to the Isili carbonate shelf sedimentary system that was sampled along a transect located from the carbonate factory (Punta Trempu section in Cherchi et al. 2000), to the marginal sectors, where displaced sediments were deposited as parallel to cross-stratified sedimentary bodies (tributary belt in the Isili village area). Submarine channel bodies crop out distally from the marginal sectors (the Isili Channel System in Vigorito et al. 2005). Details of the different stratigraphic localities sampled can be found in Bassi et al. (2006).

6.2.2.1 Coralline Algal and Rhodolith Assemblages

The Isili carbonate production area can be differentiated in shallow and deeper settings according to the coralline red algal deposits represented by algal debris of fruticose corallines (as "maerls" in Bassi et al. 2006), crustose coralline algal pavements (i.e., encrusting coralline bindstone with a wackestone–packstone matrix; Bassi 2005) and rhodolith beds. The coralline algal debris formed in the shallower area representing the carbonate factory, while crustose coralline algal pavement and rhodolith beds grew along the shelf margin settings contributing to the sedimentation into the tributary belt.

Fig. 6.6 (continued) characterized by a shallow-water carbonate factory, a marginal tributary channel belt and a main channel. Different sectors of the production areas contributed to feed the Isili main channel through erosive tributary conduits. Mixed rhodolith assemblages fed by the shelf-margin were distinguished in the filling sequences of the main channel (modified from Bassi et al. 2006). (**b**) The lower–middle Miocene channel network in the eastern Montagne del Matese (Pietraroia, Southern Apennines) consists of large amounts of Burdigalian–Langhian foramol/rhodalgal carbonate deposits which were repeatedly removed from the production areas and were laid down along heavily incised margins of a temperate-type open-shelf sedimentary system. The early-channelized succession is characterized by parautochthonous bryozoan floatstone and rhodolith/bryozoan floatstone representing soft substrates in a low water energy setting. These sediments were successively eroded and overlain by gravitative deposits constituted by allochthonous rhodolith floatstone/rudstone derived from shallower, high water energy settings from where the rhodoliths were re-mobilized down the channel and constituted the infilling deposits with a basin-ward progradation (modified from Bassi et al. 2010)

Unsorted fragmented fruticose thalli of unidentified coralline species (Fig. 6.5d) are abundant in the carbonate factory deposits together with fragmented coralline thin crusts that can locally occur with high amounts of silty micrite. Rhodoliths are rare, up to 2–3 cm in diameter, sub-spheroidal in shape with dense, symmetrical inner arrangements and abraded warty outer surfaces. Bioerosion is absent.

The deposits on shelf-margin areas are characterized by large sub-spheroidal rhodoliths, including serpulid worm-tubes and frequently encrusted by barnacles. The rhodolith outer growth forms are warty and lumpy with laminar, concentric inner arrangement made up of mastophoroids. In some peripheral areas of the shelf margin (section CN in Bassi et al. 2006), coralline crustose pavements developed as *in situ* encrusting thallial bindstone with sub-spheroidal and sub-discoidal asymmetrical rhodoliths that show a warty and encrusting outer surface (Fig. 6.7a). The crustose bindstone and the rhodoliths with an open inner arrangement are mainly made up of melobesioids; sporolithaceans and mastophoroids are subordinate. Delicate and fragile large bryozoan colonies can be very abundant. Encrustation and bioerosion are very common.

In the proximal tributary belt area (sections IG and IP in Bassi et al. 2006) small channels are filled in by spheroidal and sub-spheroidal rhodoliths with warty, lumpy and fruticose outer growth forms. The symmetrical inner arrangement generally shows low proportion of constructional voids (Fig. 6.7d, e) and is made up by mastophoroids and melobesioids with subordinate sporolithaceans and lithophylloids. Abrasion on the coralline thalli is very common. Fragmentation, encrustation and bioerosion are frequent.

The shelf-margin and the proximal tributary belt areas fed the distal sector (i.e., the main channel CU in Bassi et al. 2006) of the Isili sedimentary system by re-sedimented rhodoliths, which document a mixing from different source areas. From the parautochthonous algal debris of fruticose corallines and the in situ rhodolith beds, through deposits of partially to totally reworked rhodoliths, to the crustose coralline algal pavement, there is a clear tendency of (1) an increase in fine sediment, (2) a shift from coarser to finer fraction in the matrix of the rhodolith deposits, (3) a change from spheroidal to sub-discoidal rhodoliths (Fig. 6.8a, b), and (4) a reduction in mastophoroids and a concomitant increase in abundance of melobesioids. The rhodoliths found in the distal tributary belt area (CU in Bassi et al. 2006) show characteristic in common, in terms of shape, growth forms, and inner arrangement with those found in the proximal tributary belt. The well-differentiated growth phases of these rhodoliths (laminar and dense; warty, lumpy, encrusting morphologies) point to active growth during low-water energy periods leading to a delicate, laminar structure, and to high-energy events during which they were reworked and partially destroyed. Abrasion and encrustation are abundant, while fragmentation and bioerosion are common.

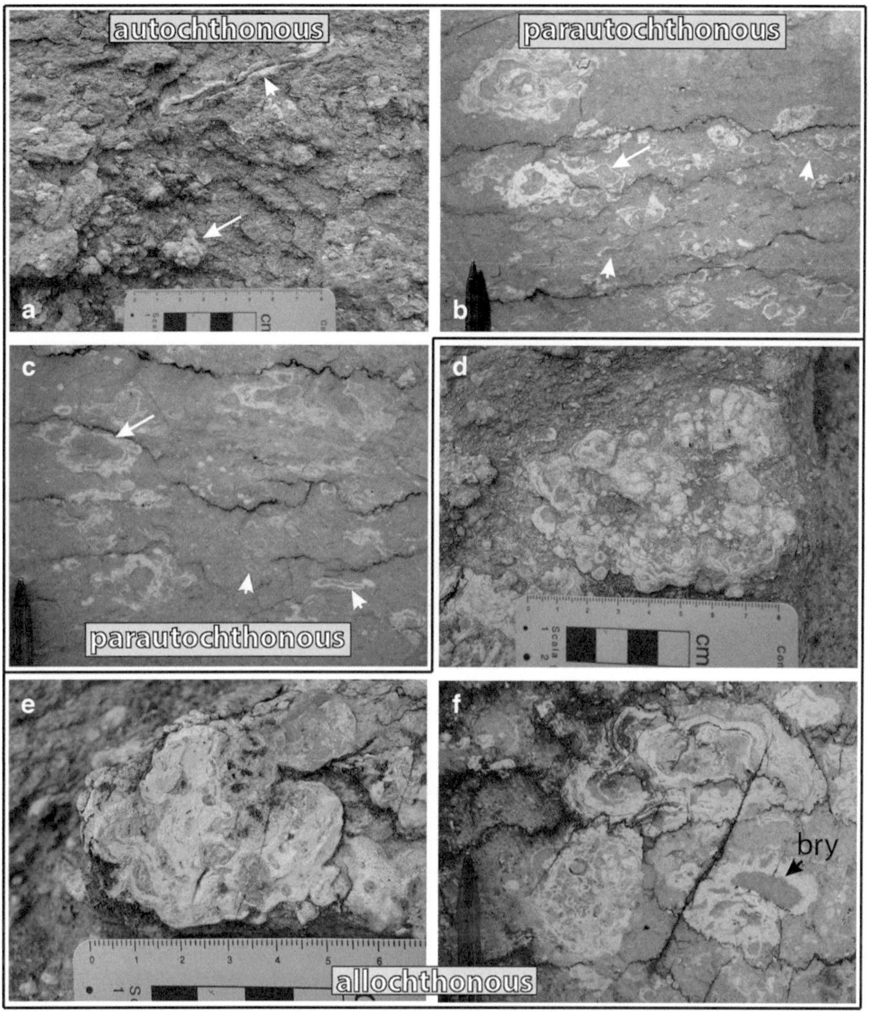

Fig. 6.7 Rhodolith deposits from shelf margin and tributary channel belt. (**a**–**c**) In protected areas of the shelf margin (**a**), autochthonous sub-spheroidal and sub-discoidal rhodoliths (*arrows*) with warty and encrusting outer surfaces occur in coralline crustose pavement characterized by encrusting coralline bindstone. In the parautochthonous sediments, bryozoan floatstone and rhodolith/bryozoan floatstone are present (**b**, **c**), sub-spheroidal and sub-discoidal rhodoliths (*arrows*) with loosely-packed inner arrangement and high constructional void spaces occur; thin encrusting superimposed corallines (*arrowheads*) coat larger bioclasts and float in the muddy matrix. (**d**–**f**) In the tributary channel belt, re-sedimented channelized deposits consist of floatstone/rudstone with allochthonous sub-spheroidal rhodoliths showing loose laminar inner arrangement and nuclei represented by massive bryozoans (bry). (**a**, **d**, **e**), Early Miocene, Isili area, southern Sardinia; (**b**, **c**, **f**), early–middle Miocene, Pietraroia, Montagne del Matese, Southern Apennines

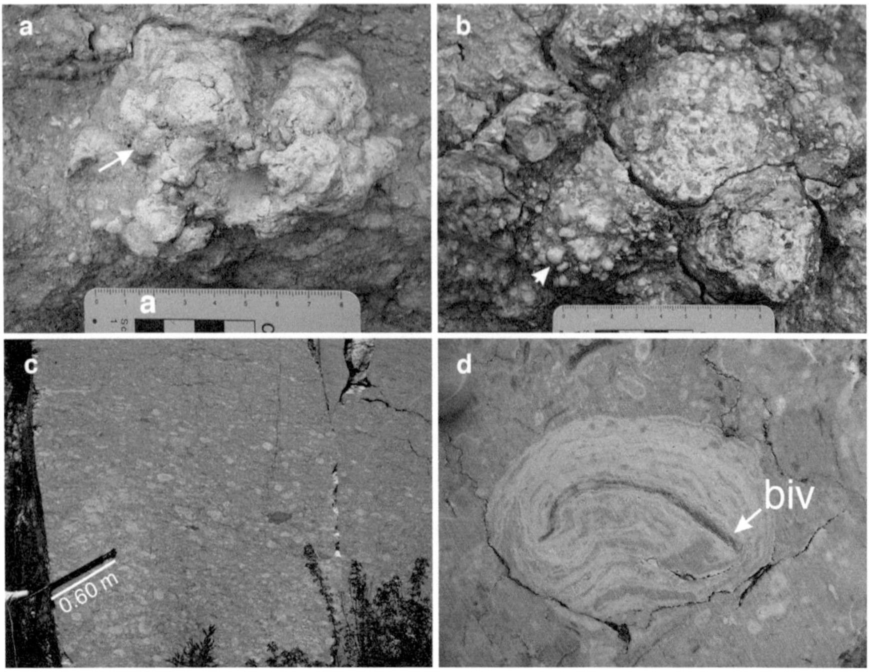

Fig. 6.8 Tributary channel belt deposits with allochthonous rhodoliths. (**a**, **b**) Sub-spheroidal rhodoliths of large sizes showing warty (*arrow*) and lumpy (*arrowhead*) outer growth forms with low proportion of constructional voids in the inner arrangement; early Miocene, Isili area, southern Sardinia. (**c**, **d**) Graded rhodolith floatstone/rudstone showing the large variability in the rhodolith size (**c**); the dominating sub-spheroidal rhodoliths show dense/open inner arrangements and their nuclei can be constituted of large bivalve shells (biv) or massive bryozoan colonies; early–middle Miocene, Pietraroia, Montagne del Matese, Southern Apennines

6.2.3 Lower–Middle Miocene Channel Networks on Tectonically Modelled Substrate (Burdigalian–Langhian, Montagne del Matese, Southern Apennines)

The Montagne del Matese area is part of the thrust and fold complex of the Central–Southern Apennines (Fig. 6.1), deformed since the late Miocene (D'Argenio et al. 1973; Sgrosso 1998). The southeastern part of the Matese mountain group is characterized by a few thousand metres-thick carbonate succession dated as Jurassic to early–middle Miocene. The carbonate succession is overlain by Tortonian–Messinian flysch sequences that represent the first phases of clastic sedimentation in the area (Fig. 6.3; Ciampo et al. 1987).

The studied lower–middle Miocene (Burdigalian–Langhian) shallow-water carbonate succession, cropping out in the eastern margin of the south-eastern Matese area, is characterized by a white massive bryo-rhodalgal and rhodalgal limestone ('Calcari a briozoi e lithothamni' *Auct.*, up to 50 m thick; Carannante et al. 1988).

Lower Miocene bioclastic coarse grainstone with miogypsinids and *Operculina* occurs at the very base of the succession or more commonly fills cavity networks produced by bioerosion and/or fracturing of the pre-Miocene substrate. These grainstone deposits are overlain by parautochthonous bryozoan floatstone and rhodolith/bryozoan floatstone (BLL-1 in Bassi et al. 2010) formed on soft substrates in a low water energy and high turbidity setting. An early subaqueous cementation phase resulted in incipient hard-grounds at the top of the bryozoan floatstone and rhodolith/bryozoan floatstone. Then, these floatstones were eroded by impressive gravitative flows and sealed by the related shallow-water rhodolith-bearing deposits (rhodolith floatstone-rudstone as BLL-2 in Bassi et al. 2010). The top of the shallow-water bryo-rhodalgal carbonate succession is characterized by very coarse phosphatic rudstones with grainstone matrix rich in planktonic foraminifera, phosphate peloids and rare glauconitic grains (Carannante 1982). Hard-grounds and/or iron or phosphate rich crusts are present throughout and especially at the top of the bryo-rhodalgal carbonate succession. The drowning event associated with the beginning of deposition of the overlying hemipelagic marls ('Marne ad *Orbulina*' *Auct.*) was synchronous in the studied Matese area (at 13.21 Ma; Lirer et al. 2005).

This Burdigalian–Langhian succession is part of a complex broad channelized carbonate sedimentary system formed on an uneven carbonate shelf margin which extended N–S for about 15 km and is up to 6 km wide (Carannante and Vigorito 2001; Bassi et al. 2010; Fig. 6.6b). The studied stratigraphic sections are located in the Pietraroia and Regia Piana areas (Figs. 6.1 and 6.3). The outcropping portion of the Pietraroia channel represents the proximal segment of a complex submarine main channel (stem) filled in with rhodalgal skeletal and deep-water deposits. The complex submarine channel networks were initially fed by shallow-water rhodalgal skeletal deposits, then covered by hemipelagic *Orbulina* marls. These shallow-water deposits removed from the main productive areas (presumably located westwards) when they were still unlithified, reached the Pietraroia area as aggradational channel-filling deposits, arranged as stacked channel bodies. Tributary channel bodies are quite common and concentrated in a belt less than 50 m wide. Secondary channels are commonly U-shaped; V-shaped tributary channels and filled scours also occur. Hard-grounds occurring throughout the studied limestones mark the different progradational events and testify erosive and/or non-deposition phases (Bassi et al. 2010). Each channelized deposit exhibits a sharp fining and thinning upward trend which is indicative of progressive demise and abandon of the channel as sedimentary conduit (Carannante and Vigorito 2001). The Regia Piana area represents a higher hydrodynamic setting than Pietraroia. A storm-influenced wide carbonate shelf produced shallow water sediments that were removed down and accumulated in a wider and gently steepened thalweg. Both in the shelf and in the thalweg high-energy hydrodynamic events swept off the mud producing highly selected rhodolith rudstone and coralline algal debris packstone/grainstone. Details of the stratigraphic and sedimentary setting are in Carannante (1982) and Bassi et al. (2010).

6.2.3.1 Coralline Algal and Rhodolith Assemblages

In the parautochthonous bryozoan floatstone and rhodolith/bryozoan floatstone (BLL-1 in Bassi et al. 2010) melobesioids are the only corallines: *Lithothamnion* is the dominating genus recorded, occasionally associated with *Mesophyllum*. The rhodoliths range from 3 to 7 cm in mean diameter and are sub-spheroidal and sub-discoidal in shape (Fig. 6.7b, c). Small rhodoliths (2–3 cm in mean diameter) with a large bryozoan nucleus and a thin laminar coralline algal coating are frequent. High proportion of constructional voids is common and nuclei are mainly represented by bryozoans. Rhodoliths are characterized by encrusting and laminar growth forms with rare warty protuberances.

These parautochthonous sediments are overlain by gravitative (allochthonous) deposits constituted of shallower-water rhodolith floatstone/rudstone (BLL-2 in Bassi et al. 2010) formed in high-energy settings from where the rhodoliths were re-mobilized down the channel, constituting the infilling deposits. Melobesioids (*Lithothamnion*, *Mesophyllum*; up to 70 % of the coralline assemblages) dominate with subordinate mastophoroids (*Spongites*, *Neogoniolithon*; 27–38 %). *Lithoporella* and lithophylloids (8–19 %), as well as *Sporolithon* (9–18 %), are very rare. The more proximal parts of the channel, with higher proportion of gravitative deposits by shallower lateral tributary channels, are characterized by melobesioid and mastophoroid corallines associated with subordinate lithophylloids and sporolithaceans. Mastophoroids and lithophylloids increase in abundance in the deposits derived from more proximal channel areas. The deeper water gravitative deposits are characterized by melobesioids and a lower coralline diversity at genus level.

In the allochthonous rhodolith floatstone/rudstone, rhodoliths are mainly spheroidal/sub-spheroidal in shape with subordinate sub-discoidal specimens, with dense and laminar symmetrical inner arrangements. Average rhodolith diameter ranges from 2 to 15 cm. Their nuclei are composed of oyster and bryozoan fragments. Outer growth forms are usually encrusting; warty and lumpy protuberances can also be common. Taphonomic features such as abrasion and encrustation are very abundant, while fragmentation and bioerosion are common.

In the Regia Piana channel deposits, the coralline taxonomic assemblage is dominated by melobesioids (up to 66 %), associated with subordinate mastophoroids (18 %), lithophylloids (8 %) and sporolithaceans (8 %). Rhodoliths are small in size ranging from 2 to 4.5 cm in mean diameter with spheroidal and sub-spheroidal shapes. Sub-discoidal rhodoliths are rare. Rhodoliths show laminar and encrusting loose, symmetrical inner arrangement developed on nuclei made up of bryozoan fragments and/or coarse to medium grained particles. Dominating coralline growth forms are encrusting and are frequently abraded. Abraded and fragmented warty and lumpy protuberances on the outer rhodolith surfaces are rare.

6.3 Discussion: Synopsis of Examples

Although the age-range of examples considered herein spans nearly 30 million years, the coralline red algal and rhodolith assemblages in the re-sedimented successions share striking similarities with respect to sedimentological, compositional and taphonomic characteristics. Re-sedimented successions exhibit similar general lithofacies ranging from silty packstone, through floatstone to rudstone with silty packstone matrix. In this section, the compositional characteristics of the re-sedimented rhodolith deposits are discussed in terms of rhodolith sphericity, coralline taxonomic assemblages, rhodolith inner arrangement, outer growth forms, and taphonomic signatures. Changes in these rhodolith characteristics have been correlated with some palaeoenvironmental conditions (Aguirre et al., this volume). Comparisons of the coralline algal components in the re-sedimented successions with in situ counterparts provide a basis for understanding the complex network of tributary and main channel systems as well as the gradients within each channel deposit (Fig. 6.9 and Table 6.1).

6.3.1 Rhodolith Sphericity

In the study cases, parautochthonous spheroidal and sub-spheroidal rhodoliths usually of small size (<5 cm in mean diameter) occur dominating in the carbonate factory (i.e., in the middle Eocene: rhodolith and *Nummulites* rudstone facies, coralline debris and larger *Nummulites* rudstone facies, larger foraminiferal packstone facies; in the early Miocene of Isili: coralline algal debris; Fig. 6.9). These rhodolith shapes were found as allochthonous in the proximal (i.e., early Miocene of Isili) and distal (i.e., early–middle Miocene of the Montagne del Matese: rhodolith floatstone/rudstone BLL-2) tributary belt erosive channels. These rhodolith shapes have been related to more frequent turning events in higher energy environments (Bosence 1983a, b; Steller and Foster 1995). In the Gulf of California, Recent spheroidal rhodoliths occur in shallower, wave influenced sediments (Foster et al. 1997; Johnson et al. 2011, 2012). In Sesoko-jima (Okinawa, Japan), sub-spheroidal rhodoliths were mainly found in water depths from 50 to 70 m, influenced by multidirectional water currents such as those produced by the combination of storm-induced and tide induced currents (Bassi et al. 2009).

In peripheral areas of the shelf margin, sub-spheroidal with a tendency to sub-discoidal rhodoliths usually occur in bindstone facies and, secondarily, in muddy coralline encrusting facies (crustose coralline algal pavement; Bassi 2005; Fig. 6.9a and Table 6.1). Overlying algal debris of fruticose corallines, these rhodolith beds are characterized by delicate bryozoan colonies indicating reduced hydrodynamic energy setting and a higher substrate stability, although episodic high-energy events could occur. This is mirrored by the sub-discoidal rhodoliths, which document occasional overturning. Slow laminar bottom currents swept the crustose coralline

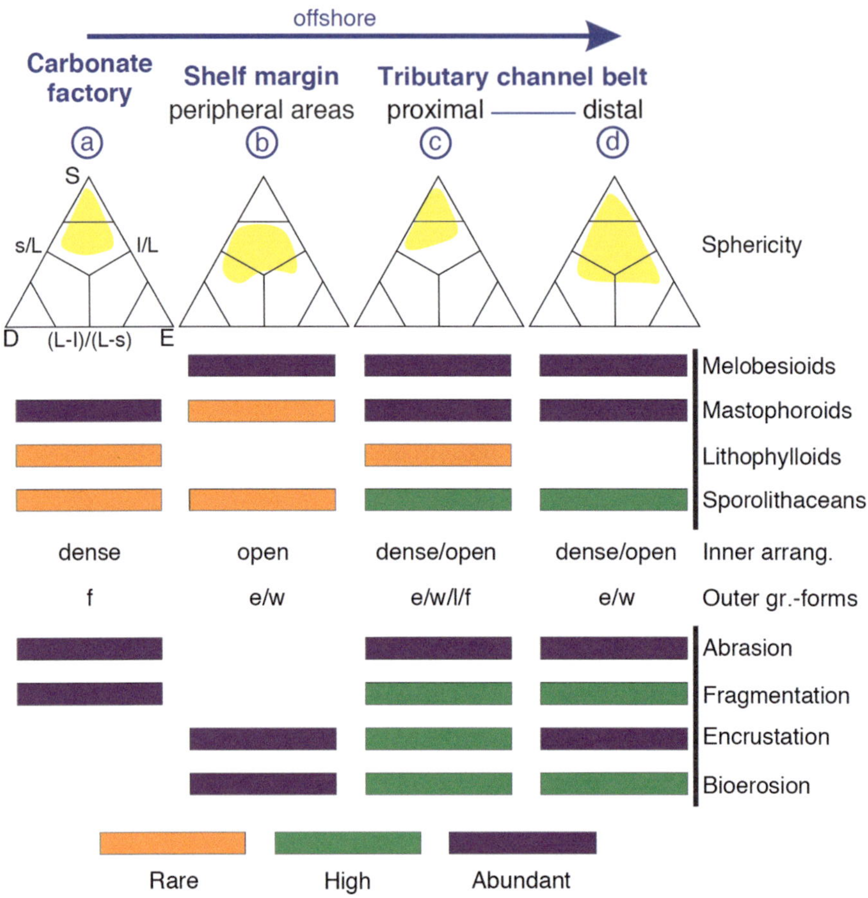

Fig. 6.9 Comparative characters of re-sedimented rhodoliths from the middle Eocene of northeast Italy, early Miocene of southern Sardinia and early–middle Miocene of southern Apennines. Summary model for rhodolith deposits and taphonomic gradients in carbonate factory, protected areas on the shelf margin, proximal and distal tributary channel belt areas. *S* spheroidal, *D* discoidal, *E* ellipsoidal, *L* long axis, *I* intermediate axis, *s* short axis, *arrang*. arrangement, *gr.-forms* growth forms, *Inner arrang*. inner arrangement, *f* fruticose, *e* encrusting, *w* warty, *l* lumpy. Coralline algal taxonomic composition is shown for the families Hapalidiaceae (subfamily Melobesioideae), Corallinaceae (subfamilies Mastophoroideae, Lithophylloideae) and Sporolithaceae. Dense inner arrangement points to low proportion of constructional voids

algal pavement hampering encrustation by bryozoans and thin encrusting coralline algae (Bassi et al. 2006).

The highest shape diversity is found in the distal zone of the tributary belt where sub-spheroidal and sub-ellipsoidal rhodoliths occur (i.e., Bassi et al. 2006, 2010; Fig. 6.9). Sub-discoidal rhodoliths are subordinate. The sphericity of these rhodoliths shows characteristics in common with the parautochthonous rhodoliths originated from the carbonate factory and the shelf margin documenting a mixing from

Table 6.1 Comparative aspects of re-sedimented rhodoliths from the middle Eocene of northeast Italy, early Miocene of southern Sardinia and early–middle Miocene of southern Apennines

Region, age, references, sedimentary system	Palaeoenvironmental setting/interpretation	Rhodolith characteristics	Corall. taxon. assemblages	Texture and lithology
Monti Berici, northeast Italy; middle Eocene; Bassi et al. (2013) Carbonate ramp	Off-shore re-deposition on a ramp	<4 cm diam., sub-spheroid., laminar inner arrang., warty/lumpy outer g-f	mastophoroids D; melobesioids, sporolithaceans S	rudstone with packst matrix
Southern Sardinia, Italy; early Miocene; Bassi et al. (2006)	Carbonate factory, parautochthonous	Fragmented fruticose; <3 cm diam., sub-spheroidal, dense inner arrang., warty outer g-f	Undetermined	rudstone with packst. matrix
Erosive tributary channels and deeper main channel	Shelf margin			
	Parautochthonous	Spheroidal, laminar concentric inner arrang., warty/lumpy outer g-f	mastophoroids	rhodolith rudstone
	Autochthonous	Encrusting binding corallines, sub-spheroid./sub-discoid., loose inner arrang., warty/encrust. outer g-f	melobesioids D; sporolithaceans, mastophoroids S	crustose coralline algal pavement, bindstone
	Prox tributary channel belt, allochthonous	Spheroid., sub-spheroid., dense inner arrang., warty/lumpy/fruticose outer g-f	mastophoroids, sporolithaceans; melobesioids, mastophoroids	rhodolith rudstone
	Distal tributary channel belt, allochthonous	Mixing algal characters from marginal sectors and tributary channels	melobesioids, mastophoroids	rhodolith rudstone
Montagne del Matese, southern Apennines, Italy; early–middle Miocene; Bassi et al. (2010)	Parautochthonous (BLL-1)	3–7 cm diam., sub-spheroidal, sub-discoidal, loosely packed inner arrang., encrusting outer g-f, rare warty	melobesioids	rhodolith floatstone
Channel networks on tectonically modelled substrate	Channel, allochthonous deposits (BLL-2)	2–15 cm diam., spheroidal/sub-spheroidal, dense and laminar inner arrang., encrusting outer g-f (rare warty, lumpy)	melobesioids D; mastophoroids S, sporolithaceans S, lithophylloids S	rhodolith floatstone/ rudstone

diam diameter, *spheroid* spheroidal, *discoid* discoidal, *arrang* arrangement, *g-f* growth form, *D* dominant, *S* subordinate, *packst* packstone, *prox* proximal different source areas. These rhodoliths were re-sedimented during different filling stages of the main channel in a distal sector of the tributary belt. Occasionally storms and gravitative events have periodically removed the rhodoliths from the original sites and deposited them in the tributary zone.

Sub-spheroidal and sub-ellipsoidal rhodoliths were generated where overturning rates on the seafloor and coralline species composition allowed for a thick coralline thallial development (e.g., present-day, northeast Australia, Lund et al. 2000; late Pliocene, south-western Spain, Aguirre et al. 1993). After re-sedimentation, the rhodoliths were generally buried and this developmental condition ceased.

6.3.2 Coralline Taxonomic Assemblages

Melobesioid coralline algae dominate in the shelf margin and in the tributary channel areas, showing locally higher species diversities than the associated mastophoroids. The latter occur commonly in the carbonate factory and in the distal tributary channel belt (Fig. 6.9 and Table 6.1). In the Early–Middle Miocene case study, mastophoroids can be highly diversified at species level (Bassi et al. 2010; Checconi et al. 2010). Lithophylloids can be common in the carbonate factory and in the proximal tributary channel belt, but were neither identified in the protected shelf margin areas nor in the distal tributary zone. Sporolithaceans are present in all the studied settings, but are subordinate to melobesioids and mastophoroids (Fig. 6.9).

Some taxonomic groups of coralline algae show particular depth distribution (e.g., Adey and Macintyre 1973; Adey et al. 1982; Adey 1986; Minnery et al. 1985; Bosence 1991; Aguirre et al. 2000, 2012). Several mastophoroid and lithophylloid species live in the upper, best-illuminated 10 m of present-day reefs (Adey et al. 1982; Iryu et al. 1995; Littler and Littler 2003). In modern tropical reef environments, mastophoroids and subordinate lithophylloids are most abundant in shallow water, melobesioids become more abundant below 10–20 m, and sporolithaceans are abundant at greater depths, although they also occupy cryptic habitats in shallow reef ecosystems. Present-day melobesioids dominate the coralline algal assemblages at deep tropical and shallow to deep temperate environments, usually reaching depths of 110–120 m (e.g., Adey 1986; Adey et al. 1982; Minnery et al. 1985; Iryu et al. 1995; Lund et al. 2000; Braga and Aguirre 2004). Off Fraser Island, non-tropical Australia, rhodoliths are dominated by melobesioids and sporolithaceans below 60 m water depths (Lund et al. 2000; Bassi et al. 2013). Rhodoliths off the Ryukyu Islands also show a possible increase in melobesioids and sporolithaceans with depth (Iryu et al. 1995; Matsuda and Iryu 2011). The rare occurrence of lithophylloids in the case studies, however, which are restricted to the carbonate factory and the proximal tributary channel belt, is in good accordance with their shallow water ecological preference which limits their re-working and taphonomic preservation into the distal and deeper settings (e.g., Gischler and Pisera 1999; Lund et al. 2000; Braga and Aguirre 2004). The high coralline diversity found in the proximal tributary channel belt area is due to the contributions from different source areas of allochthonous rhodoliths.

6.3.3 Inner Arrangement

The studied fossil re-sedimented rhodoliths show common symmetrical as well as uncommon asymmetrical patterns of accretion. This may be due to the relative turning frequency of individual rhodoliths. Asymmetric algal growth has been suggested as an indicator for extended periods of stable position for rhodoliths (e.g., Bosence 1983b; Aguirre et al. 1993; Zuschin and Piller 1997; Bassi et al. 2009). Rhodoliths from protected areas of the self-margin show asymmetrical inner arrangement confirming their occasional overturning (Bassi et al. 2006), while re-sedimented rhodoliths in proximal and distal tributary zones have a symmetrical pattern of accretion being originated from settings with higher energy (Fig. 6.9).

In shallow-water environments, laminar dense inner arrangement is developed when rhodoliths are frequently overturned (i.e., Bosence 1983a, b; Steller and Foster 1995; Perry 2005), while highly branched small rhodoliths occur in coralline algal debris facies (Nebelsick et al. 2005; Hinojosa-Arango et al. 2009). In the studied cases, re-sedimented rhodoliths with laminar internal structure were found to be autochthonous in the carbonate factory and re-sedimented in the tributary channel belt areas (Fig. 6.9).

Small rhodoliths with open inner arrangements (i.e., high proportion of constructional voids) represent moderate water energy conditions, while the larger discoidal rhodoliths with a high percentage of constructional voids suggest lower energy (e.g., Bassi 1998, 2005; Marrack 1999; Rasser and Piller 2004; Bassi and Nebelsick 2010; Nitsch et al. 2015). These are the cases of the studied rhodoliths respectively found in the parautochthonous bryozoan floatstone and rhodolith/bryozoan floatstone (BLL-1; Bassi et al. 2010) and in the crustose coralline algal pavement of the shelf margin (Bassi et al. 2006).

6.3.4 Outer Growth-Forms

In the studied re-sedimented rhodoliths, encrusting and warty outer growth-forms dominate with minor occurrence of lumpy and fruticose morphologies from the carbonate factory, through the shelf margin to the distal tributary channel areas (Fig. 6.9). Fruticose growth-forms were only recognized in the carbonate factory and rarely in the proximal tributary channel belt.

A number of studies (e.g., Penrose 1991; Penrose and Woelkerling 1991; Woelkerling and Harvey 1992) have shown that considerable variations in growth-form can occur within a species. Differences in growth-form are helpful in specimen identification in a particular geographic region (Braga et al. 2009a, b), but unfortunately this does not imply that such differences are of diagnostic value (Bosence 1976; Woelkerling and Irvine 1986; Woelkerling et al. 1993). According to some authors, differences in crustose coralline morphology are primarily due to environment (water energy, nature of the substrate) and not to differences in species (e.g., Bosence 1983a; Foster et al. 1997; Rasser and Piller 2004). Other studies, however, did not find a relationship between water energy and rhodolith outer morphology (e.g., Steller and Foster 1995; Gischler and Pisera 1999; Lund et al. 2000). Bathymetry and light intensity themselves do not seem to affect coralline growth-forms.

Encrusting and warty growth forms have been described from fossil rhodoliths on soft mobile substrates in middle ramp and talus slope settings (e.g., Bassi 2005; Braga et al. 2009a, b; Bassi and Nebelsick 2010). In the Brazilian Fernando de Noronha Archipelago, rhodolith beds observed at depths from 10 to 100 m are monospecific and possess warty or lumpy growth forms (Amado-Filho et al. 2012). In deep fore-reef to shelf areas in the Ryukyu Islands rhodolith-forming algae are mainly thin, encrusting (and less commonly, warty to lumpy) forms (Matsuda and Iryu 2011). The spheroidal rhodolith shape consisting of encrusting thalli with a dense and symmetrical pattern of accretion suggests a frequent multiple-directional overturning of the rhodoliths (e.g., Reid and Macintyre 1988; Bassi 1995, 2005; Ballantine et al. 2000; Rasser and Piller 2004) as recognized in the studied re-sedimented specimens found in the tributary channel belt areas.

6.3.5 Taphonomic Signatures

The observed taphonomic features show a characteristic distribution with respect to the water energy of the setting and the residence time on the sediment-water interface before the burial. Abrasion is very common in all the distinguished settings, being absent in the rhodoliths of the protected area on the shelf margin. Abrasion is especially evident in rhodoliths on coarse (sandy) mobile substrates on which transport and sediment friction are important factors (Fig. 6.9).

Fragmentation is widespread in the shallower carbonate factory rhodoliths, which contribute to the sediment matrix as fragments of fruticose coralline branches. In the tributary channel belt, fragmentation can be locally high. Encrustation occurs abundantly in the crustose rhodolith pavement on the self-margin and in the distal tributary channel belt area. Encrustation is the biological process driving to the rhodolith formation by corallines and other components. Nonetheless, it is also an important factor affecting preservation and is thus discussed with respect to its taphonomic connotations. This is especially the case where the process of encrustation protects the encrusted organisms from destruction by abrasion (e.g., Nebelsick and Bassi 2000; Nebelsick et al. 2011). Complex encrustation sequences are usually performed by corallines, bryozoans and encrusting foraminifera (mainly acervulinids); serpulids can be very common. These sequences have positive effects protecting thin algal crusts from destruction through abrasion by preserving vegetative (e.g. epithallial cells) and reproductive (e.g., conceptacles) features, which are diagnostic characters for the taxonomic identification.

Bioerosion can be readily recognised as sediment-filled holes, which often continue through the coralline algal thalli. Bioerosion can be very rare in the smallest rhodoliths and becomes common in the largest ones. In the middle Eocene rhodolith and *Nummulites* rudstone facies, bioerosion is present only in the larger rhodoliths as *Trypanites* and rarely small-sized *Gastrochaenolites*. In the early and middle Miocene studied cases, bioerosion is very abundant only in the autochthonous coralline crustose pavement on the self margin in the Isili area, while it is common in rhodoliths found in the proximal and distal tributary channel belt areas. Bioerosion is performed by bivalves, sponges, polychaetes, barnacles, algae, fungi and bacteria (Bassi et al. 2013). The smaller and shallower the traces are, the more prone to abrasion and scarcely preserved they are. This could explain the common high preservation state of the deepest borings (i.e., *Gastrochaenolites*, *Entobia*, *Trypanites*, *Maeandropolydora* and *Caulostrepsis*) in the rhodoliths both within high- (tributary channels) and low water energy (protected areas) settings (Checconi et al. 2010).

Rhodoliths in different environments proceed along different taphonomic pathways. Taphonomic attributes (abrasion, fragmentation, encrustation, bioerosion) of the studied rhodoliths track environmental energy and sedimentation rate, in particular its effects on the stability and reworking of rhodoliths at the sediment surface. The lower the sedimentation rates, the more bioreosion and encrustation can take place. In the autochthonous coralline crustose pavement facies of the protected areas on the self-margin, encrustation and bioerosion are abundant, whilst in parautochthonous deposits of the carbonate factory abrasion and fragmentation dominate. An abrasion/fragmentation-dominated pathway characterizes the carbonate factory. An encrustation/bioerosion-dominated pathway characterizes the protected areas on the shelf-margin. The allochthonous rhodolith deposits, re-sedimented into the proximal and distal tributary channel belt areas, show all four taphonomic signatures though with different intensities. Bioerosion increases and fragmentation decreases from the proximal to distal parts of the tributary channel belt, according to the increasing stability of the substrate (i.e., reduction in re-sedimentation events). Contrasting pathways produce distinct proportions of taphonomic attributes, proportions that delineate taphofacies (e.g., Meldahl and Flessa 1990; Brachert et al. 1998; Rico-García et al. 2008; Yesares-García and Aguirre 2004; Nebelsick et al. 2011).

6.4 Concluding Remarks

The constructional rhodolith characteristics (coralline taxonomic assemblages, rhodolith sphericity, inner arrangement, outer growth-forms) and the taphonomic signatures recorded in the rhodolith growth stages reflect the depositional processes that led to formation of re-sedimented channelized deposits.

Distinguishing *in situ* (autochthonous) and re-sedimented (allochthonous) rhodoliths in terms of rhodolith and taphonomic characteristics in carbonate and mixed siliciclastic-carbonate, channelized depositional systems appears to support palaeoenvironmental reconstruction, combining palaeobiological with physical (sedimentology and facies geometry) analyses. In the studied shelves, loose biogenic sediments only partly remain in the production area (autochthonous/parautochthon ous sediments) being generally removed offshore and re-sedimented (allochthonous) by means of extensive gravity flows through complex channel networks (Bassi et al. 2010). In the channels, after final deposition and under favourable environmental conditions (e.g., water energy, light, substrates, sedimentation rate; see Aguirre et al. this volume), coralline red algae may continue to grow and thus further contribute to the overall carbonate skeletal production.

The Middle Eocene case study illustrates middle-ramp sediments that were removed offshore before lithification. These rhodoliths then underwent early lithification after being deposited in a proximal outer ramp setting.

The Lower–Middle Miocene case studies show rhodolith beds from a temperate-type open shelves *sensu* (Carannante et al. 1988) continuously feeding complex channel networks. The absence of reef-builder assemblages and of significant early diagenetic processes produced mainly loose carbonate deposits consisting of unlithified skeletal debris that are prone to re-mobilization. The re-sedimented rhodoliths can potentially be remobilized after deposition and can continue to play an important ecological role as rhodolith beds in deeper settings (Checconi et al. 2010). Thus, combining facies analysis, mapping of the three-dimensional facies architecture, analysis of the constructional rhodolith characteristics and the taphonomic signatures allows the complex relationship patterns of autochthonous/parautochtho nous and allochthonous rhodolith deposits to be assessed. In submarine channelized carbonate settings, these patterns were governed by the interplay of changes in environmental factors such as water energy, light, nature of the substrate, and surface residence time on the sediment-water interface.

Acknowledgements DB has been funded by local research grants (FAR) at the University of Ferrara. Thoroughly constructive comments by J.C. Braga and J. Aguirre are greatly appreciated.

References

Adey WH (1986) Coralline algae as indicators of sea-level. In: Van de Plassche O (ed) Sea level research: a manual for the collection and evaluation of data. Free Univ Geo Book, Norwich, pp 229–280

Adey WH, MacIntyre IG (1973) Crustose coralline algae: a re-evaluation in the geological sciences. Geol Soc Am Bull 84:883–904

Adey WH, Townsend RA, Boykins WT (1982) The crustose coralline algae (Rhodophyta: Corallinaceae) of the Hawaiian Islands. Smithson Contrib Mar Sci 15:1–74

Aguirre J, Braga JC, Martín JM (1993) Algal nodules in the upper Pliocene deposits at the coast of Cadiz (S Spain). In: Barattolo F, De Castro P, Parente M (eds) Studies on fossil benthic algae. Boll Soc Paleontol Ital 1:1–7

Aguirre J, Riding R, Braga JC (2000) Diversity of coralline red algae: origination and extinction patterns from the Early Cretaceous to the Pleistocene. Paleobiology 26:651–667

Aguirre J, Braga JC, Martín JM, Betzler C (2012) Palaeoenvironmental and stratigraphic significance of Pliocene rhodolith beds and coralline algal bioconstructions from the Carboneras Basin (SE Spain). Geodiversitas 34:115–136

Amado-Filho GM, Pereira-Filho GH, Bahia RG, Abrantes DP, Veras PC, Matheus Z (2012) Occurrence and distribution of rhodolith beds on the Fernando de Noronha Archipelago of Brazil. Aquat Bot 101:41–45

Ballantine DL, Bowden-Kerby A, Aponte NE (2000) *Cruoriella* rhodoliths from shallow-water back reef environments in La Parguera, Puerto Rico (Caribbean Sea). Coral Reefs 19:75–81

Bassi D (1995) Crustose coralline algal pavements from Late Eocene – Colli Berici of Northern Italy. Riv Ital Paleont Strat 101:81–92

Bassi D (1998) Coralline algal facies and their palaeoenvironments in the Late Eocene of Northern Italy (Calcare di Nago). Facies 39:179–202

Bassi D (2005) Larger foraminiferal and coralline algal facies in an Upper Eocene storm influenced, shallow water carbonate platform (Colli Berici, north-eastern Italy). Palaeogeogr Palaeoclimatol Palaeoecol 226:17–35

Bassi D, Nebelsick JH (2005) Coralline red algae in the Middle-Upper Eocene shallow water carbonates of the Colli Berici, north-eastern Italy. In: Field excursion Guide Book of the 5th Reg Symp Int Fossil Algae Association, 1–3/09/2005. St Trent Sci Nat Acta Geol suppl 80(2003):7–14

Bassi D, Nebelsick JH (2010) Components, facies and ramps: redefining Upper Oligocene shallow water carbonates using coralline red algae and larger foraminifera (Venetian area, northeast Italy). Palaeogeogr Palaeoecol Palaeoclimatol 295:258–280

Bassi D, Cosovic V, Papazzoni CA, Ungaro S (2000) The Colli Berici. In: Bassi D (ed) Shallow water benthic communities at the middle–upper eocene boundary. Southern and North-Eastern Italy, Slovenia, Croatia, Hungary. Field Trip Guidebook of the 5th Meeting IGCP 393 IUGS-UNESCO, Ann Univ Ferrara Suppl Sci Terra, pp 43–57

Bassi D, Carannante G, Murru M, Simone L, Toscano F (2006) Rhodalgal/bryomol assemblages in temperate type carbonate, channelised depositional systems: the Early Miocene of the Sarcidano area (Sardinia, Italy). In: Pedley HM, Carannante G (eds) Cool-water carbonates: depositional systems and palaeoenvironmental control. Special Publication 255. Geological Society London, London, pp 35–52

Bassi D, Bianchini G, Mietto O, Nebelsick JH (2008) Southern Alps in Italy: Venetian pre-Alps. In: McCann T (ed) The geology of Central Europe. Geological Society London, London, pp 56–62

Bassi D, Nebelsick JH, Checconi A, Hohenegger J, Iryu Y (2009) Present-day and fossil rhodolith pavements compared: their potential for analysing shallow-water carbonate deposits. Sediment Geol 214:74–84

Bassi D, Carannante G, Checconi A, Simone L, Vigorito M (2010) Sedimentological and palaeoecological integrated analysis of a Miocene canalized coralline red algal carbonate margin (Matese Mountains, Central-Southern Apennines, Italy). Sediment Geol 230:105–122

Bassi D, Iryu I, Nebelsick JH (2012) To be or not to be a fossil rhodolith? Analytical methods for studying fossil rhodolith deposits. J Coast Res 28:288–295

Bassi D, Nebelsick JH, Puga-Bernabéu Á, Luciani V (2013) Middle Eocene larger nummulites palaeohighs and their offshore re-deposition: a case study from the middle Eocene of the venetian area, northeast Italy. Sediment Geol 297:1–15

Bassi D, Iryu Y, Braga JC, Takayanagi H, Tsuji T (2013) Bathymetric distribution of ichnocoenoses from recent subtropical algal nodules off Fraser Island, eastern Australia. Palaeogeogr Palaeoecol Palaeoclimatol 369:58–66

Basso D, Tomaselli V (1994) Palaeoecological potentiality of rhodoliths: a Mediterranean case history. In: Matteucci L, Carboni MG, Pignatti JS (eds) Studies on ecology and paleoecology of benthic communities. Boll Soc Paleontol Ital spec vol 2:17–27

Bosence DWJ (1976) Ecological studies on two unattached coralline algae from Western Irela1nd. Palaeontology 19:365–395

Bosence DWJ (1983a) The occurrence and ecology of recent rhodoliths – a review. In: Peryt TM (ed) Coated grains. Springer, Berlin, pp 225–242

Bosence DWJ (1983b) Description and classification of rhodoliths (rhodoids, rhodolites). In: Peryt TM (ed) Coated grains. Springer, Berlin, pp 217–224

Bosence DWJ (1991) Coralline algae: mineralisation, taxonomy, and palaeoecology. In: Riding R (ed) Calcareous algae and stromatolites. Springer, Berlin, pp 98–113

Brachert TC, Betzler C, Braga JC, Martín JM (1998) Microtaphofacies of a warm-temperate carbonate ramp (uppermost Tortonian/lowermost Messinian, Southern Spain). Palaios 13:459–475

Braga JC (2003) Application of botanical taxonomy to fossil coralline algae (Corallinales, Rhodophyta). Acta Micropal Sin 20:47–56

Braga JC, Aguirre J (2004) Coralline algae indicate Pleistocene evolution from deep, open platform to outer barrier reef environments in the northern Great Barrier Reef margin. Coral Reefs 23:547–558

Braga JC, Martín JM, Wood JL (2001) Submarine lobes and feeder channels of redeposited, temperate carbonate and mixed siliciclastic–carbonate platform deposits (Vera Basin, Almería, southern Spain). Sedimentology 48:99–116

Braga JC, Martín JM, Betzler C, Aguirre J (2006) Models of temperate carbonate deposition in Neogene basins in SE Spain: a synthesis. In: Pedley HM, Carannante G (eds) Cool-water carbonates: depositional systems and palaeoenvironmental controls. Special Publication 255. Geological Society London, London, pp 121–135

Braga JC, Aguirre J, Esteban J (2009a) Calcareous algae of Cabo de Gata–Níjar Nature Park. Field Guide, ACUAMED-Consejería Medio Ambiente. Junta de Andalucía, Madrid (Spanish text 143 pp, English text 63 pp)

Braga JC, Vescogni A, Bosellini FR, Aguirre J (2009b) Coralline algae (Corallinales, Rhodophyta) in western and central Mediterranean. Messinian reefs. Palaeogeogr Palaeoclimatol Palaeoecol 275:113–128

Braga JC, Bassi D, Piller WE (2010a) Palaeoenviromental significance of Oligocene–Miocene coralline red algae – a review. In: Mutti M, Piller WE, Betzler C (eds) Carbonate systems during the Oligocene–Miocene climatic transition. Special Publication 42. International Association of Sedimentologists, Wiley-Blackwell, pp 165–182

Braga JC, Martín JM, Aguirre J, Baird CD, Grunnaleite I, Jensen NB, Puga-Bernabéu A, Sælen G, Talbot MR (2010b) Middle–Miocene (Serravallian) temperate carbonates in a seaway connecting the Atlantic Ocean and the Mediterranean Sea (North Betic Strait, S Spain). Sediment Geol 225:19–33

Carannante G (1982) Modello deposizionale e diagenetico di un livello fosfatico del Miocene carbonatico dell'Appennino Campano. Rend Soc Geol It 5:15–20

Carannante G, Vigorito M (2001) A channelised temperate-type carbonate margin: geometries and controlling factors. Géol Mediterr 28:41–44

Carannante G, Esteban M, Milliman JD, Simone L (1988) Carbonate lithofacies as paleolatitude indicators: problems and limitations. In: Nelson CS (ed) Nontropical shelf carbonates – modern and ancient. Sediment Geol 60:333–346

Casula G, Cherchi A, Montadert L, Murru M, Sarrja E (2001) The Cenozoic grabens system of Sardinia (Italy): geodynamic evolution from new seismic and field data. Mar Petrol Geol 18:863–888

Checconi A, Bassi D, Monaco P, Carannante G (2010) Re-deposited rhodoliths in the Middle Miocene hemipelagic deposits of Vitulano (Southern Apennines, Italy): coralline assemblage characterization and related trace fossils. Sediment Geol 225:50–66

Cherchi A, Montadert L (1982) The Oligo-Miocene rift of Sardinia and the early history of the western Mediterranean basin. Nature 298:736–739
Cherchi A, Murru M, Simone L (2000) Miocene carbonate factories in the syn-rift Sardinia Graben sub-basins (Italy). Facies 43:223–240
Ciampo G, Sgrosso I, Ruggero Taddei E (1987) L'inizio della sedimentazione terrigena nel Matese, nei Monti del Casertano e nei Monti di Suio. Boll Soc Geol Ital 106:323–330
D'Argenio B, Pescatore T, Scandone P (1973) Schema geologico dell'Appennino Meridionale (Campania, Lucania). Acc Naz Lincei 183:49–72
Davaud E, Septfontaine M (1995) Post-mortem onshore transportation of epiphytic foraminifera: recent example from the Tunisian coastline. J Sediment Res 65:136–142
Fabiani R (1915) Il Paleogene del Veneto. Mem Ist Geol Regia Univ Padova 3:1–336
Foster MS, Riosmena-Rodriguez R, Steller DS, Woelkerling WJ (1997) Living rhodolith beds in the Gulf of California and their implications for palaeoenvironmental interpretation. In: Johnson ME, Ledesma-Vázquez J (eds) Pliocene carbonates and related facies flaking the Gulf of California, Baja California. Special paper 318. Geological Society of America, Boulder, pp 127–139
Funedda A, Oggiano G, Pasci S (2000) The Logudoro basin: a key area for the tectono-sedimentary evolution of North Sardinia. Boll Soc Geol Ital 119:31–38
Gischler E, Pisera A (1999) Shallow water rhodoliths from Belize reefs. N Jb Geol Paläont (Abh) 214:71–93
Graham DJ, Midglay NG (2000) Graphical representation of particle shape using triangular diagrams: an excel spreadsheet method. Earth Surf Proc Land 25:1473–1477
Harvey AS, Broadwater ST, Woelkerling WJ, Mitrovski PJ (2003) *Choreonema* (Corallinales, Rhodophyta): 18S rDNA phylogeny and resurrection of the Hapalidiaceae for the subfamilies Choreonematoideae, Austrolithoideae, and Melobesioideae. J Phycol 39:988–998
Hinojosa-Arango G, Maggs CA, Johnoson MP (2009) Like a rolling stone: the mobility of maerl (Corallinaceae) and the neutrality of the associated assemblages. Ecology 90:517–528
Iryu Y, Nakimori T, Matsuda S, Abe O (1995) Distribution of marine organisms and its geological significance in the modern reef complex of the Ryukyu Islands. Sediment Geol 99:243–258
Iryu Y, Bassi D, Woelkerling W (2012) Typification and reassessment of seventeen species of coralline red algae (Corallinales and Sporolithales, Rhodophyta) described by W. Ishijima during 1954–1978. J Syst Palaeontol 10:171–209
Johnson ME, da Silva CM, Santos A, Baarli BG, Cachão M, Mayoral EJ, Rebelo AC, Ledesma-Vázquez J (2011) Rhodolith transport and immobilization on a volcanically active rocky shore: middle Miocene at Cabeço das Laranjas on Ilhéu de Cima (Madeira Archipelago, Portugal). Palaeogeogr Palaeoclimatol Palaeoecol 300:113–127
Johnson ME, Baarli GB, Cachão M, da Silva CM, Ledesma-Vázquez J, Mayoral EJ, Ramalho RS, Santos A (2012) Rhodoliths, uniformitarianism, and Darwin: Pleistocene and Recent carbonate deposits in the Cape Verde and Canary archipelagos. Palaeogeogr Palaeoclimatol Palaeoecol 329–330:83–100
Lirer F, Persico D, Vigorito M (2005) Calcareous plankton biostratigraphy and age of the middle Miocene deposits of Longano Formation (eastern Matese Mountains, southern Apennines). Riv Ital Paleontol Strat 111:91–108
Littler DS, Littler MM (2003) South Pacific reef plants. A divers' guide to the plant life of South Pacific coral reefs. OffShore Graphics, Washington, DC
Lund M, Davies PJ, Braga JC (2000) Coralline algal nodules off Fraser Island, eastern Australia. Facies 42:25–34
Marrack EC (1999) The relationship between water motion and living rhodolith beds in the southwestern Gulf of California, Mexico. Palaios 14:159–171
Márton E, Zampieri D, Kázmér M, Dunkl I, Frisch W (2011) New Paleocene–Eocene paleomagnetic results from the foreland of the Southern Alps confirm decoupling of stable Adria from the African plate. Tectonophysics 504:89–99

Matsuda S, Iryu Y (2011) Rhodoliths from deep fore-reef to shelf areas around Okinawa-jima, Ryukyu Islands, Japan. Mar Geol 282:215–230

Meldahl KH, Flessa KW (1990) Taphonomic pathways and comparative biofacies and taphofacies in a recent intertidal/shallow shelf environment. Lethaia 23:43–60

Minnery GA, Rezak R, Bright TJ (1985) Depth zonation and growth form of crustose coralline algae: Flower Garden Banks, northwestern Gulf of Mexico. In: Toomey DF, Nitecki MH (eds) Paleoalgology. Springer, Berlin, pp 237–247

Murru M, Simone L, Vigorito M (2001) Carbonate channel network in the Miocene syn-rift Sardinia basins. Géol Mediterr 28:133–137

Murru M, Bassi D, Simone L (2015) Displaced/re-worked rhodolith deposits infilling parts of a complex Miocene multistorey submarine channel: a case history from the Sassari area (Sardinia, Italy). Sediment Geol 326:94–108

Nebelsick J, Bassi D (2000) Diversity, growth forms and taphonomy: key factors controlling the fabric of coralline algal dominated shelf carbonates. In: Insalaco E, Skelton P, Palmer T (eds) Carbonate platform systems: components and interactions. Special Publication 178. Geological Society London, London, pp 89–107

Nebelsick JH, Rasser M, Bassi D (2005) Facies dynamics in Eocene to Oligocene circumalpine carbonates. Facies 51:197–216

Nebelsick JH, Bassi D, Rasser MW (2011) Microtaphofacies: exploring the potential for taphonomic analysis in carbonates. In: Allison P, Bottjer DJ (eds) Taphonomy: process and bias through time, vol 32, 2nd edn, Topics in Geobiology. Springer Science+Business, Dordrecht, pp 337–377

Nelson WA, Sutherland JE, Farr TJ, Hart DR, Neill KF, Kim HJ, Yoon HS (2015) Multi-gene phylogenetic analyses of New Zealand coralline algae: *Corallinapetra novaezelandiae* gen. et sp. nov. and recognition of the Hapalidiales ord. nov. J Phycol 51:454–468

Nitsch F, Nebelsick JH, Bassi D (2015) Constructional and destructional patterns – void classification of rhodoliths from Giglio Island, Italy. Palaios 30:680–691

Papazzoni CA, Sirotti A (1995) *Nummulite* biostratigraphy at the Middle/Upper Eocene boundary in the northern Mediterranean area. Riv Ital Paleontol Strat 101:63–80

Payros A, Pujalte V (2008) Calciclastic submarine fans: an integrated overview. Earth-Sci Rev 86:203–246

Penrose D (1991) The genus *Spongites* (Corallinaceae, Rhodophyta): *S. fruticulosus* the type species, in southern Australia. Phycologia 30:438–448

Penrose D, Woelkerling WJ (1991) *Pneophyllum fragile* in southern Australia: implications for generic concepts in the Mastophoroideae (Corallinaceae, Rhodophyta). Phycologia 30:495–506

Perry CT (2005) Morphology and occurrence of rhodoliths in siliciclastic, intertidal environments from a high latitude reef setting, southern Mozambique. Coral Reefs 24:201–207

Rasser MW, Piller WE (2004) Crustose algal frameworks from the Eocene Alpine Foreland. Palaeogeogr Palaeoclimatol Palaeoecol 206:21–39

Reid RP, MacIntyre IG (1988) Foraminiferal–algal nodules from the Eastern Caribbean: growth history and implications on the value of nodules as paleoenvironmental indicators. Palaios 3:424–435

Reijmer JJG, Everaars JSL (1991) Carbonate platform facies reflected in carbonate basin facies (Triassic, Northern Calcareous Alps, Austria). Facies 25:253–278

Renema W, Beaman RJ, Webster JM (2013) Mixing of relict and modern tests of larger benthic foraminifera on the Great Barrier Reef shelf margin. Mar Micropaleontol 101:68–75

Rico-García A, Aguirre J, González-Delgado JA (2008) Taphonomy and taphofacies models of the Pliocene deposits of Vejer de la Frontera (Cádiz, SW Spain). Geobios 41:543–558

Rögl F (1998) Palaeogeographic considerations for Mediterranean and Paratethys seaways (Oligocene to Miocene). Ann Naturhist Mus Wien 99A:279–310

Schweighauser J (1953) Mikropaläontologische und stratigraphische Untersuchungen im Paleocaen und Eocaen des Vicentin (Norditalien). Schweiz Paläontol Abh 70:1–97

Sgrosso I (1998) Possibile evoluzione cinematica nell'orogene Centro-Sud-Appenninico. Boll Soc Geol It 117:679–724

Simone L, Bassi D, Carannante G, Cherchi A (2012) Rudist-bearing rhodalgal facies in the post-Turonian recovery of the periTethyan carbonate systems: the case history from the Nurra Region (northwestern Sardinia, Italy). Geodiversitas 34:167–187

Sneed ED, Folk RL (1958) Pebbles in the Colorado river, Texas: a study in particle morphogenesis. J Geol 66:114–150

Steller DL, Foster MS (1995) Environmental factors influencing distribution and morphology of rhodoliths in Bahía Concepción, B.C.S., México. J Exp Mar Biol Ecol 194:201–212

Ungaro S (1969) Étude micropaléontologique et stratigraphique de l'Éocène supérieur (Priabonien) de Mossano (Colli Berici). Mém Bull Rech Geol Min 69, Colloque sur l'Éocène 3:267–280

Vardabasso S (1963) Die ausseralpine Taphrogenese im Kaledonisch-variszisch konsolidierten Sardischen Vorlande. Sonderdruck Geol Rundsch 53:613–630

Vigorito M, Murru M, Simone L (2005) Anatomy of a channel system and related fan in a foramol/rhodalgal carbonate sedimentary setting: the case history from the Miocene syn-rift Sardinia Basin, Italy. Sediment Geol 174:1–30

Vigorito M, Murru M, Simone L (2006) Architectural patterns in a multistorey mixed carbonate–siliciclastic submarine channel, Porto Torres Basin, Miocene, Sardinia, Italy. Sediment Geol 186:213–236

Vigorito M, Murru M, Simone L (2010) Carbonate production in rift basins: models for platform inception, growth and dismanting, and for shelf to basin sediment transport, Miocene Sardinia Rift Basin, Italy. In Mutti M, Piller W, Betzler C (eds) Carbonate systems during the Oligocene-Miocene climatic transition. Special Publication 42. International Association of Sedimentologists, Wiley-Blackwell, pp 257–282

Woelkerling WJ, Harvey A (1992) *Mesophyllum incisum* (Corallinaceae, Rhodophyta) in southern Australia: implications for generic and specific delimitation in the Melobesioideae. Brit Phycol J 27:381–399

Woelkerling WJ, Irvine LM (1986) General characteristics of nongeniculate Corallinaceae. In: The Coralline Red algae: an analysis of the genera and subfamilies of nongeniculate corallinacee. British Museum (Natural History), London. xi + 268 pp

Woelkerling WJ, Irvine LM, Harvey AS (1993) Growth-forms in non-geniculate coralline red algae (Corallinales, Rhodophyta). Aust Syst Bot 6:277–293

Wonham JP, Jayr S, Mougamba R, Chuilon P (2000) 3D sedimentary evolution of a canyon fill (Lower Miocene-age) from the Mandorove Formation, offshore Gabon. Mar Petrol Geol 17:175–197

Yesares-García J, Aguirre J (2004) Quantitative taphonomic analysis and taphofacies in lower Pliocene temperate carbonate–siliciclastic mixed platform deposits (Almería-Níjar basin, SE Spain). Palaeogeogr Palaeoclimatol Palaeoecol 207:83–103

Yordanova EK, Hohenegger J (2002) Taphonomy of larger foraminifera: relationships between living individuals and empty tests on flat reef slopes (Sesoko Island Japan). Facies 46:169–204

Zuschin M, Piller WE (1997) Gastropod shells recycled – an example from rocky tidal flat in the northern Red Sea. Lethaia 30:127–134

Chapter 7
Neogene Rhodoliths in the Mediterranean Basins

Juan C. Braga

7.1 Introduction

7.1.1 Palaeogeographic and Climatic Context

Geodynamic evolution and emergence of extensive areas led to dramatic palaeoceanographic changes during the Neogene (from ~23 to ~2.6 million years ago). The long-lasting circum-tropical ocean was divided by landmasses in several steps that altered global geography and major current patterns to reach the configuration of modern oceans. The connections of the Mediterranean Basin to the Indian Ocean were definitively interrupted in the middle Miocene (Rögl 1998) when the suture of the Arabian margin cut off the deep-marine circulation (Robertson et al. 2012). Within the Mediterranean region, the Paratethys was individualized as a large inland sea during the middle Miocene as well (Rögl 1998). At the western end, uplift of the Betic-Rifean chain closed off seaways between the Atlantic Ocean and the Mediterranean Sea, causing a temporary desiccation of the latter during the Messinian Salinity Crisis in the late Miocene (Hsü et al. 1977). The opening of the Gibraltar Straits resumed the connection of the open ocean and the Mediterranean Sea, which remains a partially isolated, evaporitic body of water requiring additional inflow to maintain its level (Mariotti et al. 2002). As a result, Mediterranean marine ecosystems evolved independently from the rest of the world oceans and Neogene rocks in the Mediterranean region reflect a unique and complex history, part of which is recorded in rhodolith-rich deposits.

A brief glaciation interval during the earliest Miocene was followed by a relatively warm period that ended with the most recent warm phase of the Earth, the

J.C. Braga (✉)
Departamento de Estratigrafía y Paleontología, Universidad de Granada,
Campus Fuentenueva, 18002 Granada, Spain
e-mail: jbraga@ugr.es

R. Riosmena-Rodríguez et al. (eds.), *Rhodolith/Maërl Beds: A Global Perspective*, Coastal Research Library 15, DOI 10.1007/978-3-319-29315-8_7

Mid-Miocene Climatic Optimum, about 15 My ago. Since then, the global climate has been cooling with minor reversals towards the modern ‘cold house’ climate (Zachos et al. 2001). The large-scale glaciation on Antarctica, starting at the end of the Eocene, became permanent in the late Miocene, predating the beginning of Arctic ice-sheet formation in the latest Miocene.

During the early and middle Miocene, sea-level changes of 50–60 m on the million-year scale were presumably related to expansion and shrinkage of the ice sheets in Antarctica, probably triggered by the long planetary tilt cycle (Miller et al. 2011). These major changes were modulated by minor superimposed oscillations controlled mainly by orbital obliquity (41,000-year cycles, Westerhold et al. 2005). The tilt cycle dominated sea-level cycles of meters to several tens of meters of amplitude for the rest of the Neogene (Miller et al. 2011).

7.1.2 Coralline Algae

According to the literature, the Neogene saw a remarkable change in coralline diversity (Aguirre et al. 2000). The species richness of nongeniculate corallines peaked in the evolutionary history of the group during the Aquitanian. About 250 coralline species have been reported from this basal stage of the Miocene. The number of listed species decreased with small fluctuations to a minimum of 43 in the late Pliocene. Rarefaction analysis of the raw literature numbers indicates that the early Miocene diversity maximum is probably not just a simple artifact of the intense attention paid to coralline algae of that period, reflected in large number of publications dedicated to these algae. By contrast, the marked decrease in raw species richness throughout the Neogene may be the result of the scarcity of publications on Pliocene coralline algae (Aguirre et al. 2000). Although Aguirre et al. (2000) analyzed global data, most publications on fossil coralline algae refer to the Mediterranean basins, and therefore the suggested global patterns of coralline diversity should also be expected in Mediterranean assemblages. Many fossil-species descriptions, however, have traditionally been based on a few characters of doubtful diagnostic value (Bosence 1983a, 1991; Braga et al. 1993; Braga and Aguirre 1995; Aguirre and Braga 1998) and, consequently, the species lists compiled from the literature cannot generally be equated with species inventories based on identifications in fossil assemblages following a modern taxonomic approach. However, very few accounts of Neogene coralline species have been carried out in the last two decades and they are scattered in separate regions and geological intervals. The lack of updated inventories prevents accurate testing of whether the patterns observed in the literature reliably reflect the diversity patterns in fossil assemblages. Partial and indirect evidence nevertheless seems to confirm the diversity decline among coralline algae in the Mediterranean during the Neogene. In this sense, the number of species within a genus recorded in a basin by an author or a group of co-authors presumably reflects the local morphological diversity within the genus representatives, and can be a proxy of minimum species richness in the Mediterranean not affected by varying diagnostic criteria or taxonomic procedures. The intra-basin diversity of *Sporolithon* slightly increased from the early to the middle Miocene and then declined until the Pliocene (Braga and Bassi 2007), following the pattern of global diversity according to published species lists. In one of the few available modern species accounts, Checconi (2006) studied the corallines from lower and middle Miocene rocks of the Southern Apennines (Italy). The total number of Burdigalian coralline species identified by this author is significantly higher than the one in the upper Miocene reefs in Salento (S Italy) and upper Miocene and Pliocene deposits in SE Spain (Braga et al. 2009; Aguirre et al. 2012). The 35 coralline species recorded in the Burdigalian from the Apennines (Checconi 2006) nearly double the total of 21 species found in the Messinian reefs in Salento and Sorbas Basin together (Braga et al. 2009). The species richness in the Pliocene deposits in the Carboneras Basin (19–20 species) is similar to that in the upper Miocene reefs (Aguirre et al. 2012). The decrease from the early Miocene to the Pliocene is even more marked within some genera, such as *Spongites*, encompassing 13 species during the Burdigalian (Checconi 2006), 5 during the Messinian in southern Italy (Braga et al. 2009) and only 2 during the Pliocene in southern Spain (Aguirre et al. 2012).

This Neogene decrease in diversity in Mediterranean corallines took place within a global pattern of eastward displacement of maximum diversity centers from the Mediterranean region that was the diversity hotspot during the Eocene to the Indo-Australian Archipelago (Renema et al. 2008). The decline in the number of coralline species parallels the greater isolation of the Mediterranean from oceans and the falling global temperatures after the Middle Miocene Climatic Optimum. According to the available evidence, diversity diminished more markedly in groups of tropical affinity, such as *Sporolithon* (Braga and Bassi 2007) and mastophoroids (*sensu* Harvey et al. 2003), such as *Spongites*.

A biogeographical differentiation is recorded in the Miocene for the first time in the evolutionary history of coralline algae (Braga et al. 2010a). The components of shallow-water Neogene Mediterranean algal assemblages differ from their pantropical counterparts and are characterized by the genera *Spongites*, *Neogoniolithon*, and *Lithophyllum*. Although these genera also occur in Neogene tropical deposits, other genera such as *Porolithon* and *Aethesolithon* are distinctive of the coralline assemblages (Johnson 1964; Johnson and Kaska 1965; Rösler et al. 2015). The separation of a Mediterranean domain was probably due the isolation of the Mediterranean from the Indian Ocean, complete by the middle Miocene (Rögl 1998). Beyond the reduction in species richness, the scarcity of modern accounts of species prevents an accurate assessment of component variations in fossil assemblages in the Mediterranean domain during the Neogene. The few available data suggest a change through time, leading to a coralline flora similar to that of the present-day Mediterranean by the Pliocene. In this sense, the species lists of early and middle Miocene corallines published in the last two decades include few extant Mediterranean species (Piller 1993; Fravega et al. 1993; Bucur and Filipescu 1994; Vannucci et al. 1994a) whereas the inventories from latest Miocene fossil assemblages comprise many species living in the Mediterranean and Lusitanian provinces

(Braga et al. 2009). More than 90 % of the species identified in lower Pliocene deposits of southern Spain (Aguirre et al. 2012) and 88 % of the species reported in the Pliocene of Tuscany (Italy) (Checconi et al. 2007) are known in the modern algal communities of these biogeographical provinces.

7.2 Rhodolith Deposits

Rhodoliths are common components in Neogene marine platform Mediterranean carbonates and are also frequent in shallow-water siliciclastic deposits (Halfar and Mutti 2005; Braga et al. 2010a). Rhodoliths and fragments of coralline algae also occur as re-deposited particles in deeper water sediment gravity flow materials (Bassi et al., this volume). Rhodolith-dominated facies are widespread in the Neogene of the Mediterranean and Paratethys regions and can be volumetrically more important than carbonates produced by coral reefs (Esteban 1996; Halfar and Mutti 2005). In the circum-Mediterranean region, Neogene rhodolith-rich deposits continue the Oligocene record (Brandano, this volume) and have been reported from the Aquitanian (early Miocene) to the Piacenzian (late Pliocene) (Figs. 7.1 and 7.2). They seem to be most widespread in the Langhian and late Tortonian-early Messinian intervals but decline in the Pliocene (Fig. 7.1). No rhodolith-rich late Messinian deposits appear because of the Mediterranean-wide alterations of marine conditions related to the Messinian Salinity Crisis. The global abundance of rhodalgal carbonate occurrences from the late Burdigalian to the early Serravallian coincides with the so-called Monterey event, a prominent and long-lasting (17.5–13.5 Ma) middle Miocene carbon-isotope excursion to higher $\delta^{13}C$ values, and with the Middle-Miocene Climatic Optimum (MMCO) (Halfar and Mutti 2005). These authors suggest that the carbon-isotope excursion reflects increased productivity in the oceans that would have triggered the dominance of rhodalgal facies over coral-reef carbonates due to a lower tolerance of zooxanthellate corals to high nutrient levels. According to the same authors, the high productivity would later have been increased by greater continental weathering and subsequent nutrient input to shallow waters during the MMCO, while global cooling after the climatic optimum would have sustained the dominance of red algae, preventing the recovery of coral reefs. More recent research, however, suggests that the middle Miocene carbon-isotope excursion is not related to greater export productivity (Diester-Haass et al. 2009), except in high latitudes (Diester-Haass et al. 2013). On the other hand, the Miocene expansion of rhodalgal facies does not coincide with a global coral-reef decline. On the contrary, the number of coral-reef sites and the volume of reef carbonates peak in the middle Miocene and maintain high values in the late Miocene (Perrin and Kiessling 2010), which is the other interval (late Tortonian-early Messinian) of maximum expansion of rhodalgal facies. The number of coral-reef sites in the Mediterranean is higher from the Langhian to the early Messinian than in older and younger Neogene intervals (Perrin and Bosellini 2012) and rhodalgal facies coexist in some platforms with coral reefs (Pomar et al. 2012). In the

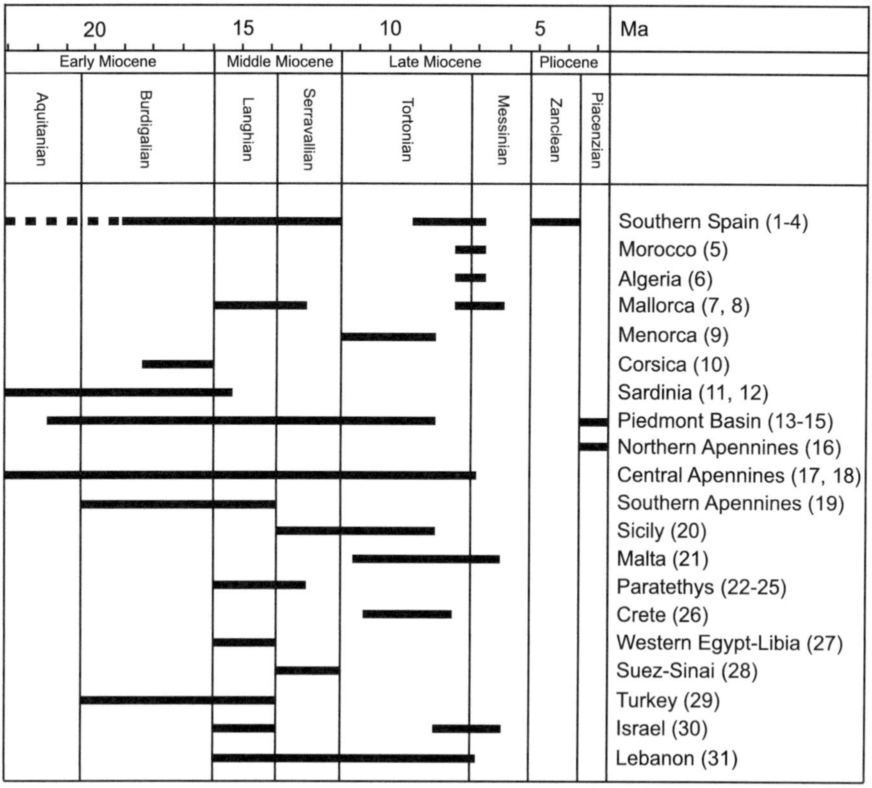

Fig. 7.1 Stratigraphic distribution of Neogene rhodolith concentrations in Mediterranean basins (Modified after Halfar and Mutti (2005). Data from *1* Geel et al. (1992), *2* Geel (2000), *3* Braga and Aguirre (2001), *4* Braga et al. (2010b), *5* Saint Martin and Corneé (1996), *6* Saint Martin (1996), *7* Pomar and Rodríguez-Perea (1983), *8* Pomar et al. (1996), *9* Brandano et al. (2005), *10* Tomassetti and Brandano (2013), *11* Benisek et al. (2010), *12* Vigorito et al. (2010), *13* Fravega et al. (1993), *14* Vannucci et al. (1994b), *15* Vannucci et al. (1996), *16* Checconi et al. (2007), *17* Brandano et al. (2010a, b), *18* Reuter et al. (2013), *19* Bassi et al. (2010), *20* Ruchonnet and Kindler (2010), *21* Pedley (1996), *22* Dullo (1983), *23* Studencki (1988), *24* Randazzo et al. (1999), *25* Bucur and Filipescu (1994), *26* Kroeger et al. (2006), *27* El-Shahat (1993), *28* Scott and Govean (1985), *29* Bassant et al. (2005), *30* Buchbinder et al. (1993), *31* BouDagher-Fadel and Clark (2006)

Mediterranean, the prevalence of extensive rhodolith-rich deposits does not seem triggered by the Monterey event but appears to be related to a high number and extension of carbonate platforms, as coralline algae are common to abundant components in one to several facies belts in most Miocene carbonate deposits. The growth of carbonate platforms is controlled by several interacting factors, including the geotectonic context, levels of siliciclastic supply, substrate topography, relative sea-level change, climate, and oceanographic conditions (Bosence 2005). The tectonic evolution of the Mediterranean region from the Burdigalian to the early Messinian gave rise to numerous shallow-water platforms around relatively small

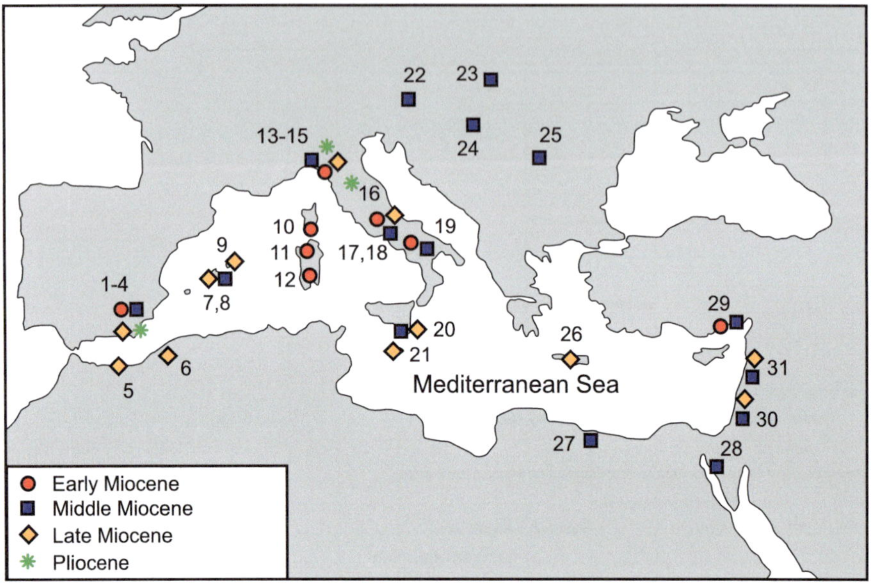

Fig. 7.2 Location of rhodolith concentrations in Mediterranean basins (Data from sources in Fig. 7.1)

islands (Popov et al. 2004; Perrin and Bosellini 2012), some made up mainly of Mesozoic to Paleogene carbonates. In the climatic context of the Mediterranean during the Miocene, these platforms were affected only by reduced amounts of siliciclastic input and, consequently, were suitable places for deposition of carbonates, including lithofacies dominated by coralline algae. The dramatic reduction of emerged examples of Pliocene rhodolith-rich deposits (Figs. 7.1 and 7.2) might be related to the limited number of onshore outcropping carbonates from this epoch in the Mediterranean. Most outcropping marine Pliocene deposits in the region comprise siliciclastics from the large reliefs that surround the Mediterranean Sea.

7.3 Palaeoenvironments of Rhodolith Formation

In the Neogene Mediterranean shallow-water palaeoenvironments, rhodolith-rich deposits formed primarily in inner to outer zones of carbonate ramps (Pedley and Carannante 2006; Pomar et al. 2012). In these ramps, they also accumulated in rocky-shores and at the toe of submarine cliffs, and in areas sheltered by sandspits or other features (Braga et al. 2006) (Fig. 7.3). They also thrived on fore-reef slopes, and locally in lagoons and on reef tops, of rimmed platforms (Pomar et al. 1996; Sola et al. 2013). Rhodoliths can also be locally abundant in siliciclastic systems, usually associated with interrupted or reduced sediment influx (Braga and Martín 1988).

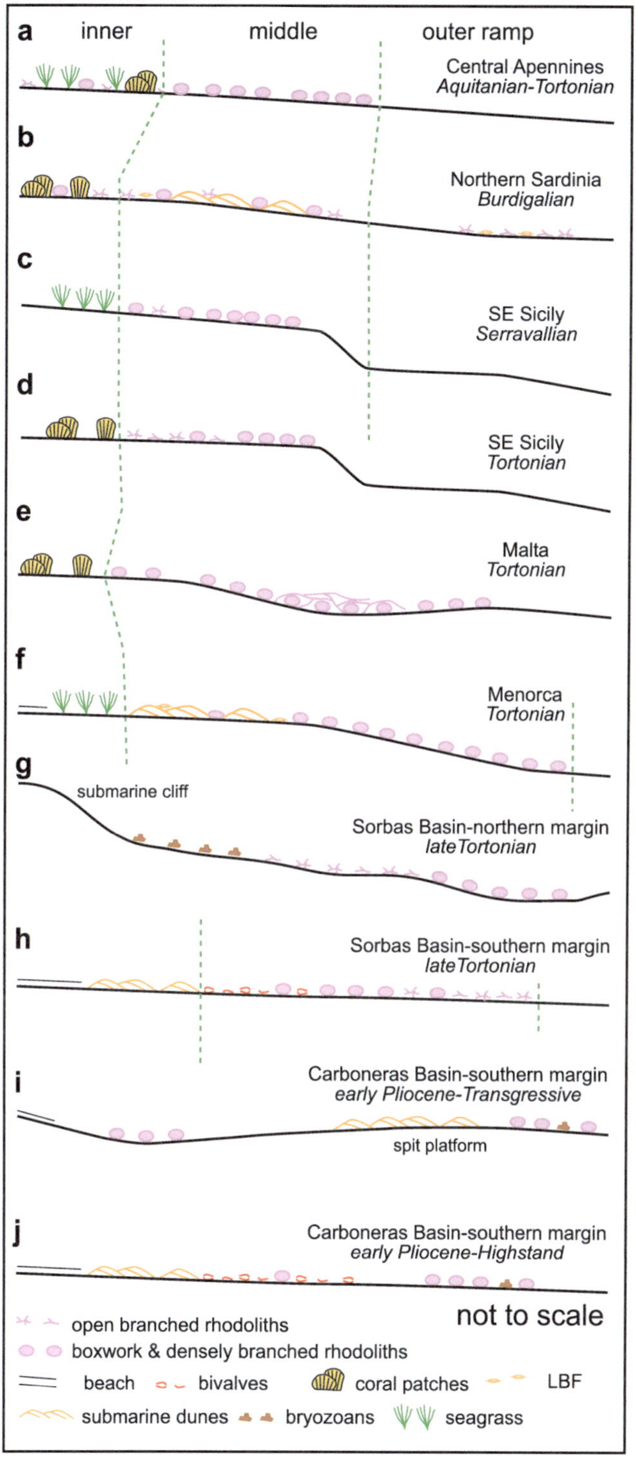

Fig. 7.3 Sketch of palaeoenvironmental distribution of rhodolith concentrations and predominant rhodolith morphologies in examples of ramp deposits from Mediterranean localities (Data from (**a**) Brandano et al. (2010a, b), (**b**) Benisek et al. (2010), (**c, d**) Ruchonnet and Kindler (2010), (**e**) Pedley (1996), (**f**) Brandano et al. (2005), (**g, h**) Puga-Bernabéu et al. (2007a, b), and (**i, j**) Aguirre et al. (2012))

7.3.1 Ramps

Rhodoliths in the Aquitanian-Langhian carbonates of the Central Apennines occur in the inner ramp deposits but were important carbonate producers mainly in the middle ramp, seawards of seagrass meadows and growing among *Porites* patches (Brandano et al. 2010a, b). They are the dominant components in the middle-ramp sediment and continue as minor scattered components in the bryozoan-bivalve floatstones in the proximal outer ramp.

In the Burdigalian platform of northern Sardinia, rhodoliths and branching corallines have been documented in inner-ramp carbonates associated with corals (Benisek et al. 2010). Rhodoliths and branching corallines also occur in cross-bedded rudstones and floatstones in the middle ramp while branching corallines are dominant in the outer ramp deposits.

In the Serravallian carbonates of the Ragusa Platform in SE Sicily, rhodoliths are the main components in wackestones to rudstones formed in the middle ramp basinwards of wackestones to packstones with foraminifers and red algal fragments interpreted as seagrass-related sediments (Ruchonnet and Kindler 2010). No significant coralline algal remains occur in the outer-ramp marly limestones and marls (Ruchonnet and Kindler 2010).

The inner ramp in the lower Tortonian packages of the Ragusa system is characterized by dispersed coral (*Porites*) colonies and coral patches in mudstone/wackestone, also including *Halimeda* and coralline fragments. Branching coralline algae and *Halimeda* are the main components in packstones/floatstones formed in the proximal middle ramp, which gradually pass basinwards into rhodolith floatstones (Ruchonnet and Kindler 2010).

In the Maltese Tortonian ramp, rhodoliths are dispersed in bioclastic wackestones locally including patch reefs. Basinwards of this deposits, a coralline algal biostrome, several kilometers across and up to 20 m in thickness, developed on a subsiding narrow corridor. The biostrome is made up of coralline crusts and rhodoliths (Bosence and Pedley 1982; Pedley 1996).

Rhodoliths in the lower Tortonian carbonates in Menorca occur mainly in rudstones and floatstones formed on the ramp slope (Brandano et al. 2005; Pomar et al. 2012), seawards of subaqueous dunes developed on the middle ramp. The inner-ramp deposits comprise poorly sorted, structureless packstones with molluscs and foraminifers, interpreted as seagrass-related sediments (Pomar et al. 2002) while wackestones and packstones with foraminifers were the main lithofacies in the outer ramp limestones (Pomar et al. 2012).

In the upper Tortonian-lower Messinian Azagador limestone in the Sorbas Basin in SE Spain, corallines are the main components in middle-ramp rudstones-floatstones, formed seawards of inner-ramp shoals and proximal middle-ramp rudstones-floatstones dominated by bivalves (oysters and pectinids). The relative abundance of branching corallines in relation to rhodoliths increased to the outer ramp (Puga-Bernabéu et al. 2007a, b).

In depositional settings with a ramp profile in the Pliocene carbonates in the Carboneras Basin in SE Spain, rhodoliths are dominant components in mid-to-outer ramp deposits that formed seawards of bivalve-dominated rudstones to floatstones, and cross-bedded, inner-ramp packstones to rudstones (Martín et al. 2004; Aguirre et al. 2012). Coralline algal/invertebrate (bryozoan/bivalve or serpulid) buildups are dispersed in finer-grained sediments accumulated in the distal outer-ramp environments.

In addition to these ramp settings, coralline algal concentrations formed at the toe of submarine cliffs at the northern margin of the Sorbas Basin during the late Tortonian-early Messinian. Branching corallines and minor rhodoliths were dominant bioclasts in rudstones to floatstones deposited basinwards of robust-bryozoan rudstones originated by the fall of bryozoans colonizing the cliff walls (Puga-Bernabéu et al. 2007a). Floatstones to rudstones with rhodoliths and bryozoans also formed close to the southern shore of the Carboneras Basin during the Pliocene. In this case, rhodoliths grew on the seafloor sheltered by a spit platform built by longshore currents (Braga et al. 2003; Aguirre et al. 2012).

7.3.2 *Reef-Rimmed Shelves*

Rhodoliths are generally less abundant and less significant components in Miocene reef-rimmed shelves than in carbonate ramps. They have been reported chiefly in fore-reef talus and proximal slope rudstones to packstones in Messinian reefs in Salento (Bosellini et al. 2001, 2002; Braga et al. 2009), Mallorca (Pomar 1991; Pomar et al. 1996), and SE Spain (Dabrio et al. 1981; Riding et al. 1991; Mankiewicz 1996; Franseen and Goldstein 1996; Braga and Aguirre 2001; Braga et al. 2009; Sola et al. 2013). Rhodoliths are abundant in distal-slope silty floatstones in the lower Messinian reefs in the Poniente Basin in SE Spain, together with coralline algae indicative of seagrass beds (Sola et al. 2013). They also occur among corals in reef cores (Braga et al. 2009), in reef-flat/reef-crest deposits (Mallorca, Pomar et al. 1996) and in lagoon sediments (Braga et al. 2009).

7.3.3 *Siliciclastic Systems*

Rhodolith beds in Neogene siliciclastic systems have been documented in several localities. In the lower Tortonian deposits in the Almanzora Corridor (SE Spain), they occur associated with delta sandstones and conglomerates, although they probably accumulated in subenvironments temporarily abandoned by the delta (Braga and Martín 1988). Rhodoliths with siliciclastic nuclei have been reported in shoreface sandstones and conglomerates laterally equivalent to the deltaic sediments (Braga and Martín 1988). In Pliocene siliciclastic shelf deposits, rhodolith concentrations occur in relatively thin bioclastic beds that probably reflect reduced terrigenous influx. Piacenzian rhodolith beds within sand and silt have been described from NW Italy (Vannucci et al. 1994b) and Zanclean examples have been reported in SE Spain (Aguirre 1998).

7.3.4 Summary of Context of Rhodolith-Bed Formation

Although the concepts underlying the internal division of ramps in the listed cases differ from each other, these examples show that rhodoliths can be abundant and main components of carbonate factories on ramps from inner to outer zones, with maximum concentrations, forming the so-called rhodolith beds, in middle to proximal outer-ramp settings (Fig. 7.3).

In some cases, rhodoliths thrived among coral patches (mainly *Porites*) on the inner ramp (Pedley 1996; Benisek et al. 2010; Brandano et al. 2010a, b) but rhodolith concentrations commonly formed seawards of coral colonies and seagrass beds on the middle and proximal outer ramp (Pedley 1996; Ruchonnet and Kindler 2010; Brandano et al. 2005, 2010a, b). In other cases, rhodolith factories developed on the middle to proximal outer ramp seawards of submarine dune fields (shoals). A facies belt rich in bivalves can be found between shoals and rhodolith beds in western Mediterranean examples (Puga-Bernabéu et al. 2007a; Martín et al. 2004; Braga et al. 2006). Basinwards of rhodolith concentrations, outer ramp sediments are fine-grained packstones to wackestones, locally rich in bryozoans (Brandano et al. 2010a), which can intercalate sediment gravity flow deposits sourced from shallower areas on the ramp (Martín et al. 2004; Pomar et al. 2012; Puga-Bernabéu et al. 2014).

In the non-deformed Pliocene ramps in the Carboneras Basin (SE Spain) in which palaeodepth of rhodolith-bed formation can be directly estimated measuring the elevation difference between rhodolith-rich facies and coeval beach/shore-line deposits, the estimated palaeobathimetry ranges from 10 to several tens of metres (less than 50 m) (Martín et al. 2004; Aguirre et al. 2012). In reef settings, rhodoliths can be abundant in very shallow subenvironments (reef crest/reef flat in the Tortonian-Messinian reef in Mallorca, Pomar et al. 1996) and in deeper contexts, below the reef core and talus to proximal slope facies dominated by coral growth and debris. In the Messinian Cariatiz reef in the Sorbas Basin in SE Spain, in which the present-day vertical distribution of facies throughout the outcrop reflects their depositional depth, rhodoliths are most common in slope deposits formed at palaeodepths of 30–40 m (Braga et al. 2009). In a similar situation, in the Messinian reefs of the Poniente Basin (SE Spain) rhodoliths are common in bivalve/coralline algal packstone to rudstone and silty coralline algal floatstone formed on slope areas about 30–40 m deeper than *Porites* frameworks (Sola et al. 2013).

Estimates based on actualistic comparisons with the depth distribution of living rhodoliths comprising similar coralline algal assemblages suggest palaeodepths of rhodolith growth ranging from 10–20 to 70–100 m in the Tortonian ramp in Menorca (Balearic Islands) (Brandano et al. 2005). Applying a similar approach,

Puga-Bernabéu et al. (2007a) inferred palaeodepths of 15–40 m for the branching coralline facies and greater depths for the rhodolith facies in the upper Tortonian-early Messinian carbonates at the northern margin of the Sorbas Basin.

These wide depth ranges of profuse rhodolith growth in Miocene platforms agree with the modern distribution of rhodolith beds. In the Gulf of California, rhodolith beds occur at depths of 2–40 m (Marrack 1999; Steller et al. 2009); on the Brazilian shelf they extend from the intertidal zone to 120 m (Amado-Filho et al. 2007); on the eastern Australian shelf, off southern Queensland, rhodoliths occur from about 30 to 120 m but are the most common components in bioclastic gravels from about 50 to 110 m (Marshall et al. 1998; Lund et al. 2000); in Sesoko-jima in the Ryukyu Islands (Japan), branching growths occur down to only 20 m; rhodoliths are scarce from 20 to 50 m depth, concentrate in a rhodolith bed from 50 to 70 m, and diminish in number down to 100 m (Bassi et al. 2009). Rhodolith concentrations in the Mediterranean Sea can be found from 40 to 90 m on the Mallorca shelf (Fornos and Ahr 1997, 2006), from 35 to about 60 m in the northern Tyrrhenian Sea (Toscano et al. 2006), and from about 30 to 100 m in the Pontian Islands at the western Italian shelf (Basso 1998). Rhodoliths are the main components of surface sediment from 30 to 70 m of water depth on the Alboran Ridge, whereas in the Oran Bay they prevail from 35 to 65 m (Betzler et al. 2011). A rhodolith bed in a small bay in Almuñecar (southern Spain) extends from 10 to 20 m, while rhodoliths concentrate between 20 and 50 m offshore Cabo de Gata in areas sheltered from terrigenous influx.

The distribution of rhodolith beds both in the ancient and present-day examples suggests that the occurrence of rhodolith concentrations is not indicative of any particular palaeodepth range within shallow-water platform settings (0 to about 120 m) although they are more common at depths of several tens of meters.

The attempts to characterize the depositional setting of rhodolith beds following the bathymetric/light zones of Pomar (2001) amended by Pomar et al. (2012) are questionable. These zones, the names of which do not correspond to the common use in oceanography, are based on the deepest records of green algae (chlorocline) and red algae (rhodocline) that separate the 'euphotic' from the 'oligophotic' zones and the latter from the 'aphotic' zone, respectively. Since red algae are most common in the shallowest waters, where their depth range coincides with that of green algae (Pomar et al. 2012, fig. 19), the separation of the 'euphotic' from the 'oligophotic' zone relies on the deepest occurrence of green algae. In Neogene deposits, this means that the separation of these two zones depends on the in situ presence/absence of dasycladaleans and *Halimeda*. In modern oceans, however, living *Halimeda* has a wide depth range and is common from the intertidal zone to depths of 120 m. This alga has a depth range to 150 m on the slope of the Eniwetok Atoll (Hillis-Colinvaux 1986); it can be found to 130 m on an unnamed seamount (Littler et al. 1986) and down to 150 m at San Salvador in the Bahamas (Blair and Norris 1988). In the Hawaiian Islands, *Halimeda* forms extensive meadows between 65 and 120 m deep (Webster et al. 2006). On the Brazilian continental margin where the shelf is extensively covered by rhodolith beds (Amado-Filho et al. 2012), *Halimeda* is found from the intertidal to 166 m depth (Bandeira-Pedrosa et al.

2004). In Miocene examples from SE Spain, *Halimeda* occurs from reef crests (Dabrio et al. 1981) to relatively deep ramp bioherms, at an estimated palaeodepth of 20–65 m (Martín et al. 1997). Consequently, the mere presence of *Halimeda* is not indicative of a narrow depth range and in situ *Halimeda* can be more abundant in relatively deep water than in shallow settings within a platform. Thus, the recognition of an 'euphotic' zone sensu Pomar (2001) has hardly any palaeoenvironmental/palaeodepth meaning. On the other hand, the modern geographic distribution of calcifying *Halimeda* is delimited by the 25 °C sea-surface isothere (same mean summer temperature) although weakly to not calcifying representatives can be found in colder seas, such as the Mediterranean (Hillis 1991). Most living dasycladalean species inhabit waters with minimum annual temperatures higher than 20 °C (Berger and Kaever 1992) and the few species occupying habitats with lower temperatures show minimal stem calcification (for example, *Dasycladus vermicularis* (Scopoli) Krasser in the Mediterranean) or their calcification is restricted to reproductive discs (*Acetabularia* species) (Elliott 1977). As the environmental requirements of post-Palaeozoic green algae are thought to have been similar to those of extant representatives (Elliott 1984), green algae did not grow on Neogene platforms with waters cooler than the above-mentioned limiting temperatures or, if they did, their preservation potential was very low due to weak calcification. In these cases, 'euphotic'-zone deposits cannot be directly identified and any kind of carbonate containing red algae has to be assigned to the 'oligophotic' zone. With the application of this scheme, all carbonates formed on the late Tortonian-early Messinian (Azagador limestone) and Pliocene ramps in SE Spain, which lack any remains of green algae, would have accumulated in the 'oligophotic' zone or in the deeper 'aphotic' zone, from the beach facies to the basin.

7.4 Rhodolith Morphology and Composition

In rhodolith beds of the Serravallian Ragusa ramp in SE Sicily, Ruchonnet and Kindler (2010) have observed within the middle ramp a basinwards replacement of spherical rhodoliths by loose branching coralline growths. The abundance of *Spongites* decreases basinwards and only *Lithothamnion* and *Mesophyllum* have been identified in the deepest coralline assemblages (Ruchonnet and Kindler 2010). By contrast, in the overlying Tortonian ramp, the most proximal coralline algal facies are dominated by loose branching thalli that are replaced by small spherical rhodoliths in the deeper algal bed within the middle ramp. *Spongites*, *Lithothamnion*, and *Sporolithon* occur in algal assemblages from all palaeodepths (Ruchonnet and Kindler 2010).

In the basal sequence of the Burdigalian ramp of northern Sardinia, Benisek et al. (2010) have described rhodoliths and branching corallines in the inner and middle ramp changing basinwards to dominant coralline branches in the outer ramp. In the overlying sequence, Benisek et al. (2010) recognise a decrease in size (from

10 to 5 cm) and sphericity (the proportion of ellipsoidal rhodoliths increases) downslope of decameter-scale clinoforms developed on the ramp slope.

Studencki (1988) suggested a pattern in the palaeobathymetrical distribution of coralline algal growth forms in the Badenian limestones from the Holy Cross Mountains in Poland. The shallowest bed comprises mostly spheroidal rhodoliths with an internal succession of branching to encrusting to columnar thalli. Rhodoliths and branching corallines, together with bryozoans, were the main components at intermediate palaeodepths, while unattached branching thalli predominate in the deepest coralline facies. *Lithothamnion*, *Mesophyllum*, *Spongites* (as *Lithophyllum*), and *Sporolithon* were the major components in the two shallowest palaeoenvironments. In contrast, the deepest branching-coralline assemblages were dominated by *Lithothamnion* and *Sporolithon* (Studencki 1988).

Rhodoliths from the Tortonian distally steepened ramp in Menorca, studied in detail by Brandano et al. (2005), show no significant changes in nodule shape (dominantly ellipsoidal) and size (from 2 to 8 cm) from the middle ramp to the toe of the slope. Minor discoidal nodules occur in rhodolith pockets between cross-bedded grainstones in the middle-ramp deposits. In these pockets, most rhodoliths have a skeletal nucleus and comprise an inner part of encrusting thalli overgrown by lumpy growth forms. Up to 18 % of nodules are made up of fruticose and lumpy thalli. An outward succession from encrusting to lumpy thalli is the most common in rhodoliths from the slope and toe of the slope. In the latter setting, however, up to 45 % of nodules are exclusively fruticose and lumpy thalli (Brandano et al. 2005). In terms of taxonomic composition, the relative proportion of *Spongites* generally decreases downslope from more than 40 % on the shallowest middle ramp to about 1 % at the toe of the slope. Melobesiods (*Mesophyllum* and *Lithothamnion*) concurrently increase, whereas *Lithophyllum* and *Sporolithon* are minor components in all palaeoenvironments (Brandano et al. 2005).

In the upper Tortonian mixed carbonate-siliciclastic ramp in the Almanzora Corridor in SE Spain, rhodoliths in the shallowest coastal palaeoenvironments are small and made up of thick encrusting to warty plants belonging mainly to *Lithophyllum* (Fig. 7.4) and minor *Spongites* (Fig. 7.5). Coralline thalli grew around a relatively large nucleus that determined the rhodolith shape. By contrast, in deeper settings, rhodoliths are generally larger, ellipsoidal to spheroidal in shape, with thick algal covers in relation to nuclei. Rhodoliths comprise thin encrusting, warty, fruticose and lumpy thalli belonging to several species, intergrown with epizoans such as foraminifers and bryozoans. *Mesophyllum* (Fig. 7.4) and *Lithothamnion* are the main components with minor *Spongites* (as *Lithophyllum*), and *Lithophyllum*, the latter genus usually occurring in the first stages of rhodolith growth (Braga and Martín 1988).

In the same region, in an upper Tortonian carbonate unit in the Sorbas Basin, rhodoliths and unattached branching coralline algae are the main components of sediments accumulated at intermediate palaeodepths on the middle ramp at the southern margin of the basin. Basinwards the branching growths (Fig. 7.6) become predominant together with small concentrations of fragments of foliose thalli (Puga-Bernabéu et al. 2007a, b). These changes in the predominant growth form are

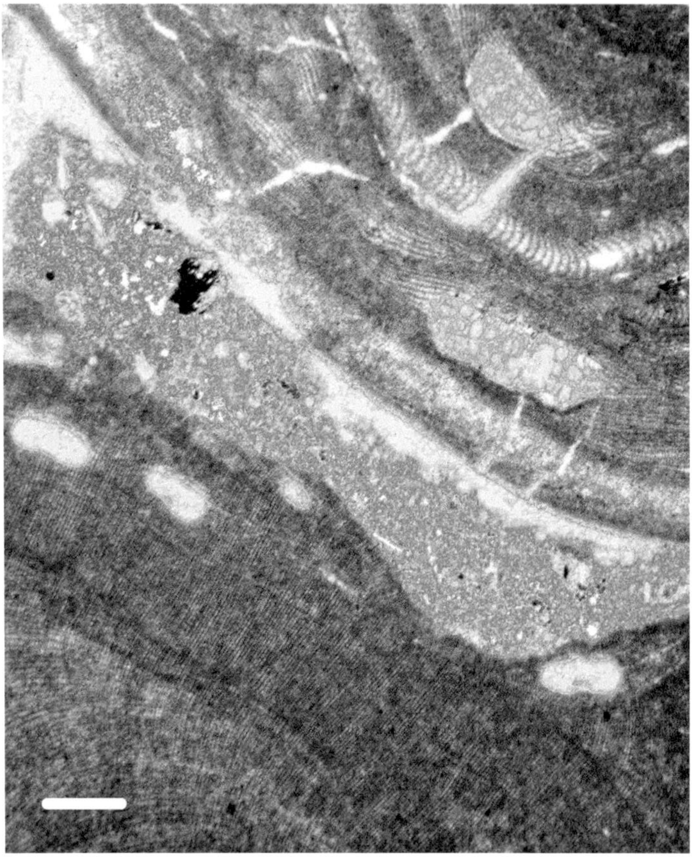

Fig. 7.4 Microphotograph of *Lithophyllum* (*bottom left*) and *Mesophyllum* (*upper right*) thalli in a rhodolith from Tortonian deposits in the Almanzora Corridor, SE Spain. Scale bar = 250 μm

accompanied by a decline in *Lithophyllum* (*L. incrustans* Philippi and *L. dentatum* (Kützing) Foslie) in the algal assemblages. Contrarily, at the northern margin of the basin unattached branching corallines dominate the most proximal algal facies at the foot of submarine palaeocliffs while rhodoliths formed in deeper settings (Puga-Bernabéu et al. 2007a). Here, the compositional change is more marked as *Lithophyllum* (*L. incrustans* and *L. dentatum*) is recorded only in the branching-algal facies while rhodoliths are made up only of melobesiods.

Rhodolith morphology and coralline assemblages do not significantly change within the narrow lateral extension of individual rhodolith beds in the Pliocene carbonates in the Carboneras Basin. However, in the shallowest context, the sheltered leeside of a spit platform at the southern margin, rhodoliths comprise a proportion of *Lithophyllum* and mastophoroids (*sensu* Harvey et al. 2003) (about 40 % of the total coralline algae) higher than the one in nodules from deeper environments (22–25 %).

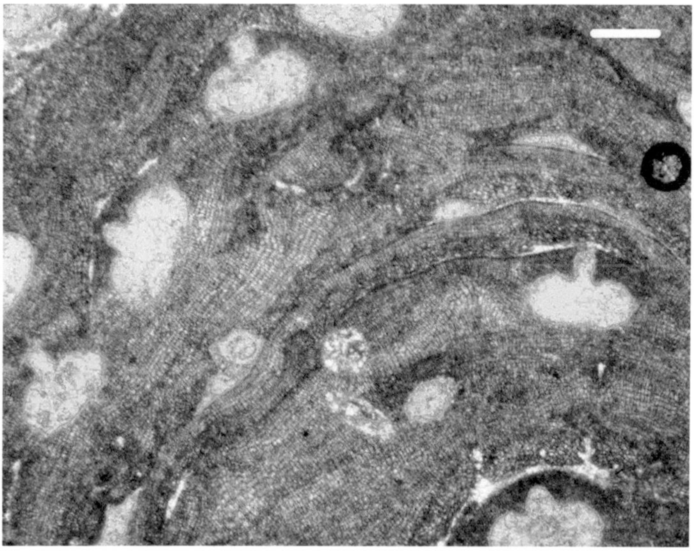

Fig. 7.5 Microphotograph of a *Spongites* thallus in a rhodolith from Tortonian deposits in the Almanzora Corridor, SE Spain. Scale bar = 250 μm

Fig. 7.6 Open branched rhodoliths. Distal middle ramp deposits. Tortonian (late Miocene), southern margin of the Sorbas Basin, SE Spain. Coin diameter 23 mm

Fig. 7.7 Spheroidal to ellipsoidal rhodoliths. Middle ramp deposits. Early Pliocene, Carboneras Basin, SE Spain. Coin diameter 23 mm

Spheroidal rhodoliths (Fig. 7.7) are also found more in the shallowest bed than in the deeper ones in which nodules are ellipsoidal to spheroidal (Aguirre et al. 2012).

In summary, in Mediterranean Miocene deposits no clear-cut patterns are discerned in rhodolith morphology or internal growth forms along depth gradients. Open-branching rhodoliths (unattached branching corallines) can be the dominant morphologies in settings both shallower (Tortonian, Ragusa ramp, Ruchonnet and Kindler 2010, Tortonian, northern Sorbas margin, Puga-Bernabéu et al. 2007a) and deeper (Burdigalian, Sardinia, Benisek et al. 2010, Serravallian, Ragusa ramp, Ruchonnet and Kindler 2010, Tortonian, southern Sorbas margin, Puga-Bernabéu et al. 2007a, b) than the ones with spheroidal to ellipsoidal boxwork/densely branched rhodoliths (Fig. 7.3). However, the latter pattern, i.e. loose branching growths characterizing the deeper rhodoliths, is the most common (Fig. 7.3). Rhodolith shape does not significantly change along proximal-distal (depth) gradients in examples of ramps studied in detail (Tortonian, Menorca, Brandano et al. 2005). Shapes between the ellipsoidal and spheroidal end morphologies are the most widespread and only in a few cases has a certain dominance of spheroidal rhodoliths been recorded in the shallower settings (Badenian, Holy Mountains, Studencki 1988, Burdigalian, Sardinia, Benisek et al. 2010, Pliocene, Carboneras Basin, Aguirre et al. 2012). In contrast, discoidal shapes dominated in the shallowest and most turbulent environments of rhodolith formation (pockets between cross-bedded grainstones) on the Tortonian ramp in Menorca.

The shape of the nucleus determines the general morphology of nodules with relatively thin algal cover. This kind of rhodoliths can be characteristic of turbulent shallow-water environments, such as beach-shore faces and rocky shores, where coralline algae settle on pebble- to cobble-sized clasts (Braga and Martín 1988; Aguirre et al. 1993). Growth of the algal cover can be limited by abrasion, and nodules can be buried after a short period of development before coralline thalli determine their shape. Large, densely branched rhodoliths, however, can dominate very shallow-water rhodolith concentrations, such as the one in the reef-flat/reef-crest of the upper Miocene reef of Mallorca (Pomar et al. 1996).

The growth forms of the corallines making up rhodoliths show no clear pattern along depth gradients, either. Branching lumpy to fruticose corallines can be the most common both in the shallowest settings, such as in the Mallorca example mentioned above, and in the deepest ones, as is the case of rhodoliths in the Tortonian Menorca ramp studied by Brandano et al. (2005). Rhodoliths in intermediate and sometimes in deeper palaeoenvironments tend to exhibit more complex combinations of coralline growth forms than the shallower nodules. The cases studied in detail show that nodules that grew at several tens of meters usually have a higher diversity of coralline taxa with different types of growth forms and, consequently, the rhodoliths are made up of varying combinations of encrusting, warty, foliose, lumpy, and fruticose corallines (Braga and Martín 1988; Brandano et al. 2005; Checconi et al. 2010).

In contrast to the absence of clear-cut patterns of rhodolith morphology and algal growth forms along depth gradients, the taxonomic composition of algal assemblages involved in rhodolith formation usually shows changes with palaeobathimetry in the Mediterranean Neogene cases studied. *Spongites*, together with *Neogoniolithon* in some cases, or *Lithophyllum* dominated or were abundant in the shallower rhodoliths on ramps or in reef frameworks and proximal slopes in reef-rimmed shelfs, whereas they were less common or absent in deeper water assemblages, which included mainly species of *Mesophyllum*, *Lithothamnion*, and *Sporolithon* (Braga and Martín 1988; Studencki 1988; Braga and Aguirre 2001; Brandano et al. 2005; Puga-Bernabéu et al. 2007a, b; Ruchonnet and Kindler 2010; Aguirre et al. 2012). This pattern of gradual replacement of mastophoroids and lithophylloids (*sensu* Harvey et al. 2003) by melobesioids and sporolithales along depth gradients has been pointed out in a number of studies on modern and Neogene-Quaternary rhodoliths (Adey and Macintyre 1973; Bosence 1983b; Bosence 1991; Braga and Martín 1988; Lund et al. 2000; Webster et al. 2009; Braga et al. 2010a, b).

As shown by the examples outlined above, rhodolith beds and local rhodolith concentrations can be lateral equivalents to coral reefs or develop in temperate platform settings in which no reef corals or calcareous green algae thrived at any palaeodepth. In the early and middle Miocene ramps of the central and western Mediterranean coral patch reefs associated to rhodolith beds are made up primarily of *Porites* (Pedley 1996; Ruchonnet and Kindler 2010; Benisek et al. 2010; Brandano et al. 2010a, b) and coral diversity in the region is generally low (Bosellini and Perrin 2008; Perrin and Bosellini 2012). This suggests that these ramps developed at the transition from the warm-temperate to the subtropical realm (Benisek et al. 2010). According to coral generic richness, minimum sea-surface palaeotemperatures remained below 20 °C during the early and middle Miocene in the Mediterranean basins and no temperature increase is recorded by coral diversity during the Mid-Miocene Climatic Optimum (Bosellini and Perrin 2008). Estimated palaeotemperatures in the Mediterranean decreased during the late Miocene. The low numbers of coral genera suggest sea-surface temperatures of 16–17 °C (Bosellini and Perrin 2008, 2010). In the western Mediterranean, reefs with very low coral diversity (and *Halimeda*) alternate in time with carbonates formed only by heterozoan associations rich in rhodalgal deposits (Martín and Braga 1994; Braga and Aguirre 2001; Brachert et al. 1996; Martín et al. 2010). This alternation has been interpreted as the result of climatic oscillations involving sea-level changes. Oxygen isotope values and planktonic foraminiferal assemblages in laterally equivalent basinal marls indicate sea-surface temperatures higher during the intervals of reef development than during deposition of carbonates with only heterozoan associations (Sánchez-Almazo et al. 2001, 2007; Martín et al. 2010). Oxygen isotope values of the planktonic foraminifera suggest a further temperature decrease by 4.9 Ma in the early Pliocene (Zanclean) (Martín et al. 2010) and Mediterranean Pliocene carbonates only comprise heterozoan lithofacies. These palaeotemperature changes are reflected in the coralline algal assemblages forming rhodoliths. *Spongites* is abundant in shallow-water rhodoliths coeval with coral reefs (Studencki 1988; Ruchonnet and Kindler 2010; Braga and Aguirre 2001) whereas *Lithophyllum* is the main component in shallow-water rhodoliths in temperate ramps (Braga and Aguirre 2001; Puga-Bernabéu et al. 2007a, b; Aguirre et al. 2012). Melobesiods dominate the deeper water assemblages in all palaeoclimatic contexts (Braga and Aguirre 2001).

7.5 Concluding Remarks

Rhodoliths are common in Neogene shallow-water marine deposits of the Mediterranean and Paratethys regions. Rhodolith-rich carbonates are most widespread in the Langhian and late Tortonian-early Messinian intervals and become scarce in the Piacenzian (late Pliocene). The existence of numerous small islands surrounded by shallow-water platforms with limited siliciclastic sedimentation combined with the Miocene subtropical to warm-temperate climate in the Mediterranean promoted the deposition of carbonates rich in coralline algae.

Rhodolith beds typically occur in ramp deposits, mainly from middle to proximal outer zones. They have been recorded in inner-ramp deposits among coral patches, in rocky-shore facies and in the leeside sediments of spit platforms. Most rhodolith beds, however, developed seawards of corals, seagrass facies or submarine dune fields (shoals) on middle and proximal outer ramps. Rhodoliths also occur in lagoon, reef crest, and reef-slope deposits from rimmed platforms. Rhodolith concentrations in siliciclastic systems seem to reflect temporarily interrupted or reduced sediment influx in the area where they accumulated. Consequently, rhodolith beds are not characteristic of any palaeodepth range within the platform although rhodolith concentrations are most common in deposits from intermediate palaeodepths of several tens of meters.

Changes in rhodolith shape and internal growth forms of coralline thalli are generally not indicative of palaeodepth gradients. Loose branching growths can predominate in settings both shallower and deeper than the ones rich in complex spheroidal to ellipsoidal boxwork and branching rhodoliths. In most well-known examples, however, loose branching growths occur in the deepest coralline facies. Rhodoliths from intermediate palaeodepths tend to include more complex combinations of coralline growth forms than shallower and deeper nodules.

By contrast, the taxonomic composition of corallines making up rhodoliths can be used to identify proximal-distal gradients and to roughly estimate palaeobathimetry by comparisons with modern depth distribution of taxa. *Spongites* and/or *Lithophyllum* are abundant in the shallower rhodoliths and some extant species, such as *S. fruticulosus* Kützing and *L. incrustans*, occur mostly in the top 40–50 m. These genera are less common or absent in deeper water assemblages consisting mainly of several species of *Mesophyllum*, *Lithothamnion*, and *Sporolithon*.

Acknowledgements This work was funded through the project CGL2010-20857 of the Ministerio de Ciencia e Innovación of Spain. I am very grateful to Marco Brandano and an anonymous reviewer by their constructive comments on the first version of the manuscript.

References

Adey WH, Macintyre IG (1973) Crustose coralline algae: a re-evaluation in the geological sciences. Geol Soc Am Bull 84:883–904

Aguirre J (1998) El Plioceno del SE de la Península Ibérica (provincia de Almería). Síntesis estratigráfica, sedimentaria, bioestratigráfica y paleogeográfica. Rev Soc Geol España 11:297–315

Aguirre J, Braga JC (1998) Redescription of Lemoine's (1939) types of coralline algal species from Algeria. Palaeontology 41:489–507

Aguirre J, Braga JC, Martín JM (1993) Algal nodules in the upper Pliocene deposits at the coast of Cadiz (S Spain). Bolletino Soc Paleontol Ital 1:1–7

Aguirre J, Riding R, Braga JC (2000) Diversity of coralline red algae: origination and extinction patterns from the Early Cretaceous to the Pleistocene. Paleobiology 26:651–667

Aguirre J, Braga JC, Martín JM, Betzler C (2012) Palaeoenvironmental and stratigraphic significance of Pliocene rhodolith beds and coralline algal bioconstructions from the Carboneras Basin (SE Spain). Geodiversitas 34:115–136

Amado-Filho GM, Maneveldt G, Manso RCC, Marins-Rosa BV, Pacheco MR, Guimarães SMPB (2007) Structure of rhodolith beds from 4 to 55 meters deep along the southern coast of Espírito Santo State, Brazil. Cienc Mar 33:399–410

Amado-Filho GM, Moura LM, Bastos AC, Salgado LT, Sumida PY, Guth AZ, Francini-Filho RB, Pereira-Filho GH, Abrantes DP, Brasileiro PS, Bahia RG, Leal RN, Kaufman L, Kleypas JA, Farina M, Thompson FL (2012) Rhodolith beds are major $CaCO_3$ bio-factories in the tropical south west Atlantic. Plos One. doi:10.1371/journal.pone.0035171

Bandeira-Pedrosa ME, Pereira SMB, Oliveira EC (2004) Taxonomy and distribution of the green algal genus *Halimeda* (Bryopsidales, Chlorophyta) in Brazil. Rev Brasil Bot 27:363–377

Bassant P, van Buchem FSP, Görür N (2005) The stratigraphic architecture and evolution of the Burdigalian carbonate–siliciclastic sedimentary systems of the Mut Basin, Turkey. Sediment Geol 173:187–232

Bassi D, Nebelsick JH, Checconi A, Hohenegger J, Iryu Y (2009) Present-day and fossil rhodolith pavements compared: their potential for analysing shallow-water carbonate deposits. Sediment Geol 214:74–84
Bassi D, Carannante G, Checconi A, Simone L, Vigorito M (2010) Sedimentological and palaeoecological integrated analysis of a Miocene channelized carbonate margin, Matese Mountains, Southern Apennines, Italy. Sediment Geol 230:105–122
Basso D (1998) Deep rhodolith distribution in the Pontian Islands, Italy: a model for the paleoecology of a temperate sea. Palaeogeogr Palaeoclimatol Palaeoecol 137:173–187
Benisek MF, Marcano G, Betzler C, Mutti M (2010) Facies and stratigraphic architecture of a Miocene warm-temperate to tropical fault-block carbonate platform, Sardinia (Central Mediterranean Sea). In: Mutti M, Piller WE, Betzler C (eds) Carbonate systems during the Oligocene-Miocene climatic transition. IAS Spec Publ vol 42, pp 129–148
Berger S, Kaever MJ (1992) Dasycladales. An illustrated monograph of a fascinating algal order. Georg Thieme, Stuttgart
Betzler C, Braga JC, Jaramillo-Vogel D, Römer M, Hübscher C, Schmiedl G, Lindhorst S (2011) Late Pleistocene and Holocene cool-water carbonates of the Western Mediterranean Sea. Sedimentology 58:643–669
Blair SM, Norris IN (1988) The deep-water species of *Halimeda* Lamouroux (Halimedaceae, Chlorophyta) from San Salvador Island, Bahamas: species composition, distribution and depth records. Coral Reefs 6:227–236
Bosellini FR, Perrin C (2008) Estimating Mediterranean Oligocene-Miocene sea-surface temperatures: an approach based on coral taxonomic richness. Palaeogeogr Palaeoclimatol Palaeoecol 258:71–88
Bosellini FR, Perrin C (2010) Coral diversity and temperature: a palaeoclimatic perspective for the Oligo-Miocene of the Mediterranean region. In: Mutti M, Piller WE, Betzler C (eds) Carbonate systems during the oligocene-miocene climatic transition, vol 42, IAS Special Publication., pp 227–242
Bosellini FR, Russo A, Vescogni A (2001) Messinian reef – building assemblages of the Salerno Peninsula (southern Italy): palaeobathymetric and palaeoclimatic significance. Palaeogeogr Palaeoclimatol Palaeoecol 175:7–26
Bosellini FR, Russo A, Vescogni A (2002) The Messinian reef complex of the Salerno Peninsula (southern Italy): stratigraphy, facies and paleoenvironmental interpretation. Facies 47:91–112
Bosence DWJ (1983a) Coralline algae from the Miocene of Malta. Palaeontology 26:147–173
Bosence DWJ (1983b) The occurrence and ecology of recent rhodoliths. In: Peryt TM (ed) Coated grains. Springer, Berlin, pp 225–242
Bosence DWJ (1991) Coralline Algae: mineralisation, taxonomy, and palaeoecology. In: Riding R (ed) Calcareous algae and stromatolites. Springer, Berlin, pp 98–113
Bosence DWJ (2005) A genetic classification of carbonate platforms based on their basinal and tectonic settings in the Cenozoic. Sediment Geol 175:49–72
Bosence DWJ, Pedley HM (1982) Sedimentology and palaeoecology of a Miocene coralline algal biostrome from the Maltese Islands. Palaeogeogr Palaeoclimatol Palaeoecol 38:9–43
BouDagher-Fadel M, Clark GN (2006) Stratigraphy, paleoenvironment and paleogeography of Maritime Lebanon: a key to Eastern Mediterranean Cenozoic history. Stratigraphy 3:81–118
Brachert TC, Betzler C, Braga JC, Martín JM (1996) Record of climatic change in neritic carbonates: turnover in biogenic associations and depositional models (Late Miocene, southern Spain). Geol Rundschau 85:327–337
Braga JC, Aguirre J (1995) Taxonomy of fossil coralline algal species: Neogene Lithophylloideae (Rhodophyta, Corallinaceae) from Southern Spain. Rev Paleobot Palynol 86:265–285
Braga JC, Aguirre J (2001) Coralline algal assemblages in upper Neogene reef and temperate carbonates in Southern Spain. Palaeogeogr Palaeoclimatol Palaeoecol 175:27–41
Braga JC, Bassi D (2007) Neogene history of *Sporolithon* Heydrich (Corallinales, Rhodophyta) in the Mediterranean region. Palaeogeogr Palaeoclimatol Palaeoecol 243:189–203

Braga JC, Martín JM (1988) Neogene coralline-algal growth-forms and their palaeoenvironments in the Almanzora River Valley (Almeria, S.E. Spain). Palaeogeogr Palaeoclimatol Palaeoecol 67:285–303

Braga JC, Bosence DWJ, Steneck RS (1993) New anatomical character in fossil coralline algae and their taxonomic implications. Palaeontology 36:535–547

Braga JC, Betzler C, Martín JM, Aguirre J (2003) Spit-platform temperate carbonates: the origin of landward-downlapping beds along a basin margin (Lower Pliocene, Carboneras Basin, SE Spain). Sedimentology 50:553–563

Braga JC, Martín JM, Betzler C, Aguirre J (2006) Models of temperate carbonate deposition in Neogene basins in SE Spain: a synthesis. In: Pedley HM, Carannante G (eds) Cool-water carbonates: depositional systems and palaeoenvironmental controls, vol 225. Geological Society, London, pp 121–135, Spec Publ

Braga JC, Vescogni A, Bosellini F, Aguirre J (2009) Coralline algae (Corallinales, Rhodophyta) in western and central Mediterranean Messinian reefs. Palaeogeogr Palaeoclimatol Palaeoecol 275:113–128

Braga JC, Bassi D, Piller WE (2010a) Palaeoenvironmental significance of Oligocene-Miocene coralline red algae – a review. In: Mutti M, Piller WE, Betzler C (eds) Carbonate systems during the Oligocene-Miocene climatic transition. IAS Spec Publ vol 42, pp 165–182

Braga JC, Martín JM, Aguirre J, Baird CD, Grunnaleite I, Bo Jensen N, Puga-Bernabéu A, Sælen G, Talbot MR (2010b) Middle–Miocene (Serravallian) temperate carbonates in a seaway connecting the Atlantic Ocean and the Mediterranean Sea (North Betic Strait, S Spain). Sediment Geol 225:19–33

Brandano M, Vannucci G, Pomar L, Obrador A (2005) Rhodolith assemblages from the lower Tortonian carbonate ramp of Menorca (Spain): environmental and paleoclimatic implications. Palaeogeogr Palaeoclimatol Palaeoecol 226:307–323

Brandano M, Wetsphal H, Mateu-Vicens G (2010a) The sensitivity of a tropical foramol-rhodalgal carbonate ramp to relative sea-level change: Miocene of the central Apennines, Italy. In: Mutti M, Piller WE, Betzler C (eds) Carbonate systems during the Oligocene-Miocene climatic transition. IAS Spec Publ vol 42, pp 89–106

Brandano M, Corda L, Castorina F (2010b) Facies and sequence architecture of a tropical foramol-rhodalgal carbonate ramp: Miocene of the central Apennines (Italy). In: Mutti M, Piller WE, Betzler C (eds) Carbonate systems during the Oligocene-Miocene climatic transition. IAS Spec Publ vol 42, pp 107–128

Buchbinder B, Martinotti GM, Siman-Tov R, Zilberman E (1993) Temporal and spatial relationships in Miocene reef carbonates in Israel. Palaeogeogr Palaeoclimatol Palaeoecol 101:97–116

Bucur II, Filipescu S (1994) Middle Miocene red algae from the Transylvanian Basin (Romania). Beitr Paläontol 19:39–47

Checconi A (2006) Le associazioni ad alghe calcaree corallinacee (Corallinales, Rhodophyta) delle successioni carbonatiche d'acqua bassa del Miocene Inferiore-Medio dell'Appennino Meridionale (Formazione di Cusano): analisi paleoecologica e paleoambientale. PhD Thesis, Università degli Studi di Perugia, Perugia (Italy)

Checconi A, Bassi D, Passeri L, Rettori R (2007) Coralline red algal assemblage from the Middle Pliocene shallow-water temperate carbonates of the Monte Cetona (Northern Apennines, Italy). Facies 53:57–66

Checconi A, Bassi D, Carannante G, Monaco P (2010) Re-deposited rhodoliths in the middle Miocene hemipelagic deposits of vitulano (Southern Apennines, Italy): coralline assemblage characterization and related trace fossils. Sediment Geol 225:50–66

Dabrio CJ, Esteban M, Martín JM (1981) The coral reef of Níjar, Messinian (uppermost Miocene), Almería Province, S.E. Spain. J Sed Petrol 51:521–539

Diester-Haass L, Billups K, Gröcke DR, François L, Lefebvre V, Emeis KC (2009) Mid Miocene paleoproductivity in the Atlantic Ocean and implications for the global carbon cycle. Paleoceanography 24, PA1209. doi:10.1029/2008PA001605

Diester-Haass L, Billups K, Jacquemin I, Emeis KC, Lefebvre V, François L (2013) Paleoproductivity during the middle Miocene carbon isotope events: a data-model approach. Paleoceanography 28:334–346. doi:10.1002/palo.20033
Dullo WC (1983) Fossildiagenese im miozänen Leitha-Kalk der Paratethys von Österreich: Ein Beispiel für Faunenverschiebung durch Diagenesunterschiede. Facies 8:1–112
Elliott GF (1977) Inferred isocrymal distribution of Jurassic dasycladacean Algae in Europe, north Africa and southwestern Asia. J Geol Soc Lond 133:363–373
Elliott GF (1984) Climatic tolerance of some aragonitic green algae of the post-Palaeozoic. Palaeogeogr Palaeoclimatol Palaeoecol 48:163–169
El-Shahat A (1993) Middle Miocene carbonates from the Northern Plateau of the Western Desert (Egypt): petrography and geochemistry. Facies 28:67–76
Esteban M (1996) An overview of Miocene reefs from Mediterranean areas: general trends and facies models. In: Franseen EK, Esteban M, Ward WC, Rouchy J (eds) Models for Carbonate Stratigraphy from Miocene Reef Complexes of Mediterranean Region. SEPM Concepts in Sedimentology and Paleontology, Tulsa, U.S.A. vol 5, pp 3–53
Fornos JJ, Ahr WM (1997) Temperate carbonates on a modern, low-energy, isolated ramp: the Balearic platform, Spain. J Sediment Res 67:364–373
Fornos JJ, Ahr WM (2006) Present-day temperate carbonate sedimentation on the Balearic platform, western Mediterranean: compositional and textural variation along a low-energy isolated ramp. In: Pedley HM, Carannante G (eds) Cool-water carbonates: depositional systems and palaeoenvironmental controls, vol 225. Geological Society, London, pp 71–84, Spec Publ
Franseen EK, Goldstein RH (1996) Paleoslope, sea-level and climate controls on Upper Miocene platform evolution, Las Negras area, southeastern Spain. In: Franseen EK, Esteban M, Ward WC, Rouchy J (eds) Models for Carbonate Stratigraphy from Miocene Reef Complexes of Mediterranean Region. SEPM Concepts in Sedimentology and Paleontology, Tulsa, vol 5, pp 159–176
Fravega P, Piazza M, Vannucci G (1993) Importance and significance of the rhodolith bodies in the miocenic sequences of Tertiary Piedmont Basin In: De Castro P, Barattolo F, Parente M (eds) Studies on fossil benthic algae. Boll Soc Paleont Ital Spec, vol 1, pp 197–210
Geel T (2000) Recognition of stratigraphic sequences in carbonate platform and slope deposits: empirical models based on microfacies analysis of Palaeogene deposits in southeastern Spain. Palaeogeogr Palaeoclimatol Palaeoecol 155:211–238
Geel T, Roep TB, Ten Kate W, Smit J (1992) Early-Middle Miocene stratigraphic turning points in the Alicante region (SE Spain): reflections of Western Mediterranean plate-tectonic reorganization. Sediment Geol 75:223–239
Halfar J, Mutti M (2005) Global dominance of coralline red algal facies: a response to Miocene oceanographic events. Geology 33:481–484
Harvey AS, Broadwater ST, Woelkerling WJ, Mitrovski PJ (2003) *Choreonema* (Corallinales, Rhodophyta): 18S rDNA phylogeny and resurrection of the Hapalidiaceae for the subfamilies Choreonematoideae, Austrolithoideae, and Melobesioideae. J Phycol 39:988–998
Hillis L (1991) Recent calcified halimedaceae. In: Riding R (ed) Calcareous algae and stromatolites. Springer, Berlin, pp 167–188
Hillis-Colinvaux L (1986) Deep water populations of *Halimeda* in the economy of an atoll. Bull Mar Sci 38:155–169
Hsü KJ, Montadert L, Bernoulli D, Cita MB, Erickson A, Garrison RE, Kidd RB, Mèlierés F, Müller C, Wright R (1977) History of the Messinian salinity crisis. Nature 267:399–403
Johnson JH (1964) Fossil and recent calcareous algae from Guam. US Geol Surv Prof Pap 403-G:G1–G40
Johnson JH, Kaska HV (1965) Fossil algae from Guatemala. Prof Contr Colorado Sch Min 1:1–152
Kroeger KF, Reuter M, Brachert TC (2006) Palaeoenvironmental reconstruction based on non-geniculate coralline red algal assemblages in Miocene limestone of central Crete. Facies 52:381–409

Littler MM, Littler DS, Blair SM, Norris JN (1986) Deep-water plant communities from an uncharted seamount off San Salvador Island, Bahamas: distribution, abundance, and primary productivity. Deep Sea Res 33:881–892
Lund MJ, Davies PJ, Braga JC (2000) Coralline algal nodules off Fraser Island, eastern Australia. Facies 42:25–34
Mankiewicz C (1996) The middle to upper Miocene carbonate complex of Níjar, Almería Province, southeastern Spain. In: Franseen EK, Esteban M, Ward WC, Rouchy J (eds) Models for carbonate stratigraphy from Miocene reef complexes of Mediterranean Region. SEPM concepts in sedimentology and paleontology, Tulsa, vol 5, pp 141–157
Mariotti A, Strugli MV, Zeng N, Lau KM (2002) The hydrological cycle in the Mediterranean region and implications for the water budget of the Mediterranean Sea. J Clim 15:1674–1690
Marrack E (1999) The relationship between water motion and living rhodolith beds in the southwestern Gulf of California, México. Palaios 14:159–171
Marshall JF, Tsuji Y, Matsuda H, Davies PJ, Iryu Y, Honda N, Satoh Y (1998) Quaternary and Tertiary subtropical carbonate platform development on the continental margin of southern Queensland, Australia. In: Camoin GF, Davies PJ (eds) Reefs and carbonate platforms in the Pacific and Indian oceans. IAS Spec Publ, London, vol 25, pp 163–195
Martín JM, Braga JC (1994) Messinian events in the Sorbas Basin in southeastern Spain and their implications in the recent history of the Mediterranean. Sediment Geol 90:257–268
Martín JM, Braga JC, Riding R (1997) Late Miocene *Halimeda* alga-microbial segment reefs in the marginal Mediterranean Sorbas Basin, Spain. Sedimentology 44:441–456
Martín JM, Braga JC, Aguirre J, Betzler C (2004) Contrasting models of temperate carbonate sedimentation in a small Mediterranean embayment: the Pliocene Carboneras Basin, SE Spain. J Geol Soc Lond 161:387–399
Martín JM, Braga JC, Sánchez-Almazo IM, Aguirre J (2010) Temperate and tropical carbonate-sedimentation episodes in the Neogene Betic basins (southern Spain) linked to climatic oscillations and changes in Atlantic-Mediterranean connections: constraints from isotopic data. In: Mutti M, Piller WE, Betzler C (eds) Carbonate systems during the Oligocene-Miocene climatic transition. IAS Spec Publ, vol 42, pp 49–69
Miller KG, Mountain GS, Wright JD, Browning JV (2011) A 180-million-year record of sea level and ice volume variations from continental margin and deep-sea isotopic records. Oceanography 24:40–53
Pedley HM (1996) Miocene reefs facies of the Pelagian Region (Central Mediterranean). In: Franseen EK, Esteban M, Ward WC, Rouchy J (eds) Models for carbonate stratigraphy from Miocene Reef complexes of Mediterranean region. SEPM concepts in sedimentology and paleontology, Tulsa, vol 5, pp 247–259
Pedley HM, Carannante G (2006) Cool-water carbonate ramps: a review. In: Pedley HM, Carannante G (eds) Cool-water carbonates: depositional systems and palaeoenvironmental controls. Geological Society, London, Spec Publ, vol 225, pp 1–9
Perrin C, Bosellini F (2012) Paleobiogeography of scleractinian reef corals: changing patterns during the Oligocene–Miocene climatic transition in the Mediterranean. Earth Sci Rev 111:1–24
Perrin C, Kiessling W (2010) Latitudinal trends in Cenozoic reef patterns and their relationship to climate. In: Mutti M, Piller WE, Betzler C (eds) Carbonate systems during the Oligocene-Miocene climatic transition. IAS Spec Publ, vol 42, pp 17–34
Piller, WE (1993) Facies development and coralline algae in the Vienna and Eisenstadt Basins (Miocene) In: Höfling R, Moussavian E, Piller WE (eds) Facial development of algae-bearing carbonate sequences in the Eastern Alps. Field trip guidebook, Alpine Algae '93, Munich-Vienna, pp 1–24
Pomar L (1991) Reef geometries, erosion surfaces and high frequency sea-level changes, Upper Miocene Reef Complex, Mallorca, Spain. Sedimentology 38:243–269
Pomar L (2001) Types of carbonate platforms: a genetic approach. Basin Res 13:313–334
Pomar L, Rodríguez-Perea A (1983) El Neogeno inferior de Mallorca: Randa. In: Pomar L, Obrador A, Fornos JJ, Rodríguez-Perea A (eds) Terciario de las Baleares, Guía de las

Excursiones del X Congreso Nacional de Sedimentología Palma de Mallorca, Universitat de Palma de Mallorca, pp 115–137
Pomar L, Ward WC, Green DG (1996) Upper Miocene reef complex of the Llucmajor area, Mallorca, Spain. In: Franseen EK, Esteban M, Ward WC, Rouchy J (eds) Models for carbonate stratigraphy from Miocene reef complexes of Mediterranean region. SEPM concepts in sedimentology and paleontology, Tulsa, vol 5, pp 191–225
Pomar L, Obrador A, Westphal H (2002) Sub-wavebase cross-bedded grainstones on a distally steepened carbonate ramp, upper Miocene, Menorca, Spain. Sedimentology 49:139–169
Pomar L, Bassant P, Brandano M, Ruchonnet C, Janson X (2012) Impact of carbonate producing biota on platform architecture: Insights from Miocene examples of the Mediterranean region. Earth Sci Rev 113:186–211
Popov SV, Rögl F, Rozanov AY, Steininger FF, Shcherba IG, Kováè M (2004) Lithological–paleogeographic maps of paratethys 10 maps late Eocene to Pliocene. Cour Forsch Senckenberg 250:1–46
Puga-Bernabéu A, Martín JM, Braga JC (2007a) Tsunami-related deposits in temperate carbonate ramps, Sorbas Basin, southern Spain. Sediment Geol 199:107–127
Puga-Bernabéu A, Braga JC, Martín JM (2007b) High-frequency cycles in Upper-Miocene ramp-temperate carbonates (Sorbas Basin, SE Spain). Facies 53:329–345
Puga-Bernabéu A, Martín JM, Braga JC, Aguirre J (2014) Offshore remobilization processes and deposits in low-energy temperate-water carbonate-ramp systems: examples from the Neogene basins of the Betic Cordillera (SE Spain). Sediment Geol 304:11–27
Randazzo AF, Müller P, Lelkes G, Juhasz E, Hámor T (1999) Cool-water limestones of the Pannonian basinal system, Middle Miocene, Hungary. J Sediment Res 69:283–293
Renema W, Bellwood DR, Braga JC, Bromfield K, Hall R, Johnson KG, Lunt P, Meyer CP, McMonagle LB, Morley RJ, O'Dea A, Todd JA, Wesselingh FP, Wilson MEJ, Pandolfi JM (2008) Hopping hotspots: global shifts in marine biodiversity. Science 321:654–657
Reuter M, Piller WE, Brandano M, Harzhauser M (2013) Correlating Mediterranean shallow water deposits with global Oligocene–Miocene stratigraphy and oceanic events. Glob Planet Chang 111:226–236
Riding R, Martín JM, Braga JC (1991) Coral-stromatolite reef framework, Upper Miocene, Almería, Spain. Sedimentology 38:799–818
Robertson AHF, Parlak O, Ustaömer T (2012) Overview of the Palaeozoic-Neogene evolution of Neotethys in the Eastern Mediterranean region (southern Turkey, Cyprus, Syria). Pet Geosci 18:381–404
Rögl F (1998) Palaeogeographic considerations for Mediterranean and Paratethys seaways (Oligocene to Miocene). Ann Natur Mus Wien 99:279–310
Rösler A, Pretkovic V, Novak V, Renema W, Braga JC (2015) Coralline algae from the Miocene Mahakam Delta (East Kalimantan, SE Asia). Palaios 30:83–93
Ruchonnet C, Kindler P (2010) Facies models and geometries of the Ragusa Platform (SE Sicily, Italy) near the Serravallian–Tortonian boundary. In: Mutti M, Piller WE, Betzler C (eds) Carbonate systems during the Oligocene-Miocene climatic transition. IAS Spec Publ, vol 42, pp 71–88
Saint Martin J-P (1996) Messinian coral reefs of western Algeria. In: Franseen EK, Esteban M, Ward WC, Rouchy J (eds) Models for Carbonate stratigraphy from Miocene reef complexes of Mediterranean region. SEPM concepts in sedimentology and paleontology, Tulsa, vol 5, pp 239–246
Saint Martin J-P, Corneé J-J (1996) The Messinian reef complex of Melilla, northeastern Rif, Morocco. In: Franseen EK, Esteban M, Ward WC, Rouchy J (eds) Models for carbonate stratigraphy from Miocene reef complexes of Mediterranean region. SEPM Concepts in Sedimentology and Paleontology, Tulsa, vol 5, pp 227—238
Sánchez-Almazo IM, Spiro B, Braga JC, Martín JM (2001) Constraints of stable isotope signatures on the depositional palaeoenvironments of upper Miocene reef and temperate carbonates in the Sorbas Basin, SE Spain. Palaeogeogr Palaeoclim Palaeoecol 175:153–172

Sánchez-Almazo IM, Braga JC, Dinarès-Turell J, Martín JM, Spiro B (2007) Palaeoceanographic controls on reef deposition: the Messinian Cariatiz reef (Sorbas Basin, Almería, SE Spain). Sedimentology 54:637–660
Scott RW, Govean FM (1985) Early depositional history of a rift basin: Miocene in western Sinai. Palaeogeogr Palaeoclimatol Palaeoecol 52:143–158
Sola F, Braga JC, Aguirre J (2013) Hooked and tubular coralline algae indicate seagrass beds associated to Mediterranean Messinian reefs (Poniente Basin, Almería, SE Spain). Palaeogeogr Palaeoclimatol Palaeoecol 374:218–229
Steller DL, Riosmena-Rodríguez R, Foster MS (2009) Living rhodolith bed ecosystems in the Gulf of California. In: Johnson ME, Ledesma-Vázquez J (eds) Atlas of coastal ecosystems in the western Gulf of California. The University of Arizona Press, Tucson, pp 72–82
Studencki W (1988) Red algae from the Pinczów Limestone (Middle Miocene; Swietokrzyskie Mts, Central Poland). Acta Palaeont Pol 33:4–57
Tomassetti L, Brandano M (2013) Sea level changes recorded in mixed siliciclastic-carbonate shallow-water deposits: the Cala di Labra formation (Burdigalian, Corsica). Sediment Geol 294:58–67
Toscano F, Vigliotti M, Simone L (2006) Variety of coralline algal deposits (rhodalgal facies) from the Bays of Naples and Pozzuoli (northern Tyrrhenian Sea, Italy). In: Pedley HM, Carannante G (eds) Cool-water carbonates: depositional systems and palaeoenvironmental controls. Geological Society, London, Spec Publ, vol 225, pp 85–94
Vannucci G, Piazza M, Fravega P, Arnera V (1994a) Le rodoliti del Miocene inferiore del settore SW del Bacino Terziario del Piemonte (Spigno Monferrato – Alessandria). Atti Soc Tosc Sc Nat Mem ser A 100(1993):93–117
Vannucci G, Piazza M, Fravega P, Chiesa I (1994b) Calcareous Rhodophyceae from the "Astian facies" (Pliocene), Valle San Bartolomeo, Alessandria, NW Italy. In: Matteucci R, Carboni MG, Pignatti JS (eds) Studies on ecology and palaeoecology of benthic communities. Boll Soc Paleontol Ital Spec, Vol 2, pp 351–364
Vannucci G, Piazza M, Fravega P, Abate C (1996) Litostratigrafia e paleoecologia di successioni a rodoliti della "Pietra da Cantoni" (Monferrato orientale, Italia nord-occidentale). Atti Soc Tosc Sc Nat Mem ser A 103:69–86
Vigorito M, Murru M, Simone L (2010) Carbonate production in rift basins: models for platform inception, growth and dismantling, and for shelf to basin sediment transport, Miocene Sardinia Rift Basin, Italy. In: Mutti M, Piller WE, Betzler C (eds) Carbonate systems during the Oligocene-Miocene climatic transition. IAS Spec Publ, vol 42, pp 257–282
Webster JM, Clague DA, Braga JC, Spalding H, Renema W, Kelley C, Applegate B, Smith JR, Paull CK, Moore JG, Potts D (2006) Drowned coralline algal dominated deposits off Lanai, Hawaii; carbonate accretion and vertical tectonics over the last 30 ka. Mar Geol 225:223–246
Webster JM, Braga JC, Clague DA, Gallup C, Hein JR, Potts DC, Renema W, Riding R, Riker-Coleman K, Silver E, Wallace LM (2009) Coral reef evolution on rapidly subsiding margins. Global Planet Change 66:129–148
Westerhold T, Bickert T, Röhl U (2005) Middle to late Miocene oxygen isotope stratigraphy of ODP site 1085 (SE Atlantic): new constrains on Miocene climate variability and sea-level fluctuations. Palaeogeogr Palaeoclimatol Palaeoecol 217:205–222
Zachos J, Pagani M, Sloan L, Thomas E, Billups K (2001) Trends, rhythms, and aberrations in global climate 65 Ma to present. Science 292:686–693

Chapter 8
Oligocene Rhodolith Beds in the Central Mediterranean Area

Marco Brandano

8.1 Introduction

The Oligocene is of particular interest because it represents an early "icehouse" epoch, in which the transition between Eocene and Oligocene is characterized by global cooling followed by rapid expansion of the Antarctic continental ice sheet (Zachos et al. 2001; Tripati et al. 2005), recorded as the Oi-1 glaciation. This event was linked to reorganization of the ocean/climate system, as shown by deepening of the calcite compensation depth by about 1 km (Coxall et al. 2005) and increased ocean fertility (Thomas et al. 2000; Zachos et al. 2001). Oi-1 also coincided with aridification and cooling in continental interiors (Dupont-Nivet et al. 2007; Zanazzi et al. 2007) and reduction of atmospheric CO_2 levels (DeConto and Pollard 2003). According to Allen and Armstrong (2008), reduced atmospheric CO_2 and global cooling were produced by complementary mechanisms, such as the waning of pre-collision arc magmatism, increased weathering of silicate, storage of organic carbon in Paratethyan basins, and oceanographic changes, producing upwelling. These mechanisms were induced by the early phase of the Arabian-Eurasian collision and the closure of the Tethys. The climate improved in the late Oligocene (26–24 Ma; Zachos et al. 2001), and it has been attributed to deep-sea warming, coupled with collapse of the Antarctic ice sheet (e.g., Miller et al. 1987; Zachos et al. 2001). However, Pekar et al. (2006) also showed that the late Oligocene was punctuated by extensions of ice volumes, recorded by two late Oligocene $\delta^{18}O$ events, Oi-2c (24.4 Ma) and Mi-1 (23.0 Ma). Icehouse conditions during the Oligocene have important implications for the amplitude of sea level oscillations, which have been estimated as up to 60 m (Pekar et al. 2002, 2006).

M. Brandano (✉)
Dipartimento di Scienze della Terra, La Sapienza Università di Roma,
P. A. Moro 5, I-00185 Rome, Italy
e-mail: marco.brandano@uniroma1.it

R. Riosmena-Rodríguez et al. (eds.), *Rhodolith/Maërl Beds: A Global Perspective*, Coastal Research Library 15, DOI 10.1007/978-3-319-29315-8_8

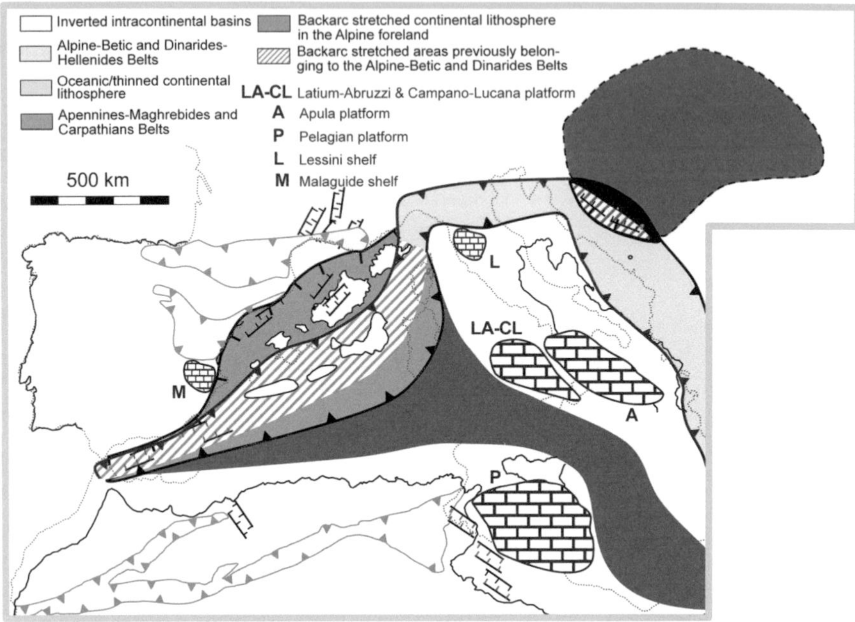

Fig. 8.1 Oligocene palaeogeographical and palaeotectonic map of western-central Mediterranean area (Modified from Carminati et al. 2010)

During the Oligocene, the Mediterrranean area underwent the final stages of the Alpine and Dinarides subduction systems and the younger NW-directed Apennine subduction system (Carminati et al. 2010) (Fig. 8.1). There were many carbonate platforms in the foreland zone of the latter: Latium-Abruzzi-Campana and Apulia in the central and southern Apennine domain (Brandano and Corda 2011; Mutti et al. 1997, 1999; Brandano et al. 2012), the Ragusa platform (Hyblea, Sicily) and Malta in the Pelagian area (Pedley 1998; Brandano et al. 2009a). In the Southern Alps, Oligocene neritic carbonate sedimentation developed on the Lessini shelf (Bassi and Nebelsick 2010; Nebelsick et al. 2013). During Oligocene, shallow-water carbonate sedimentation was established along the eastern margin of Iberian plate in the Prebetic of the Betic Cordillera (E-SE Spain) (Geel 2000; Höntzsch et al. 2013), where the Oligocene shallow-water succession was deposited in the Malaguide complex (Braga and Bassi 2011).

This chapter presents the sedimentological and paleo-ecological characteristics and significance of the rhodolith beds in the above platforms.

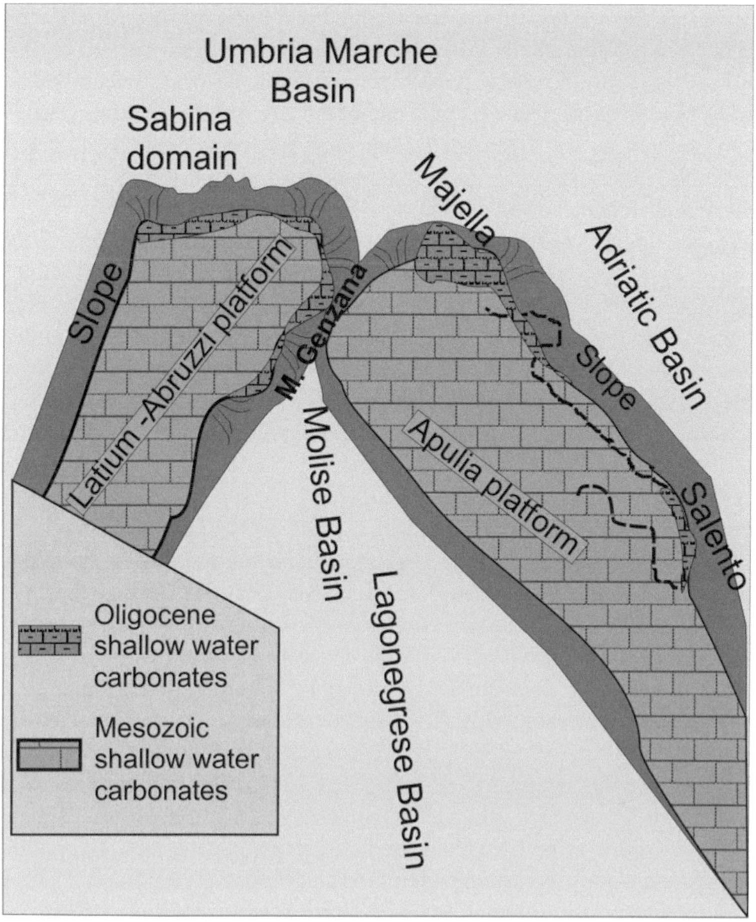

Fig. 8.2 Oligocene palaeogeographical reconstruction of the Apennines platforms

8.2 Case Histories

8.2.1 Apulia Platform

The Apennines contain two main carbonate platform domains represented by the Latium-Abruzzi-Campana Platform separated from the Apulia Platform (Mostardini and Merlini 1986; Vezzani et al. 2010) to the north by the basinal corridor of Monte Genzana and to the south by the Molisan and Lagonegrese basin (Fig. 8.2). These platforms developed along the southern margin of the Tethys (Bernoulli 2001).

Parts of the Apulia Platform are exposed in the Maiella Mountains, Gargano, Murge and Salento areas of Central and Southern Italy. The eastern part of the Apulia Platform represents the gently deformed foreland of the Apennine thrust chain, whereas the western part is downfaulted and buried underneath the foredeep and adjacent Apennine chain (Mostardini and Merlini 1986; Casero et al. 1988) (Fig. 8.1). This platform records almost exclusively shallow-marine carbonate sedimentation from the Late Triassic to the end of the Cretaceous (Bosellini and Parente 1994), represented by open-shelf deposits (Calcari del Ciolo Formation, Maastrichtian) outcropping over a limited area of the eastern Salento coastline (Bosellini and Parente 1994). Like in the adjacent Latium-Abruzzi-Campana Platform, the shallow-water sedimentation of the Paleogene is preserved along the platform margins exposed in the northern, eastern and south-eastern parts in the Maiella Mountains (Vecsei et al. 1998), the Gargano Promontory (Bosellini et al. 1999), and along the Salento peninsula (Bosellini and Parente 1994; Parente 1994; Bosellini et al. 1999; Bosellini 2006), respectively, lower Cenozoic sediments are deposited along the slope and basin of the Maiella, whereas a long hiatus is recorded on the platform top, where Paleocene to middle Eocene deposits only occur as thin, discontinuous beds. Small coral build-ups of Priabonian to Rupelian age are recorded (Pesco Falcone Limestone), followed by prograding shallow-platform sediments over the slope (Vecsei and Moussavian 1997). Lastly, a ramp developed above the former shelf and slope (Bolognano Formation) from the end of the Rupelian until the early Messinian (Vecsei and Sanders 1999; Benedetti et al. 2010; Reuter et al. 2013). The Oligocene portion of the Bolognano Formation consists of a *Lepidocyclina* calcarenite (Fig. 8.3). This unit was deposited on a carbonate ramp, tens of kilometers long, characterized by a large downslope-migrating dune field in the middle and outer ramp (Fig. 8.4a), produced by uni-directional basinward-flowing currents (Brandano et al. 2012). The inner ramp lithofacies do not crop out, probably because they were eroded. The effects of strong return currents caused by storms and winds may have produced basinward-migrating bedforms. The components of the *Lepidocyclina* limestone are distinctive of a generally oligophotic zone in a Chattian mid-ramp environment (Buxton and Pedley 1989; Pedley 1998; Pomar 2001; Brandano et al. 2009a; Bassi and Nebelsick 2010). Taphonomic analysis highlights the coexistence of both abraded and well-preserved components, the main ones represented by larger benthic foraminifera (*Nephrolepidina*, *Eulepidina*, *Amphistegina*, *Heterostegina*), echinoid plates, bryozoans and bivalve fragments. Red algae are very common, but are present as well-rounded red algal debris (Fig. 8.4b). Only in the proximal middle ramp coralline algae formed free-living branches and small rhodoliths composed of melobesioids (*Lithothamnion*) and sporolithales (*Sporolithon*) (Brandano et al. 2012).

The Oligocene deposits cropping out in the Salento peninsula are represented by the Castro Limestone and Porto Badisco Calcarenite Formations (Fig. 8.3). According to Bosellini and Russo (1992) and Bosellini et al. (1999), the Castro Limestone is attributed to the early Chattian (SBZ 22B of Chauzac and Poignant 1997) and the Porto Badisco Calcarenite to the late Chattian (SBZ 23). The Castro Limestone is extensively exposed along the eastern coast of the Salento peninsula and has been interpreted as a fringing reef complex, disconformably overlying a tectonically deformed Cretaceous–Eocene carbonate substrate (Bosellini 2006). Its geometry is clearly sigmoidal, resulting from the seaward limited progradation of

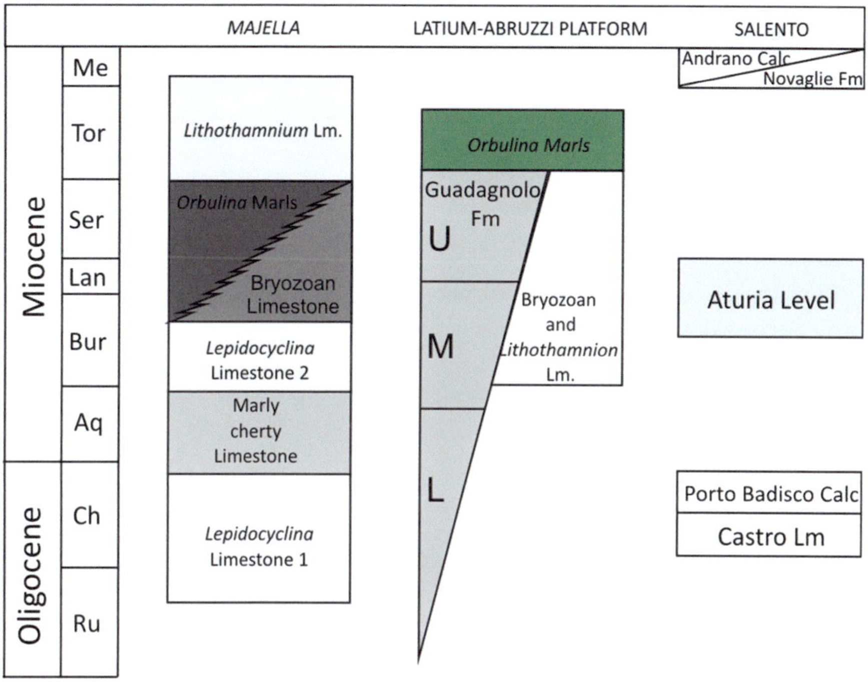

Fig. 8.3 Oligo-Miocene stratigraphic scheme of Apennine platform (After Bosellini et al. 1999; Civitelli and Brandano 2005; Brandano and Corda 2011; Reuter et al. 2013; Brandano et al. 2016); Guadagnolo Formation (Fm) subdivision: *L* Lower member, *M* Middle spongolitic member, *U* Upper calcarenitic member. *Lm* Limestones. *Calc* Calcarenites. Abbreviations of the name of the stages can be seen in Fig. 8.6 (*Ru* Rupelian)

the reef complex. Thickness, ranging from a minimum of 5 m to a maximum of 80–100 m, is controlled by accommodation space available during deposition, which was minimum on the platform and maximum along the slope (Bosellini et al. 1999; Bosellini 2006). According to Bosellini (2006), the coralline algae (*Sporolithon*, *Lithothamnion*, *Mesophyllum*), and peyssonneliaceans in the Castro Limestone bind coral colonies only with thin crusts and contribute little to reef construction.

The Porto Badisco Calcarenite has a maximum thickness of 50–60 m and is dominated by red algae and larger benthic foraminifera. In contrast to the *Lepidocyclina* calcarenite of Maiella, the Porto Badisco Calcarenite was deposited on a small carbonate ramp of only a few km, onlapping the underlying formations and infilling pre-existing depressions (Nardin and Rossi 1966). Rhodolith beds accumulated in these depressions and in the oligophotic zone of the ramp (Bosellini et al. 1999; Brandano et al. 2010).

The inherited substrate topography controlled the locus and mode of rhodolith accumulation because it favored the development of shallow channel-like depressions acting as traps for rhodoliths formed at shallower depths (Fig. 8.4c), as

Fig. 8.4 (**a**) In Majella structure the cross-bedding of the *Lepidocyclina* calcarenites is produced by migration of submarine dunes. (**b**) The skeletal assemblages is dominated by larger benthic foraminifera (*Nephroledina* and *Eulepidina*), the red algae are common as detritus, while nodules are subordinate. (**c**) The rhodoliths of the Porto Badisco Calcarenites infill the pre-existing depressions. (**d**) The columnar structures and nodules indicate a growth in deep, low-energy environment. (**e**) The central portion of the depression may act as a locus of increased water turbulence with large rhodoliths produced by continuous overturning. (**f**) In the middle ramp environment, rhodoliths formed in moderate to low energy conditions, as indicated asymmetric overgrowth on oligophotic biota such as *Eulepidina*

indicated by changes in taxonomic compositions and inner structures from the initial to the final growth stage (Brandano et al. 2010).

The abundance of mastophoroid thalli and the dominance of laminar growth forms in the inner portion of rhodoliths coming from the central part of the depression suggest initial growth in a shallow, high-energy setting. By contrast, the dominance of lumpy thalli growth forms and the more marked prevalence of melobesioids in the external portions of the rhodoliths indicate that the final stage of growth occurred in a deeper, lower-energy environment (Fig. 8.4d).

Systematic changes in the texture of the rhodolith facies (from rudstone to floatstone) and in the growth forms of rhodoliths (from laminar to columnar) from the central to the marginal parts of the lithosome, reflect variations in hydrodynamic regime across the channel section. The depression acted as a locus of increased water turbulence, in which mature rhodoliths were periodically overturned and continued growing to a larger size (up to 9 cm) (Fig. 8.4e).

The rhodolith deposits of the mid-ramp environment consist of horizontally bedded rhodolith floatstone. The rhodoliths are ellipsoidal in shape. Laminar/columnar and columnar growth forms characterize the mid-ramp rhodoliths, which are multispecific, being dominated by melobesioids, although sporolithaceans are also common. Mastophoroids are subordinate. These rhodoliths formed in moderate to low energy conditions, as indicated by the dominance of columnar structures and asymmetric overgrowth (Fig. 8.4d), which implies limited overturning in oligophotic conditions, as shown by the algal assemblages (Brandano et al. 2010).

8.2.2 Latium-Abruzzi Platform

The Latium-Abruzzi Platform represents the northern portion of the wide Latium-Abruzzi-Campana platform (Tyrrhenian platform, sensu Mostardini and Merlini 1986). Carbonate sedimentation in this platform started in the Late Triassic and ended in the late Miocene. A substantial hiatus occurred between the Early and Late Cretaceous and between the Paleocene and Eocene. The Oligocene is almost absent in the internal sector of the platform, and the Miocene carbonates directly overlie Cretaceous platform carbonates (Accordi et al. 1967). Miocene carbonates are represented by the Aquitanian-Tortonian Bryozoan and *Lithothamnion* Limestones deposited in a carbonate ramp environment (Brandano and Corda 2002; Civitelli and Brandano 2005). The Miocene carbonate sedimentation on the Latium-Abruzzi platform was terminated by the arrival of terrigenous sediments due to drowning caused by plate flexure and coeval inputs (Cipollari and Cosentino 1995; Carminati et al. 2007).

The Oligo-Miocene sediments in the platform are represented by the Guadagnolo Formation (Chattian to Serravallian). This formation overlies mainly Cretaceous strata along the platform margin and Cenozoic deposits along its slope, reaching a thickness of 500 m in some places. This formation is commonly divided into three members (Carboni et al. 1982; Brandano and Corda 2011): the basal member is Chattian to Aquitanian in age and is up to 100 m thick (Fig. 8.3). Along the northern and eastern margins of the platform, this member was deposited on a small carbonate ramp, only a few km in size, on the tectonically deformed Cretaceous to Eocene carbonate substrate. At the northern margin, larger benthic foraminifera dominate the skeletal assemblages; at the eastern margin, red algae are important contributors to carbonate production. The second member of the Guadagnolo Formation is Burdigalian to Langhian in age, reaches a thickness of 300 m, and is made up of a thick monotonous stack of alternating marl to calcareous marl and cross-bedded bioclastic calcarenite, deposited in an outer ramp environment. Planktonic foraminifers and sponge spicules dominate in the marly lithology, whereas bryozoans, echinoids, bivalves and benthic foraminifera are the main components of the calcarenite. The uppermost member is Serravallian in age, and represents an outer ramp on which coarse-grained bioclastic grainstone-packstone was deposited. The contact with the underlying spongolitic member is marked by a major unconformity.

The Chattian member of the Guadagnolo Formation at the eastern margin of the Latium-Abruzzi platform overlies an Eocene nummulitic limestone and is characterized by cross-bedded red algal rudstone to floatstone, alternating with coarse cross-bedded bioclastic packstone (Brandano and Corda 2011).

The main components of the bioclastic packstone are larger benthic foraminifera (*Amphistegina*, *Spiroclypeus*, *Heterostegina*, *Operculina*, *Nephrolepidina*, *Eulepidina dilatata*, *Miogypsinoides complanatus*, and miogypsinids). Other components are bryozoans and mollusc fragments, echinoid plates, and rotaliids such as *Neorotalia viennoti*, *Cibicides*, lageniids and serpulids (*Ditrupa*). This lithofacies is characterized by planar cross-beds 10–20 cm thick with an updip of up to 10°. In these beds, lamination forms angles of 5–10° with the bedding planes (bedding-plane discordant), producing straight-planar first-order sets. These stack in 2–3 m-thick co-sets bounded by a subhorizontal surface (Fig. 8.5a).

The red algae form rhodoliths, free-living branches, encrusting thalli and coralline debris (Fig. 8.5b–d). The rhodoliths are ellipsoidal in shape and up to 2 cm in size, most of them being non-nucleated or with red algal free-living branches as nuclei. Bryozoans and encrusting foraminifera represent an important percentage of the rhodolith components. Growth forms are laminar, encrusting, and lumpy. Free-living branches show branching density of groups I-II, sensu Bosence (1983a). Melobesioids dominate the algal assemblages, consisting primarily of *Lithothamnion*, with accessory sporolithales.

The facies and skeletal assemblages of the Chattian member of the Guadagnolo Formation are characterized by strong grain fragmentation, relatively enriched with echinoid fragments resistant to mechanical abrasion. These characteristics and the red algal association indicate oligophotic zone sedimentation (sensu Pomar 2001) of a mid-ramp environment, resulting from the combined accumulation of in situ products and sediments swept from the shallower inner ramp by waves and currents, as shown by sedimentary structures (planar cross-bedding).

8.2.3 Pelagian Province

The Pelagian Province represents the northern extension of the African Plate, comprising the Hyblean (SE Sicily) and Maltese regions, the Pelagian islands, eastern Tunisia and north-west Libya. At present, the Pelagian Province, running NW to

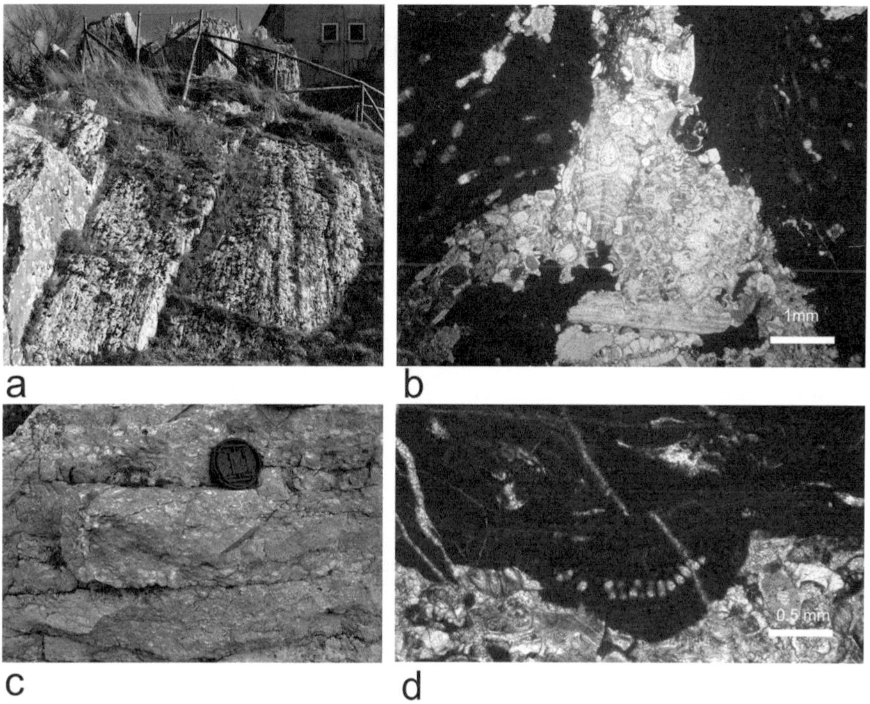

Fig. 8.5 (**a**) The Chattian ramp in the Latium-Abruzzi domain was characterized by high energy as testified by diffuse cross bedding. (**b**) Coralline algae formed small non-nucleated rhodoliths, free-living branches, encrusting thalli and coralline debris; the echinoid fragments, bryozoans, larger benthic foraminifera represent an important percentage of the skeletal assemblages. The red algae assemblages are dominated by melobesioids (**c**) and by Sporolithales (**d**)

SE, is characterized by alternating horsts and grabens separated by normal faults. It underwent extensional tectonics since the Mesozoic (Middle Jurassic–Early Cretaceous), developing a rift zone with main fault systems trending between NW-SE and E-W (Grasso et al. 1999; Klett 2001). Later, in the Late Cretaceous–Miocene interval, the northward drift of Africa induced a tectonic inversion of the faults, which had formed in the Mesozoic, with main activity during the Eocene. Lastly, Plio-Pleistocene extensional tectonics led to the development of the Sicily channel (Corti et al. 2006).

Oligocene carbonate ramps developed in the Hyblean and Malta plateaus. Since the Late Triassic, the Hyblean plateau has been subdivided into two main paleogeographic domains, represented by the SW basin (Ragusa Belt) and the NE carbonate shelf (Siracusa Belt) (Carbone et al. 1982; Bianchi et al. 1987).

The Oligocene carbonate ramp deposits overlie the Cretaceous to middle Eocene carbonates of the Amerillo Formation (Grasso et al. 1999). According to Pedley (1998), the Oligocene Hyblean ramp was detached and isolated, protected by an east-facing patch reef belt and a broad shoaling zone without an inner ramp.

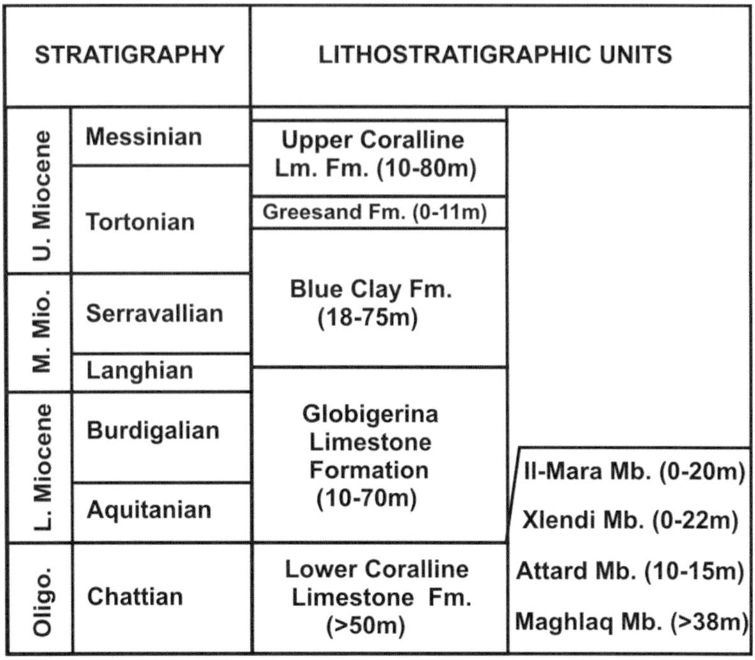

Fig. 8.6 Lithostratigraphy and age of the units outcropping on the Maltese Islands (Modified from Pedley 1978). *Lm* Limestone, *Fm* Formation, *Mb* Member, *Oligo* Oligocene, *Mio* Miocene, *L* lower, *M* middle, *U* upper

Sediment was supplied from the midramp zone, where the carbonate factory was dominated by larger benthic foraminifers and rhodoliths. Unfortunately, there are no descriptions of the red algal facies concerning the Oligocene Hyblean ramp. On the contrary rhodolith beds in the Malta Plateau are documented (Bosence 1983b; Bosence and Pedley 1982; Brandano et al. 2009a, 2012; Quaranta et al. 2012).

The Malta Plateau lies in the NE part of the Pelagian Province and has been a structurally elevated area since the Late Triassic (Bishop and Debono 1996). It consists of 4,500 m of Mesozoic and Paleogene limestone and dolomite deposited on a restricted environment of the carbonate platform. The Oligo-Miocene sequence crops out in the island of Malta (Fig. 8.6). The lowermost unit is the Lower Coralline Limestone Formation (Chattian), consisting of rudstone and packstone with skeletal fragments typical of heterozoan carbonates. This lithostratigraphic unit is overlain by the hemipelagic planktonic wackestone of the *Globigerina* Limestone Formation (Aquitanian–Langhian), in turn followed by hemipelagic clay to marlstone of the Blue Clay Formation (Serravallian to lower Tortonian). The latter formation is unconformably overlain by the shallow-water Upper Coralline Limestone Formation (Tortonian to Messinian) with the discontinuous interposition of the Green Sand Formation, of Tortonian age (Pedley 1998).

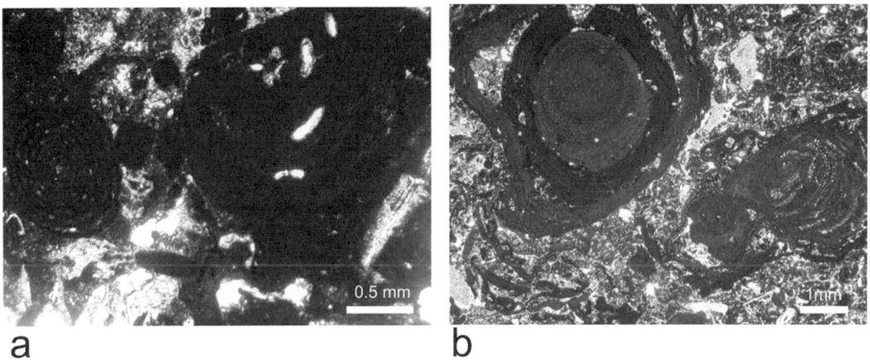

Fig. 8.7 (**a**) The inner ramp of Chattian Malta ramp is dominated by seagrass environment where larger porcelaneous foraminifera (*Borelis*) and red algal rhodoliths of mastophoroids were common. (**b**) In the middle ramp, rhodoliths constituted by genera such as *Sporolithon*, *Lithoporella* and *Lithothamnion*

The Lower Coralline Limestone Formation is subdivided into four members (Pedley 1978). The oldest is the Maghlaq Member, composed of inner-ramp wackestone and mudstone. It is overlain by the Attard Member, consisting of rhodolithic rudstone, floatstone and bioclastic packstone, generally overlying the Maghlaq Member. The Xlendi Member is a bioclastic unit, and comprises the *Scutella* beds and cross-bedded grainstone, which generally lie laterally to and above the Attard Member. The Mara Member consists of outer-ramp wackestone, including bryozoan and *Lepidocyclina* beds, which lie in outer-ramp locations, laterally and down-ramp to the Attard Member (Pedley 1998).

The rhodolith and algal assemblages of the Attard carbonate ramp were studied by Quaranta et al. (2012). The Attard ramp contains geniculate corallines, and free-living non-geniculate branches and rhodoliths in the inner and middle ramp lithofacies (Fig. 8.7a, b).

The rhodolith floatstone to rudstone of the inner ramp shows abundant intergenicula of geniculate coralline algae attributable to the Janieae tribe. The free-living branches are a minor component, and display group I–II branching density according to Bosence (1976, 1983a). Most of the rhodoliths are ellipsoidal, subordinately subspheroidal. Most of their nuclei are skeletal grains, rarely coral fragments. The rhodoliths are monospecific, with a generally laminar inner structure (50 %). Rhodoliths with a columnar/branching structure are common (33.3 %), with branching density of groups III or, rarely, group IV. Laminar/columnar structure is less common (16.7 %). The red algal assemblages are dominated by Sporolithaceae (31.61 %) and Melobesioideae (30.74 %); Mastophoroideae (19.92 %) and Peyssonneliaceae (17.73 %) are subordinate. The melobesioids are represented mainly by *Lithothamnion roveretoi* Airoldi 1932 and subordinately by *Phymatolithon*. Most of the mastophoroids are *Spongites* cf. *albanensis* (Lemoine) (Braga et al. 1993); *Lithoporella melobesioides* (Foslie) (Foslie 1909) and *Karpathia sphaerocellulosa* (Maslov 1962) are rare. Peyssonneliaceans are represented only by *Polystrata alba* (Pfender) (Denizot 1968).

Free-living branching algae are produced mainly by *Lithothamnion roveretoi* and subordinately by *Phymatolithon, Sporolithon and Spongites* cf. *albanensis*.

In the middle ramp, with red algal foraminiferal floatstone to packstone, the intergenicula of the geniculate coralline algae are still present and can be attributed to the Janieae tribe. The free-living branches are abundant and show group I–II branching density.

Most of the rhodoliths are ellipsoidal, rarely subspheroidal and discoidal, with skeletal grains and rare lumpy or fruticose coralline thalli at the nuclei. Rhodoliths are multispecific. The most common algal growth form is laminar/columnar (48 %), passing outward to columnar/branching. Examples of laminar thalli (32 %) are abundant and are commonly replaced outward by laminar/columnar ones. Rhodolith with columnar/branching algal growth forms (20 %) are subordinate, with group III branching density.

The red algal association includes dominant Sporolithaceae (50.25 %) and Melobesioideae (27.56 %) and subordinate Mastophoroideae (12.16 %) and Peyssonneliaceae, represented by *P. alba* (10.03 %). Sporolithaceans include *Sporolithon*. Melobesioids are represented by *L. roveretoi* and *Phymatolithon*. *Lithoporella melobesioides* and *Spongites* cf. *albanensis* predominate among the mastophoroids. The most abundant species and genera of algae among the non-geniculate free-living branches are *L. roveretoi*, *Phymatolithon*, *Sporolithon* and *S.* cf. *albanensis*.

8.2.4 Lessini Shelf and Venetian Basin (Southern Alps)

The Lessini Shelf is a Tertiary carbonate platform overlying the Jurassic Trento Plateau (Bosellini 1989). The eastern margin of the shelf encompasses the Lessini and Berici Hills, located in the northern part of the Adria plate involved in the Apennine (Fig. 8.1), Alpine and Dinarides subduction zone during the Cenozoic (Bosellini 1989; Cuffaro et al. 2010).

During the Rupelian (early Oligocene), shallow-water carbonates occurred over a wide area at the eastern margin of the Lessini Shelf, represented by the calcarenite of the Castelgomberto Formation (Nebelsick et al. 2013). This formation is characterized by well bedded calcarenitic units alternating with marly horizons rich in corals. Bosellini and Trevisani (1992) interpreted this alternation as a response to hydrodynamic changes produced by seal level changes. During highstand periods formed a high-energy open shelf where the currents induced by storms formed large subaqueous dunes made up of skeletal particles. During lowstands, low-energy protected environments and coral bioconstructions developed.

Nebelsick et al. (2013) distinguished four carbonate facies: (1) coralline algal facies (I) dominated by coralline algae, forming rhodoliths and encrusting thalli. The main algal assemblage consists of the genera *Lithoporella*, *Mesophyllum*,

Neogoniolithon, Spongites and Subterraniphyllum in a wackestone to packstone matrix; (2) coralline algal-coral rudstone to floatstone facies (II), in a packstone matrix with abundant coralline red algae, present in all growth forms identified and dominated by *Lithoporella* and *Spongites* and subordinate corals; (3) coralline algal larger foraminiferal facies (III) dominated by coralline algae (mostly *Spongites*, *Mesophyllum*, *Sporolithon*), forming rhodoliths and encrusting thalli, and subordinate larger benthic foraminifera in a packstone matrix; (4) coralline algal debris facies (IV) characterized by abundant coralline algal branches and fragments within a packstone to grainstone matrix.

According to facies distribution, energy conditions and the lack of build-ups, Nebelsick et al. (2013) suggest deposition on a homoclinal ramp setting. To explain the vertical staking pattern of the identified facies, two hypothesis are proposed: (1) sediments were deposited in fairly uniform conditions in a limited water depth, with small variations in ecological parameters, leading to a facies mosaic in an inner ramp environment. In low-energy conditions, soft substrates were colonized by coralline algae, especially *Neogoniolithon*. The coralline algae and foraminiferal facies developed in the seagrass environment; stable substrates created by corals led to complex encrustation sequences in the coralline and coral facies, whereas rare high-energy conditions in channels favored the formation of coralline algal debris facies. (2) The vertical variation of facies reflects sea level changes in an inner ramp setting, with marly sediments deposited in deeper water conditions, whereas shallower and more proximal environment characterized by seagrass meadows and higher-energy facies of the inner ramp.

During the Chattian in the eastern Lessini Shelf, northern Berici Hills and the adjacent Venetian foreland basin, a gently-dipping homoclinal ramp developed, characterized by increasing terrigenous inputs moving eastward (Bassi and Nebelsick 2010). Most of the benthic carbonate production occurred in distal-inner to proximal-middle ramp settings, where larger foraminiferal facies and rhodolith beds developed. Larger foraminifera, represented by two assemblages (dominated respectively by *Archaias* and *Austrotrillina*, and *Operculina*), are most diversified in the shallower settings. Coralline algae predominate in deeper areas (proximal middle ramp), constructing extensive rhodolith beds.

The rhodoliths occur in a packstone matrix dominated by coarse red algal debris. Minor frequent components are encrusting and branching bryozoans, encrusting foraminifera and rare larger foraminifera (*Nummulites*, *Operculina*, lepidocyclinids, *Sphaerogypsina*, *Amphistegina*). Bassi and Nebelsick (2010) identified three rhodolith bed assemblages: (1) the mastophoroid bed dominated by mastophoroid genera (*Spongites*, *Neogoniolithon*, *Lithoporella*) and *Sporolithon* associated with minor melobesioids (*Lithothamnion* and *Mesophyllum*). The rhodoliths are spheroidal, subordinately subellipsoidal and subdiscoidal, with a maximum diameter of up to 12 cm. The rhodoliths are composed both internally and externally by encrusting and warty thalli. (2) The mixed melobesioid/mastophoroid rhodolith bed is characterized by equal amounts of melobesioids (*Lithothamnion*, rare *Mesophyllum*) and mastophoroids (*Spongites*, *Neogoniolithon*). The rhodoliths are spheroidal in shape and up to 13.6 cm in diameter. Corallines and bryozoans dominate the rhodolith nuclei. The rhodoliths are composed of encrusting and rarely warty thalli. (3) The melobesioid rhodolith bed is dominated by melobesioids (*Mesophyllum* dominant, *Lithothamnion*). Mastophoroids (*Spongites*, *Neogoniolithon*, *Lithoporella*) and *Sporolithon* are rare. The rhodoliths are spheroidal (diameter up to 11 cm) and subordinately subellipsoidal. The nuclei are made up of bryozoans, corallines and sediment-filled voids. Encrusting thalli constitute the loosely packed, asymmetrical and symmetrical inner portions of the rhodoliths, passing outwards to encrusting and warty thalli dominated by *Mesophyllum*.

The mastophoroid rhodolith bed is interpreted as having accumulated in a shallow-water, low-turbidity setting, with high water motion and frequent to moderate rhodolith overturning in the proximal mid-ramp environment. The melobesioid/mastophoroid bed which, formed near the inner/middle ramp boundary is characterized by unidirectional bottom currents. This bed is associated with *Operculina*-rich and coralline debris facies. The melobesioid rhodolith bed is interpreted as having formed in a relatively deep-water or less well illuminated setting, below the fair-weather wave base in a hydrodynamic regime where the storm generated currents were able to continuously overturn the rhodoliths. However, high turbidity is indicated by the dominance of low-light preference of the dominating melobesioid assemblages.

8.2.5 Malaguide Complex (Spain)

Red algal rich Oligocene carbonates are scattered in the Iberian Peninsula, and only Braga and Bassi (2011) report coralline algae-rich beds from the Sierra Espuña, part of the Malaguide Complex in the Internal Zones of the Betic Cordillera, S Spain (Fig. 8.1). The Internal Zones represent a crustal wedge emplaced against the southern Iberian margin during the early Miocene. The Malaguide complex is characterized by a Paleogene to lower Miocene succession overlying Mesozoic sediments deposited along the western margin of Tethys (Geel 1973; Martín-Martín 1996).

The Oligocene carbonates were deposited on a carbonate ramp which developed on conglomerate deltaic deposits. The inner ramp was located in a low-energy environment characterized by terrigenous inputs and porcellaneous foraminifer-dominated deposits. The shallow middle ramp shows coral and coralline algal-dominated production, whereas in a deeper middle ramp, rhodoliths and loose-branching, and foliose coralline algae were the main sediment producers, together with larger hyaline benthic foraminifera. The mastophoroids *Neogoniolithon* and *Spongites* dominate the shallower coralline algal assemblages. Melobesioids (*Lithothamnion* and *Mesophyllum*) and sporolithales (*Sporolithon*) are the most abundant components in the various deeper-water assemblages.

8.3 Discussion

8.3.1 Oligocene Shallow-Water Depositional Models and Coralline Algal Palaeoecology

The transition from the warm Eocene to the Oligocene "icehouse" took place during a series of major changes in climate (Fig. 8.8), ice volume and ocean circulation, which influenced the composition and production of the carbonate factory (Nebelsick et al. 2005; Pomar and Hallock 2008; Brandano et al. 2009a, b). In particular, stepwise global cooling gave way to a period in which larger benthic foraminifera suffered a major decline, coral build-up underwent renewed development, and red algae began to dominate, causing a change in the facies association and depositional profile of the Tertiary carbonate platforms (Brandano et al. 2009b).

Classically, the depositional model of the Oligocene Mediterranean is based on the standard model of Buxton and Pedley (1989), which comprises a platform with a ramp depositional profile (Fig. 8.9). The inner ramp association is characterized by peritidal and inner lagoon facies, separated from the mid-ramp association by a grainstone barrier facies. The mid-ramp association is dominated by shallow subtidal facies, consisting of gastropod and foraminifera packstone to wackestone, and by coralline algal rhodolith facies (rhodolith rudstone to floatstone) interfingering with coralgal patch-reefs. The outer ramp association is characterized by the gradual

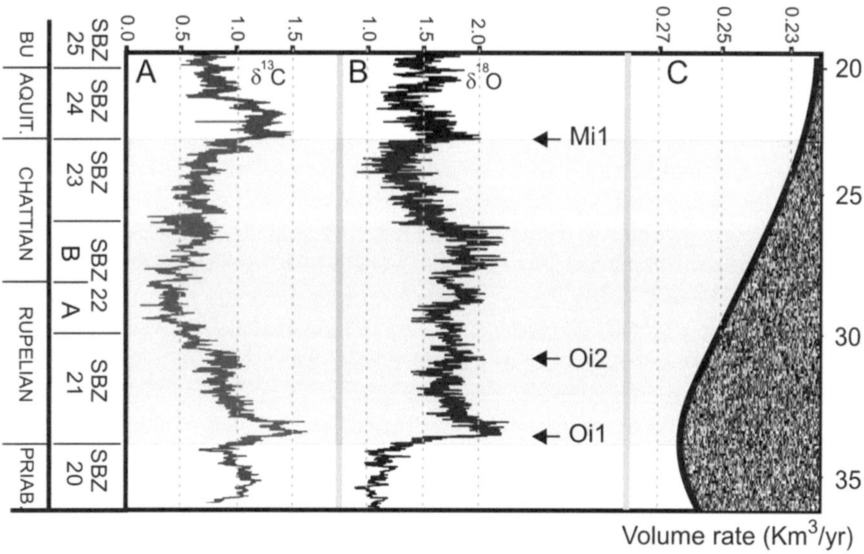

Fig. 8.8 Correlation of SBZ zones (Cahuzac and Poignant 1997) with Oligocene oxygen (**a**) and carbon (**b**) isotope curves (After Cramer et al. 2009; Vandenberghe et al. 2012). (**c**) Decreasing of volume of mantle-derived peridotite reacting with seawater during Oligocene determining a decrease of Mg/Ca ratio

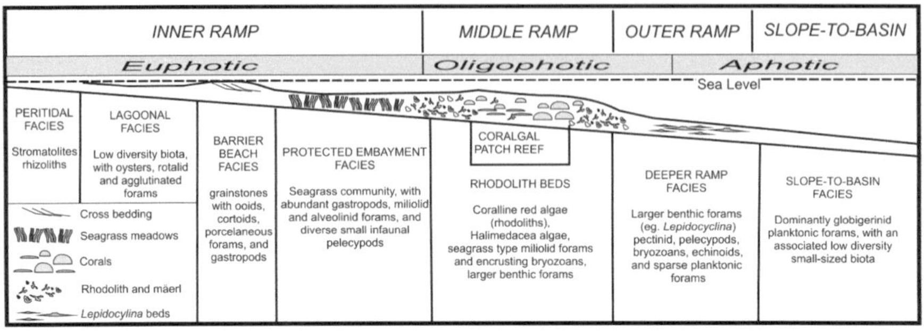

Fig. 8.9 Depositional model and facies association of Oligocene Mediterranean ramp (Modified from Buxton and Pedley 1989; Jorry et al. 2006; Brandano et al. 2009b)

passage from coralline algal to large benthic foraminifer facies (LBF wackestone to packstone) and distal pelagic marl facies (plankton-rich wackestone).

The examples presented in this chapter show some obvious variations according to the varying contexts in the Mediterranean during the Oligocene. Rhodolith deposits formed not only in pure carbonate systems (e.g., Pelagian and Apennine ramps, Rupelian Lessini) but also in mixed carbonate-siliciclastic systems (Chattian Lessini, Veneto, Malaguide). This shows that, as in modern systems, red algal-dominated carbonate production is not hampered by siliciclastic sediments, because the main coralline algal production takes place at a depth and in a depositional environment not reached or less affected by terrigenous inputs (e.g., Betzler et al. 1997; Brandano and Civitelli 2007).

One important characteristic of shallow-water sedimentation during the Oligocene was the expansion of seagrass meadows, as documented in Malta, Apulia and the Lessini (Brandano et al. 2009, 2012; Nebelsick et al. 2013). During the Oligocene, this expansion may have caused the red algal-dominated facies and production in the middle ramp to be collocated in the mesophotic and oligophotic zones. This facies occupies an environment slightly basinward of the seagrass meadows, corresponding to the circalittoral zone of the modern Mediterranean benthic biocenosis of Pérès and Picard (1964), where they form rhodolith bed facies (Bassi and Nebelsick 2010). The rhodolith bed facies generally formed in a hydrodynamic regime which could produce frequent or moderate movement of the rhodoliths, depending on water depth or current energy. Obviously, red algae can also produce sediments in shallower seagrass euphotic environments (infralittoral of Pérès and Picard 1964), as they did in Malta and on the Lessini Shelf, where they are present as geniculate coralline algae, free-living non-geniculate branches and rhodoliths.

The different taxonomic composition of algal assemblages in the various depositional settings of the Oligocene ramp is correlated to the shallow/deep and light penetration gradients existing along the ramp. In modern shallow-water settings, mastophoroids dominate red algal assemblages, with an increase in melobesioids and *Sporolithon* in areas deeper than 10–15 m, becoming dominant in the deep oligophotic setting (Adey et al. 1982; Braga and Davies 1993; Aguirre et al. 2000, this volume).

The red algal assemblages of the Mediterranean Oligocene ramps show a similar general trend, with mastophoroids abundant in shallow-water environments, whereas melobesioids and sporolithales dominate in the deep environment of the oligophotic zone. However, there are some differences. In a general review of the Oligocene coralline algae, Braga et al. (2010) note that, in the northern margin of Tethys, shallower environments are dominated by *Lithothamnion* species, whereas *Mesophyllum* prevails in deep environments. In the Attard ramp of Malta, melobesioids and sporolithales are also abundant in the inner ramp, with the genus *Lithothamnion* dominanting within the melobesioid subfamily; the mastophoroids, which are common in the inner ramp and decrease in the mid-ramp area dominated by sporolithales. Quaranta et al. (2012) interpreted this apparent anomaly as reflecting the dynamics of seagrass-colonized environments, characterized by a spatially complex mosaic in which patches of seagrass are adjacent to erosional blow-outs, each with a different balance between production and loss of sediment (cf Wright and Burgess 2005). High percentage of sediment shed from the inner to the middle ramps is testified by the presence of mastophoroids in the nuclei of rhodoliths. However, there are also some genera of mastophoroids, such as *Neogoniolithon*, which are the main components of the rhodoliths and encrusting thalli which grow on soft substrates in relatively deep, low-energy, low-light conditions, in association with larger benthic foraminifera typical of the mid-ramp oligophotic zone like *Operculina*, *Spyroclypeus* and *Nummulites*.

Another interesting feature of the Mediterranean Oligocene red algal assemblages, at least from reported examples, is the absence of the genus *Lithophyllum*. As reported by Braga and Bassi (2011), this absence may represent an early paleobiogeographic trait of the Oligocene coralline flora.

8.3.2 Climate and Trophic Resources

The biotic assemblages characterizing the Mediterranean carbonate ramp throughout the Oligocene indicate tropical waters. Coral patch reefs developed not only during the Rupelian and Chattian in the Apulia platform (Vecsei et al. 1998; Bosellini et al. 1999; Bosellini 2006) and during the Chattian in the Pelagian domain (Pedley 1998; Brandano et al. 2009a, b), but also in the northern Piedmont Tertiary basin (Quaranta et al. 2009; Silvestri 2012), in Slovenia during the Rupelian (Silvestri et al. 2011), as well as in the Prebetic (Geel 2000; Höntzsch et al. 2013) and in the Malaguide (Braga and Bassi 2011) in southern Spain. The larger Oligocene benthic foraminifers, such as *Lepidocyclina*, *Archaias* and *Borelis*, are thought to have been restricted to tropical regions (cfr. Betzler et al. 1997), with mean temperatures of 18–20 °C during the coldest months of the year (Adams et al. 1990), the vast majority of occurrences being defined by the 25 °C summer isotherm (Wright and Murray 1972). The red algal association shows abundant *Sporolithon*, *Spongites*, *Neogoniolithon* and common *Lithoporella*, mainly tropical genera (Bosence 1983a; Braga and Aguirre 2001).

The paleolatitudinal reconstruction of the Pelagian area shows that, during most of the Oligocene, the Maltese islands were mainly situated south of 30°N (Brandano et al. 2009b). This position is similar to that of the present-day northern Red Sea, where typical tropical to subtropical skeletal assemblages develop, similar to those of the Oligocene ramps (cfr. Piller and Monseur 1990; Riegl and Piller 1999).

Bassi and Nebelsick (2010) showed that the Lessini Shelf and adjacent Venetian Basin during the Chattian lacked the typical larger tropical benthic foraminifera, which are found in more southerly areas of the Tethys. According to the above authors, this is interpreted as the onset of a climatic gradient toward a tropical/warm temperate water transition from the southern to the northern Mediterranean, although the absence of tropical components such as corals or larger tropical benthic foraminifera may also be due to the lack of appropriate carbonate platform conditions.

It is also evident that trophic regimes exerted substantial control on skeletal assemblages (Carannante et al. 1988; Pomar et al. 2004; Wilson and Vecsei 2005; Bassi and Nebelsick 2010). The carbonate sedimentation in the Lessini and Malaguide regions was affected by the proximity of the Alps and the Betics, as nutrients were mainly delivered by high rates of siliciclastic supply deriving from the advancing orogenic wedge (Stefani et al. 2007; Bassi and Nebelsick 2010). Several studies (e.g., Hallock and Schlager 1986; Pomar et al. 2004) have shown that nutrients can not only inhibit the photozoan carbonate factory but also encourage the growth of larger benthic foraminifera and, in particular, of larger porcellaneous types, whereas they have a lesser effect on red coralline algae (Björk et al. 1995; Ichiki et al. 2000; Koop et al. 2001).

The Apulian and Pelagian ramps show evidence of carbonate sedimentation in oligotrophic settings. Zooxanthellate coral patch reefs and the LBF assemblages of these ramps (*Borelis*, *Archaias*, *Peneroplis*, *Sorites*, *Lepidocyclina*) argue against persistently high-nutrient levels, since these foraminifera thrive in oligotrophic waters (Langer and Hottinger 2000). In addition, the absence of low-oxygen foraminiferal assemblages (LOFAs) and the very scarce presence of filter-feeding organisms, such as barnacles and serpulids, which normally appear in environments with relatively abundant organic matter, match these trophic regimes.

8.3.3 Controlling Factors of Red Algae Diffusion During the Oligocene

The Oligocene saw the reorganization of carbonate factories, and the Eocene-Oligocene transition was associated with extinctions and evolutionary turnover and by the shift of geochemical proxies that revealed considerable reorganization in the ocean system. The extinction of various foraminifer genera (e.g., *Hantkenina*, *Discocyclina*, many species of *Nummulites*) was due to rapid environmental changes and cooling, influenced by water mass stratification (Coxall and Pearson 2007). The

Eocene-Oligocene transition was also marked by high surface ocean productivity, indicating an increase in trophic resources. Reduced atmospheric CO_2 and deepening of CCD both indicate a rise in alkalinity, which promoted carbonate production (Van Andel 1975; Pagani et al. 2005). In this context, Hallock (1996) suggested that CO_2 reduction had an important impact on the expansion of zooxanthellate corals, probably favoured by the rising Mg/Ca ratios, which enhanced hypercalcification of aragonite (Stanley and Hardie 1998).

Red algae also became important contributors to carbonate sedimentation during the Oligocene because of rising Mg/Ca ratios (Stanley and Hardie 1998). Ligi et al. (2013) show that the reactions of mantle-derived peridotite with seawater along slow-spreading mid-ocean ridge peridotite combined to river discharge contributed to the post-Cretaceous Mg/Ca increase. The above authors also show that the quantity of available mid-ocean ridge peridotite varies throughout the Cenozoic, peaking at the Eocene-Oligocene boundary and then decreasing until the early Miocene (Fig. 8.7c). As a consequence, oscillations in the Mg/Ca ratio may have influenced the type of foraminiferal-dominated carbonate production in euphotic seagrass meadows, red algae and LBF in the mesophotic and oligophotic zone, and oscillating contributions by corals.

8.4 Concluding Remarks

During the Oligocene the type and loci of carbonate production favoured the diffusion in the Mediterranean area of platform with a ramp depositional profile where the red algae are important sediment contributors of the middle ramp environment. Red algal rich deposits developed also in mixed carbonate-siliciclastic systems of the Chattian Lessini Shelf and in the Malaguide complex showing that red algal carbonate production was not hindered by siliciclastic inputs.

The expansion of seagrass during the Oligocene may have displaced the loci of carbonate production by red algae to the meso- and oligophotic zones, producing rhodolith beds.

The red algal assemblages of the Mediterranean Oligocene ramps show a similar general trend with the mastophoroids abundant in shallow-water environments, whereas melobesioids and sporolithales dominating in deep environments within the mesophotic and oligophotic zone.

The red algal assemblages in the various Mediterranean examples, together with the larger benthic foraminifera and associated coral facies, indicate that carbonate production took place in tropical to subtropical waters in a general context of oligotrophic setting. However, local trophic regimes did not prevent the expansion of red algal facies.

Lastly, during the Oligocene, reduction in atmospheric CO_2, and rising Mg/Ca ratios favoured hypercalcification of aragonite as well as red algal production.

Acknowledgments Grazia Vannucci is gratefully acknowledged for discussion and collaboration through the years. The constructive criticisms and the comments by Juan C. Braga and Julio Aguirre are greatly appreciated. This work was supported by PRIN Project 2010–2011. Gabriel Walton is thanked for the revision of the English text.

References

Accordi B, Devoto G, La Monica GB, Praturlon A, Sirna G, Zalaffi M (1967) Il Neogene nell'Appennino laziale-abruzzese. Commitee Mediterranean Neogene Stratigraphy Proc IV Session Bologna. Giorn Geol 35:235–268

Adams CG, Lee DE, Rosen BR (1990) Conflicting isotopic and biotic evidence for tropical sea-surface temperatures during the Tertiary. Palaeogeogr Palaeoclimatol Palaeoecol 77:289–313

Adey WH, Townsend RA, Boykins WT (1982) The crustose coralline algae (Rhodophyta: Corallinaceae) of the Hawaiian Islands. Smithson Contrib Mar Sci 15:1–74

Aguirre J, Riding R, Braga JC (2000) Diversity of coralline red algae: origination and extinction patterns from the Early Cretaceous to the Pleistocene. Paleobiology 26:651–667

Airoldi M (1932) Contributo allo studio delle Corallinacee del terziario italiano. I. Le Corallinacee dell'Oligocene ligure-piemontese. Palaeontogr Ital 33:55–83

Allen MB, Armstrong HA (2008) Arabia-Eurasia collision and the forcing of mid-Cenozoic global cooling. Palaeogeogr Palaeoclimatol Palaeoecol 265:52–58

Bassi D, Nebelsick JH (2010) Components, facies and ramps: redefining Upper Oligocene shallow water carbonates using coralline red algae and larger foraminifera (Venetian area, northeast Italy). Palaeogeogr Palaeoclimatol Palaeoecol 295:258–280

Benedetti A, Di Carlo M, Pignatti JS (2010) Embryo size variation in larger foraminiferal lineages: stratigraphy versus paleoecology in *Nephrolepidina praemarginata* (R. Douvillé, 1908) from the Majella Mt. (Central Apennines). J Med Earth Sci 2:19–29

Bernoulli D (2001) Mesozoic–tertiary carbonate platforms, slopes and basin of the external Apennines and Sicily. In: Vai GB, Martini IP (eds) Anatomy of an orogen: the Apennines and adjacent Mediterranean basins. Kluwer Academic Publishers, Dordrecht, pp 307–326

Betzler C, Brachert TC, Nebelsick J (1997) The warm temperate carbonate province – a review of facies, zonations, and delimitations. Cour Forschungsinst Senckenb 201:83–99

Bianchi F, Carbone S, Grasso M, Invernizzi G, Lentini F, Longaretti G, Merlini S, Mostardini F (1987) Sicilia orientale: profilo geologico Nebrodi-Iblei. Mem Soc Geol Ital 38:429–458

Bishop WF, Debono G (1996) Hydrocarbon geology of southern offshore Malta and surrounding regions. J Pet Geol 19(2):129–160

Björk M, Mohammed SM, Bjöklund M, Semesi A (1995) Coralline algae, important coral-reef builders threatened by pollution. Ambio 24:502–505

Bosellini A (1989) Dynamics of tethyan carbonate platforms. In: Crevello PD, Wilson JL, Sarc JF, Read JF (eds) Controls on carbonate platform and basin platform. S.E.P.M., spec. publ. 44:3–13, 13 Figs., Sepm (Society for Sedimentary Geology), Tulsa

Bosellini FR (2006) Biotic changes and their control on Oligocene-Miocene reefs: a case study from the Apulia platform margin (southern Italy). Palaeogeogr Palaeoclimatol Palaeoecol 241:393–409

Bosellini FR, Trevisani E (1992) Coral facies and cyclicity in the Castelgomberto Limestone (early Oligocene, eastern Lessini Mountains, northern Italy). Riv Ital Paleontol Stratigr 98(3):339–352

Bosellini A, Parente M (1994) The Apulia platform margin in the Salento peninsula (southern Italy). Giorn Geol 56:167–177

Bosellini FR, Trevisani E (1992) Coral facies and cyclicity in the Castelgomberto Limestone (early Oligocene, eastern Lessini Mountains, northern Italy). Riv Ital Paleontol Stratigr 98(3):339–352
Bosellini A, Bosellini FR, Colalongo ML, Parente M, Russo A, Vescogni A (1999) Stratigraphic architecture of the Salento coast from Capo d'Otranto to S. Maria di Leuca (Apulia, southern Italy). Riv Ital Paleontol Stratigr 105:397–416
Bosence DWJ (1976) Ecological studies on two unattached coralline algae from western Ireland. Palaeontology 19(2):365–395
Bosence DWJ (1983a) Description and classification of rhodoliths (rhodoids, rhodolites). In: Peryt TM (ed) Coated grains. Springer, Berlin, pp 217–224
Bosence DWJ (1983b) Coralline algae from the Miocene of Malta. Palaeontology 26(1):147–173
Bosence DWJ, Pedley HM (1982) Sedimentology and palaeoecology of a Miocene coralline algal biostrome from the Maltese Islands. Palaeogeogr Palaeoclimatol Palaeoecol 38:9–43
Braga JC, Aguirre J (2001) Coralline algal assemblages in Upper Neogene reef and temperate carbonates in southern Spain. Palaeogeogr Palaeoclimatol Palaeoecol 175:27–41
Braga JC, Bassi D (2011) Facies and coralline algae from Oligocene limestones in the Malaguide Complex (SE Spain). Ann Naturhist Mus Wien Ser A 113:291–308
Braga JC, Davies PJ (1993) Coralline algal distribution in one tree reef (Southern Great Barrier Reef, NE Australia). In: International Society for Reef Studies 1st European Regional Meet, Vienna, Abstr 9
Braga JC, Bosence DWJ, Steneck RS (1993) New anatomical characters in fossil coralline algae and their taxonomic implications. Palaeontology 36(3):535–547
Braga JC, Bassi D, Piller W (2010) Palaeoenviromental significance of Oligocene–Miocene coralline red algae – a review. In: Mutti M, Piller W, Betzler C (eds) Oligocene – Miocene carbonate systems. International Association of Sedimentologists, Spec Publ. Wiley-Blackwell, Chichester, 42:165–182
Brandano M, Civitelli G (2007) Non-seagrass meadow sedimentary facies of the Pontinian Islands, Tyrrhenian Sea: a modern example of mixed carbonate–siliciclastic sedimentation. Sed Geol 201:286–301
Brandano M, Corda L (2002) Nutrient, sea level and tectonics constrains for the facies architecture of Miocene carbonate ramp in Central Italy. Terra Nova 14:257–262
Brandano M, Corda L (2011) The lower Miocene spongolitic sequence of the central apennines:a record of the Burdigalian siliceous event in the central Mediterranean. Ann Naturhist Mus Wien Ser A 113:145–166
Brandano M, Cornacchia I, Raffi I, Tomassetti L (2016) The Oligocene-Miocene stratigraphic evolution of the Majella carbonate platform (Central Apennines, Italy). Sediment Geol 333:1–14
Brandano M, Frezza V, Tomassetti L, Pedley M, Matteucci R (2009a) Facies analysis and palaeoenvironmental interpretation of the late Oligocene attard member (lower coralline limestone formation), Malta. Sedimentology 56:138–1158
Brandano M, Frezza V, Tomassetti L, Cuffaro M (2009b) Heterozoan carbonates in oligotrophic tropical waters: the Attard member of the lower coralline limestone formation (Upper Oligocene, Malta). Palaeogeogr Palaeoclimatol Palaeoecol 274:54–63
Brandano M, Morsilli M, Vannucci G, Parente M, Bosellini F, Mateu-Vicens G (2010) Rhodolith-rich lithofacies of the Porto Badisco Calcarenites (upper Chattian, Salento, southern Italy). Ital J Geosci 129(1):119–131
Brandano M, Lipparini L, Campagnoni V, Tomassetti L (2012) Downslope-migrating large dunes in the Chattian carbonate ramp of the Majella Mountains (Central Apennines, Italy). Sediment Geol 255–256:29–41
Buxton MWN, Pedley HM (1989) A standardized model for Tethyan Tertiary carbonate ramps. J Geol Soc Lond 146:746–748
Cahuzac B, Poignant A (1997) Essai de biozonation de l'Oligo-Miocène dans les bassins européens à l'aide des grands foraminifères néritiques. Bull Soc Geol Fr 168:155–169

Carannante G, Esteban M, Milliam JD, Simone L (1988) Carbonate lithofacies as paleolatitude indicators: problems and limitations. Sed Geol 60:333–346
Carbone S, Grasso M, Lentini F (1982) Considerazione sull'evoluzione geodinamica della Sicilia sud-orientale dal Cretaceo al Quaternario. Mem Soc Geol Ital 24:367–386
Carboni MG, Civitelli G, Corda L, Esu D, Matteucci R, Pallini G, Schiavinotto F, Ventura G (1982) Sedimenti spongolitici del Miocene inferiore dell'Appennino centrale: un inquadramento preliminare. Geol Romana 21:529–543
Carminati E, Corda L, Mariotti G, Brandano M (2007) Tectonic control on the architecture of a Miocene carbonate ramp in the Central Apennines (Italy): insights from facies and backstripping analyses. Sediment Geol 198:233–254
Carminati E, Lustrino M, Cuffaro M, Doglioni C (2010) Tectonics, magmatism and geodynamics of Italy: what we know and what we imagine. In: The geology of Italy: tectonics and life along plate margins. Beltrando M, Peccerillo A, Mattei M, Conticelli S, Doglioni C (eds) J Virt Expl Electronic Edition 36, paper 9.doi: 10.3809/jvirtex.2010.00226
Casero P, Rour F, Moretti I, Endignoux L, Muller C, Sage L, Vially R (1988) Neogene geodynamic evolution of the Southern Appennines. Mem Soc Geol Ital 41:109–120
Cipollari P, Cosentino D (1995) Miocene unconformities in the Central Apennines: geodynamic significance and sedimentary basin evolution. Tectonophysics 252:375–389
Civitelli G, Brandano M (2005) Atlante delle litofacies e modello deposizionale dei calcari a briozoi e Litotamni nella Piattaforma carbonatica laziale-abruzzese. Boll Soc Geol Ital 124:611–643
Corti G, Doglioni C, Cuffaro M, Innocenti F, Manetti P (2006) Coexisting geodynamic processes in the Sicily Channel. In: Dilek Y, Pavlides S (eds) Postcollisional tectonics and magmatism in the Mediterranean region and Asia, Geol Soc Am Sp Paper 409. Geological Society of America, Boulder, pp 83–96. doi:10.1130/2006.2409(05)
Coxall HK, Pearson PN (2007) The Eocene-Oligocene transition. In: Williams M, Haywood AM, Gregory FJ, Schmidt DN (eds) Deep-time perspectives on climate change: marrying the signal from computer models and biological proxies, The Micropalaeontological Society Spec. Publ. The Geological Society, London, pp 351–387
Coxall HK, Wilson PA, Pälike H, Lear CH, Backman J (2005) Rapid stepwise onset of Antarctic glaciation and deeper calcite compensation in the Pacific Ocean. Nature 433:53–57
Cramer BS, Toggweiler JR, Wright JD, Katz ME, Miller KG (2009) Ocean overturning since the Late Cretaceous: inferences from a new benthic foraminiferal isotope compilation. Paleoceanography 24(4), PA4216. http://dx.doi.org/10.1029/2008pa001683
Cuffaro M, Riguzzi F, Scrocca D, Antonioli F, Carminati E, Livani M, Doglioni C (2010) On the geodynamics of the northern Adriatic plate. Rend Fis Acc Lincei 21(Suppl 1):S253–S279. doi:10.1007/s12210-010-0098-9
DeConto RM, Pollard D (2003) Rapid Cenozoic glaciation of Antarctica induced by declining atmospheric CO_2. Nature 421:245–249
Denizot M (1968) Les algues floridées encroûtantes (à l'exclusion des Corallinacées). Thèse, Laboratoire de cryptogamie, Muséum national d'Histoire naturelle, Paris, 310 p
Dupont-Nivet G, Krijgsman W, Langereis CG, Abels HA, Dai S, Fang XM (2007) Tibetan plateau aridification linked to global cooling at the Eocene–Oligocene transition. Nature 445:635–638
Foslie M (1909) Algolgiske notiser. VI. K Nor Vidensk Selsk Skr 2:1–63
Geel T (1973) The geology of the Betic of Málaga, the Subbetic, and the zone between these two units in the Vélez Rubio area (Southern Spain). GUA Pap Geol 5:1–131
Geel T (2000) Recognition of stratigraphic sequences in carbonate platform and slope deposits: empirical models based on microfacies analysis of Paleogene deposits in southeastern Spain. Palaeogeogr Palaeoclimatol Palaeoecol 155:211–238
Grasso M, Torelli L, Mazzoldi G (1999) Cretaceous–Palaeogene sedimentation patterns and structural evolution of the Tunisian shelf, offshore the Pelagian Islands (Central Mediterranean). Tectonophysics 315:235–250

Hallock P (1996) Reefs and reef limestones in Earth history. In: Birkeland C (ed) Life and death of coral reefs. Chapman and Hall, New York, pp 13–42
Hallock P, Schlager W (1986) Nutrient excess and the demise of coral reefs and carbonate platforms. Palaios 1:389–398
Höntzsch S, Scheibner C, Brock JP, Kuss J (2013) Circum-Tethyan carbonate platform evolution during the Palaeogene: the Prebetic platform as a test for climatically controlled facies shifts. Turk J Earth Sci 22:891–918
Ichiki S, Mizuta H, Yamamoto H (2000) Effects of irradiance, water temperature and nutrients on the growth of sporelings of the crustose coralline alga *Lithophyllum yessoense* Fosile (Corallinales, Rhodophyceae). Phycol Res 48:115–120
Jorry SJ, Hasler C, Davaud E (2006) Hydrodynamic behavior of *Nummulites*: implications for depositional models. Facies 52:221–235
Klett TR (2001) Total petroleum system of the Pelagian Province, Tunisia, Libya, and Malta-The Bou Dabbous-Tertiary and Jurassic-Cretaceous composite. U S Geol Surv Bull 2202-D:1–27
Koop K, Booth D, Broadbent A, Brodie J, Bucher D, Capone D, Coll J, Dennison W, Erdmann M, Harrison P, Hoegh-Guildberg O, Hutchings P, Jones GB, Larkum AWD, O'Neil J, Steven A, Tentori E, Ward S, Williamson J, Yellowlees D (2001) Encore: the effect of nutrient enrichment on coral reefs. Synthesis of results and conclusions. Mar Pollut Bull 42:91–120
Langer M, Hottinger L (2000) Biogeography of selected "larger" foraminifera. Micropaleontology 46(suppl 1):57–86
Ligi M, Bonatti E, Cuffaro M, Brunelli D (2013) Post-Mesozoic rapid increase of seawater Mg/Ca due to enhanced mantle-seawater interaction. Sci Rep 3:2752. doi:10.1038/srep02752
Martín-Martín M (1996) El Terciario del Dominio Maláguide en Sierra Espuña (Cordillera Bética Oriental, SE de España). Estratigrafía y Evolución Palogeográfica. Thesis, University of Granada, Granada, 297pp
Maslov VP (1962) Fossil red algae of USSR and their connection with facies. Trudy Inst Geol Akad Nauk SSSR 53:1–221 (in Russian)
Miller KG, Fairbanks RA, Mountain GS (1987) Tertiary oxygen isotope synthesis, sealevel history and continental margin erosion. Paleoceanography 2:1–19
Mostardini F, Merlini S (1986) Appennino centro-meridionale: sezioni Geologiche e proposta di modello strutturale. Mem Soc Geol Ital 35:177–202
Mutti M, Bernoulli D, Stille P (1997) Temperate carbonate platform drowning linked to Miocene oceanographic events: Maiella platforms margin, Italy. Terra Nova 9:122–125
Mutti M, Bernoulli D, Spezzaferri S, Stille P (1999) Lower and Middle Miocene carbonate facies in the Central Mediterranean: the impact of paleoceanography on sequence stratigraphy. In: Harris PMM et al (eds) Advances in carbonate sequence stratigraphy: application to reservoirs, outcrops and models, SEPM Spec Publ Soc Econ Paleont Miner 63. Sepm (Society for Sedimentary Geology), Tulsa, pp 371–384
Nardin M, Rossi D (1966) Condizioni strutturali della zona compresa nel Foglio Otranto (Provincia di Lecce). Mem Museo Civico St Nat 14:415–430, Verona
Nebelsick JH, Rasser M, Bassi D (2005) Facies dynamics in Eocene to Oligocene circumalpine carbonates. Facies 51:197–216
Nebelsick JH, Bassi D, Lempp J (2013) Tracking paleoenvironmental changes in coralline algal-dominated carbonates of the Lower Oligocene Calcareniti di Castelgomberto formation (Monti Berici, Italy). Facies 59:133–148. doi:10.1007/s10347-012-0349-6
Pagani M, Zachos JC, Freeman KH, Tipple B, Bohaty S (2005) Marked decline in atmospheric carbon dioxide concentrations during the Paleogene. Science 309:600–603
Parente M (1994) A revised stratigraphy of the upper Cretaceous to Oligocene units from south-eastern Salento (Apulia, southern Italy). Boll Soc Paleontol Ital 33:155–170
Pedley HM (1978) A new lithostratigraphical and palaeoenvinronmental interpretation for the coralline limestone formations (Miocene) of the Maltese Islands. Overseas Geol Miner Resour (IGS) 54:17 pp

Pedley HM (1998) A review of sediment distributions and processes in Oligo-Miocene ramps of southern Italy and Malta (Mediterranean divide). In: Wrigth VP, Burchette TP (eds) Carbonate ramps. Geol Soc London Geol Soc Spec Publ. The Geological Society, London, 149:163–179
Pekar SF, Christie-Blick N, Kominz MA, Miller KG (2002) Calibration between eustatic estimates from backstripping and oxygen isotopic records for the Oligocene. Geology 30:903–906
Pekar SF, DeConto RM, Harwood DM (2006) Resolving a late Oligocene conundrum: deep-sea warming and Antarctic glaciation. Palaeogeogr Palaeoclimatol Palaeoecol 231:29–40
Pérès JM, Picard J (1964) Nouveau manuel de bionomie benthique de la Mer Méditerranée. Rec Trav Stn Mar Endoume Marseille Fasc Hors Ser Suppl 31(47):5–138
Piller WE, Mansour AM (1990) The northern Bay of Safaga (Red Sea, Egypt): an actuopalaeontological approach II. Sediment analyses and sedimentary facies. Beitr PalaKont OG Sterr 16:1–102
Pomar L (2001) Types of carbonate platforms: a genetic approach. Basin Res 13:313–334
Pomar L, Hallock P (2008) Changes in coral-reef structure through the Miocene in the Mediterranean province: adaptive vs. environmental influence. Geology 35:899–902
Pomar L, Brandano M, Westphal H (2004) Environmental factors influencing skeletal grain sediment associations: a critical review of Miocene examples from the western Mediterranean. Sedimentology 51:627–651
Quaranta F, Piazza M, Vannucci G (2009) Climatic and tectonic control on the distribution of the Oligocene reefs of the Tertiary Piedmont Basin (NW Italy). Boll Soc Geol Ital 128(2):587–591
Quaranta F, Tomassetti L, Vannucci G, Brandano M (2012) Coralline algae as environmental indicators: a case study from the Attard member (Chattian, Malta). Geodiversitas 34(1):151–166, dx.doi.org/10.5252/g2012n1a9
Reuter M, Piller WE, Brandano M, Harzhauser M (2013) Correlating Mediterranean shallow water deposits with global Oligocene–Miocene stratigraphy and oceanic events. Glob Planet Chang 111:226–236. doi:10.1016/j.gloplacha.2013.09.018
Riegl B, Piller WE (1999) Coral frameworks revisited: reefs and coral carpets in the northern Red Sea. Coral Reefs 18:241–253
Silvestri G (2012) Paleoecological reconstruction of Cenozoic turbid-water coral assemblages through taphonomic and biofacies analysis. PhD thesis, Università degli studi di Modena e Reggio Emilia (Italy), 150 pp
Silvestri G, Bosellini FR, Nebelsick JH (2011) Microtaphofacies analysis of lower Oligocene turbid-water coral assemblages. Palaios 26:805–820
Stanley SM, Hardie LA (1998) Secular oscillations in the carbonate mineralogy of reef-building and sediment-producing organisms driven by tectonically forced shifts in seawater chemistry. Palaeogeogr Palaeoclimatol Palaeoecol 144:3–19
Stefani C, Fellin MG, Zattin M, Zuffa GG, Dalmonte C, Mancin N, Zanferrari A (2007) Provenance and palaeogeographic evolution in a multi-source foreland: the Cenozoic Venetian-Friulian Basin (NE Italy). J Sed Res 77:867–887
Thomas E, Zachos JC, Bralower TJ (2000) Deep-sea environments on a warm earth: latest Paleocene–early Eocene. In: Huber BT, MacLeod KG, Wing SL (eds) Warm climates in earth history. Cambridge University Press, New York, pp 132–160
Tripati A, Backman J, Elderfield H, Ferretti P (2005) Eocene bipolar glaciation associated with global carbon cycle changes. Nature 436:341–346
Van Andel TH (1975) Mesozoic/Cenozoic calcite compensation depth and the global distribution of calcareous sediments. Earth Planet Sci Lett 26:187–194
Vandenberghe N, Hilgen FJ, Speijer RP, Ogg JG, Gradstein FM, Hammer O, Hollis CJ, Hooker JJ (2012) Chapter 28 – The paleogene period. In: Gradstein FM, Ogg JG, Schmitz M, Ogg G (eds) The geologic time scale. Elsevier, Boston, pp 855–921. http://dx.doi.org/10.1016/B978-0-444-59425-9.00028-7

Vecsei A, Moussavian E (1997) Paleocene reefs on the Maiella platform margin, Italy: an example of the effects of the Cretaceous/Tertiary boundary events on reefs and carbonate platforms. Facies 36:123–140

Vecsei A, Sanders DGK (1999) Facies analysis and sequence stratigraphy of a Miocene warm–temperate carbonate ramp, Montagna della Maiella, Italy. Sediment Geol 123:103–127

Vecsei A, Sanders DGK, Bernoulli D, Eberli GP, Pignatti JS (1998) Cretaceous to Miocene sequence stratigraphy and evolution of the Maiella carbonate platform margin, Italy. Mesozoic Cenozoic Seq Stratigr Eur Basin: SEPM Spec Publ 60:53–74

Vezzani L, Festa A, Ghisetti F (2010) Geology and tectonic evolution of the central-southern Apennines, Italy. Geol Soc Am Spec Publ 469:1–58

Wilson MEJ, Vecsei A (2005) The apparent paradox of abundant foramol facies in low latitudes: their environmental significance and effect on platform development. Earth Sci Rev 69:133–168

Wright VP, Burgess PM (2005) The carbonate factory continuum, facies mosaics and microfacies: an appraisal of some of the key concepts underpinning carbonate sedimentology. Facies 51:17–23

Wright CA, Murray JW (1972) Comparisons of modern and Palaeogene foraminiferid distributions and their environmental implications. Mem Bur Rech Geol Min (BRGM) Fr 79:87–96

Zachos J, Pagani M, Sloan L, Thomas E, Billups K (2001) Trends, rhythms, and aberrations in global climate 65 ma to present. Science 292:686–693

Zanazzi A, Kohn MJ, MacFadden BJ, Terry DO (2007) Large temperature drop across the Eocene–Oligocene transition in central North America. Nature 445:639–642

Chapter 9
Taphonomic Range and Sedimentary Dynamics of Modern and Fossil Rhodolith Beds: Macaronesian Realm (North Atlantic Ocean)

Markes E. Johnson, Jorge Ledesma-Vázquez, Ricardo S. Ramalho, Carlos M. da Silva, Ana Cristina Rebelo, Ana Santos, B. Gudveig Baarli, Eduardo Mayoral, and Mário Cachão

9.1 Introduction

The modern collective name Macaronesia applies to clusters of volcanic islands in the northeast Atlantic Ocean off the coasts of Western Europe and Africa. This geographic realm includes (from north to south) the Azores (Portugal), Madeira and Selvagens islands (Portugal), the Canaries (Spain) and Cape Verde (an independent

M.E. Johnson (✉) • B.G. Baarli
Department of Geosciences, Williams College, Williamstown, MA 01267, USA
e-mail: markes.e.johnson@williams.edu

J. Ledesma-Vázquez
Facultad de Ciencias Marinas, Universidad Autónoma de Baja California, Ensenada, BC 22800, Mexico

R.S. Ramalho
School of Earth Sciences, University of Bristol, Wills Memorial Building, Queens's Road, Bristol BS8 1RJ, UK

Lamont-Doherty Earth Observatory at Columbia University, Comer Geochemistry Building, P.O. Box 1000, Palisades, NY 10964, USA

C.M. da Silva • M. Cachão
Departamento de Geologia e Centro de Geologia, Faculdade de Ciências da Universidade de Lisboa, Campo Grande, 1749-016 Lisbon, Portugal

A.C. Rebelo
Departamento de Biologia, Universidade dos Açores, Campus de Ponta Delgada, 9501-801 Ponta Delgada, Açores, Portugal

Staatliches Museum für Naturkunde, Rosenstein 1, Stuttgart, Germany

A. Santos • E. Mayoral
Departamento de Geodinámica y Paleontología, Facultad de Ciencias Experimentales, Universidad de Huelva, Campus de El Carmen, Avd. 3 de Marzo, s/n, 21071 Huelva, Spain

R. Riosmena-Rodríguez et al. (eds.), *Rhodolith/Maërl Beds: A Global Perspective*, Coastal Research Library 15, DOI 10.1007/978-3-319-29315-8_9

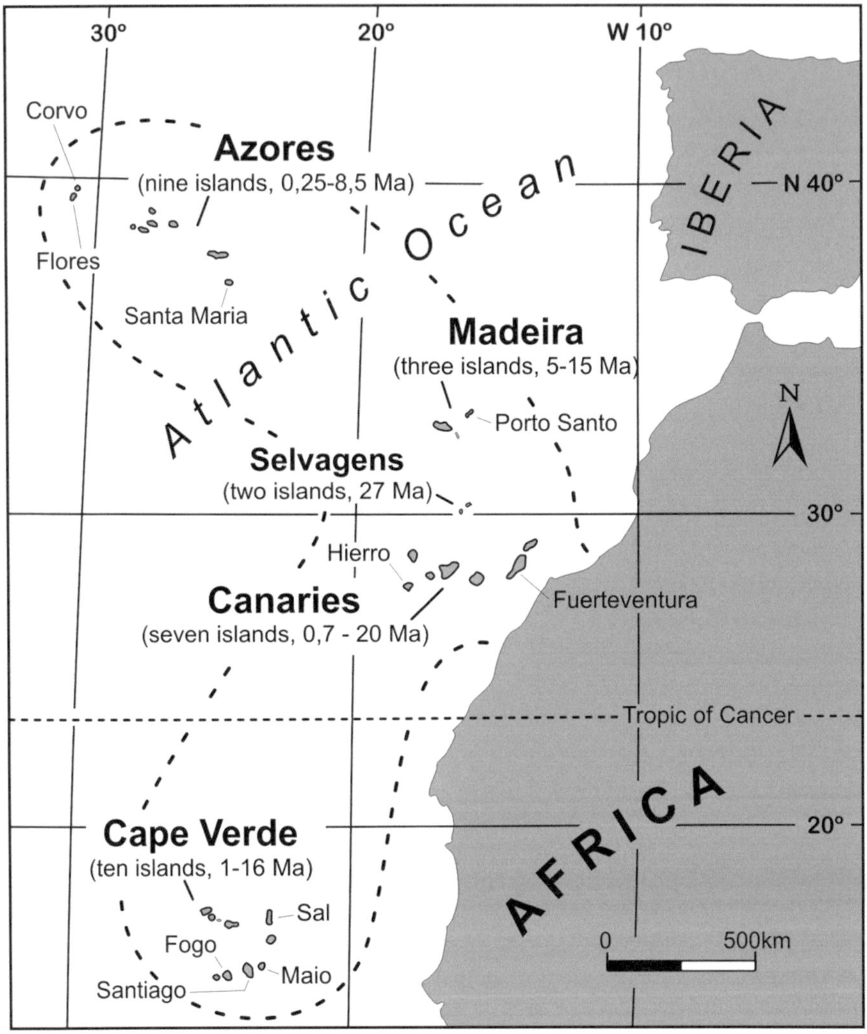

Fig. 9.1 Distribution of volcanic archipelagos in the Macaronesian realm of the North Atlantic Ocean, including the Azores, Madeira, Selvagens, Canaries, and Cape Verde islands

republic). Five archipelagos are located between 15° and 40° N latitude, with distances from the European or African continents varying from 100 km (Fuerteventura, in the Canaries) to 1,900 km (Flores and Corvo, in the Azores) and a total latitudinal span of almost 2,900 km (Fig. 9.1). Living rhodoliths are present in shallow waters around one or more islands in all island groups. Fossil rhodoliths dating from the middle Miocene to Pleistocene are preserved in carbonate strata on the oldest islands in each of these archipelagos.

This contribution is parted into two main subdivisions that feature: (1) a description of modern coastal deposits with substantial rhodolith debris as related to shelf development based on an island example drawn from each of the archipelagos in the Macaronesian realm, and (2) a review and comparison of fossil rhodolith beds representing a wide range of taphofacies reported in the literature together with new material previously unreported. Contemporary deposits from some of the older Macaronesian islands bear directly on the interpretation of biostratinomy and the extraordinary range of taphofacies to which rhodoliths conform. To the extent possible, determination of taphonomic grade conforms to the categories defined by Brandt (1989) on the basis of rhodolith breakage and corrosion. Our model follows the plan of Johnson et al. (2012), which in turn builds on the legacy begun by Darwin (1839, 1844) as one the earliest students of rhodoliths. Taphofacies vary from fossil rhodoliths preserved in growth position from a Madeira islet (Santos et al. 2012) to dune sands dominated by rhodolith debris in the Cape Verdes (Johnson et al. 2013). The specific goals of this survey are to illuminate the many intermediate taphonomic modes and sedimentary combinations between these end-points. The entire array is set in the context of physical processes related to atmospheric and oceanographic circulation patterns that developed over the Macaronesian realm from the middle Miocene to the present day, while also in the context of geomorphology favorable to the evolution of insular shelves.

9.2 Geographic Setting

With a total emerged area of 7,496 km^2, the Canary Islands are nearly twice the area of Cape Verde (4,020 km^2), almost three times that of the Azores (2,324 km^2), and about ten times larger than the Madeiran island group (824 km^2). The small Selvagens cluster registers an emerged area of only 2.7 km^2. When compared with their original size, the present-day area and topography of the Macaronesian islands is strongly determined by geological age. Younger islands, such as the main island of Madeira with an estimated age of approximately 5 Ma, tend to preserve higher reliefs (highest peak on Madeira: Pico Ruivo, 1,861 m) and narrower shelves. In comparison, older islands such as Porto Santo with an estimated age of about 15 Ma, show a more modest topography (highest peak on Porto Santo: Pico do Facho, 516 m) and a wider shelf. The fossil record of rhodoliths across Macaronesia is more extensive on older islands with well-developed shelves that reflect histories of sea-level change associated with eustasy and tectonic uplift. Recent volcanic activity is known from Cumbre Vieja (1,949 m) on La Palma and the island of El Hierro (1,500 m) in the Canary Islands, as well as Fogo (3,000 m) in the Cape Verde Islands. Modern rhodoliths are still very scarce in the coastal waters around the active volcanoes, as indicated by the prevalence of black-sand beaches.

9.3 Foundational Studies by Charles Darwin

Based on original observations made during the 1832 visit and his reappraisal during a shorter return visit in 1836, Darwin contributed to the earliest studies on former rocky shores associated with rhodoliths on volcanic islands (Herbert 2005; Johnson et al. 2012; Baarli et al. 2013). Darwin (1844) reported abundant fossil oysters, limpets, and turritellid gastropods, as well as echinoids from the white carbonate sedimentary beds bracketed stratigraphically by basalt formations. He wrote the following description regarding these coastal exposures (Darwin 1844, 4):

> Numerous white balls appearing like pisolitic concretions, from the size of a walnut to that of an apple, are embedded in this deposit; they usually have a small pebble in their centers. Although so like concretions, a close examination convinced me that they were Nulliporæ, retaining their proper forms, but with their surfaces slightly abraded (…).

The "white balls" are actually rhodoliths made up of coralline red algae, as has been verified *in loco* (Johnson et al. 2012; Baarli et al. 2013) and in the figures provided by Herbert (2005: his figs. 5.5–5.6) showing specimens collected by Darwin on Quail Island and currently deposited in the Sedgwick Museum (University of Cambridge).

9.4 Environmental Constraints on Living Rhodoliths

Due to the latitudinal extent of Macaronesia and variations in island size, topographic relief, and morphology, the realm is subject to a broad set of climatic conditions that range from a sub-Saharan climate in the eastern Canary islands, to Mediterranean-like conditions in the Madeira, Selvagens, and western Canary islands, to the typically arid-tropical climate of the Cape Verde islands. As living rhodoliths show a wide latitudinal tolerance to climate (Foster 2001), shelf extension and the bathymetric gradient around volcanic islands in the Macaronesian realm are more limiting constraints tied to variations in water turbulence associated with regional wind and wave patterns.

In terms of geomorphology, older islands with no recent volcanism have a more complex history of uplift that makes for more complex marine shelves (Ramalho et al. 2013). For example, islands such as Fuerteventura (Canaries), Sal, Boa Vista, and Maio (Cape Verdes), Porto Santo (Madeira) and Santa Maria (Azores) all exhibit wider and shallower shelves than most other younger islands belonging to the same archipelagos. Many of the Macaronesian islands are strongly impacted by the northeast trade winds. Due to differential erosion, shelf width varies in accordance with coastal exposure and increased wave energy. As such, windward shores tend to exhibit wider, sediment-stripped shelves. The persistent winds that reach these islands represent a characteristic physical trait that Darwin (1839, 1844) made explicit in his commentary regarding Santiago Island. On the whole, the Cape Verde archipelago is subjected to winds that today commonly register 5–6 on the Beaufort

Scale, which equates to wind speeds between 8 and 10.8 m/s (Brand 2011). The same wind field over the Canaries is no less significant, where persistent northeast trade winds commonly attain wind speeds of 5 on the Beaufort scale, or 8–10.7 m/s (Brand 2011). Winds are more variable over the Madeira island group, but the northeast trade winds commonly attain speeds reaching 3 on the Beaufort scale, or 5.4 m/s. In the Azores, the prevailing winds are the westerlies (Hontoria et al. 2012). According to Rusu and Soares (2012), roughly 75 % of the waves impacting the Azores arrive from the west-northwest.

9.5 Models for Contemporary Rhodolith Fecundity

A comprehensive survey of rhodolith fecundity throughout all islands in the Macaronesian realm is beyond the scope of this contribution. A general appraisal of rhodolith banks is possible, however, based on contemporary coastal deposits formed by rhodoliths with regard to certain islands selected as models for each of the five archipelagos. Our survey of modern coastal settings includes extensive rock platforms explored during low tides. Supratidal deposits with abundant rhodoliths may be separated from offshore zones with living rhodoliths by clean rock platforms that typically exceed 130 m in width. This has implications for the physical relocation of rhodoliths and their mechanical disintegration. Hence, this overview offers a practical precursor to the integration of a taphonomic model for studies in rhodolith biostratinomy.

9.5.1 Sal (Cape Verde Islands)

Sal is a windward island in the northeast sector of the Cape Verde archipelago (Figs. 9.1, 9.2, and 9.3). It is built on a basalt core complex that dates to 25 Ma in age (Tores et al. 2002). With a coastal circumference of only 83 km, it is also one of the smaller islands. The 50-m isobath is closest to shore around the north end of the island, where strong surf is constant, but expands to a distance 5 km offshore around the southern half of the island. Morri and Bianchi (1996), found corals living in the relative shelter of the northeastern coast between Palmeira and Buracona at a maximum depth of 20 m, where rhodoliths are absent. During a coastal tour in March 2013, no trace of modern rhodoliths was found washed onto beaches or rocky platforms around the northern part of the island from Palmeira to Ponta Norte (Fig. 9.3). This sector receives the strongest surf on the island due to direct exposure to the trade winds. Absence of rhodoliths may be due to the steep bathymetric gradient and narrow insular shelf seaward of the intertidal rock platform in this sector. To the northeast at Calhetinha, only a few whole rhodoltihs were found washed onto the black-sand beach and none were observed on the exposed rock platform.

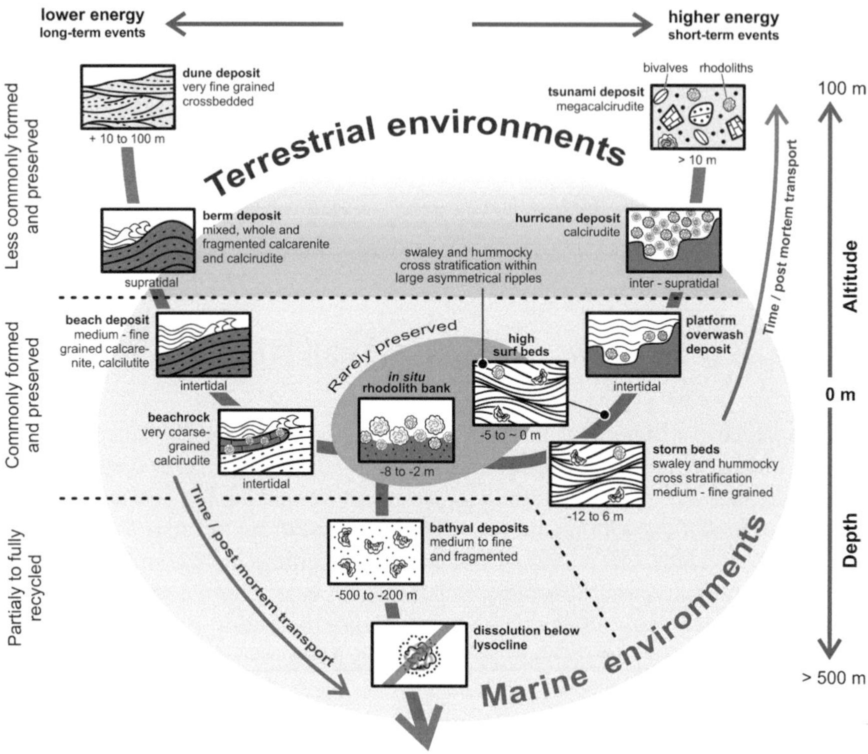

Fig. 9.2 Schematic diagram for taphofacies inferred from sedimentary dynamics in carbonate deposits dominated by rhodoliths and degraded rhodolith materials (Modified from Johnson et al. 2012)

The marine shelf is wider off the east coast than off the west coast, including a major projection from the southeast end of the island. Recent rhodoliths with a lumpy morphology are very abundant in the supra-tidal zone on the landward side of broad rock platforms along the east shore at localities such as Baía Negra (Figs. 9.3 and 9.4a). These are uniformly large rhodoliths with diameters of 10–12 cm. The deposits do not include sand, and were these rhodoliths to be cemented together, they would constitute a packstone. Very little breakage is evident and wear is only moderate. Assignment to a taphonomic grade according to the scheme of Brandt (1989) may be misleading, because relocation from the original site of living rhodoliths clearly entailed transportation across a wide, intervening rock platform without significant deterioration of the rhodoliths. Wind-driven waves that impact Sal at an oblique angle from the northeast account for the transport of the larger rhodoliths to the eastern shores potentially from as far away as the 40-m isobath. Our interview with conch divers on Sal (March, 2013) provided information on the widespread occurrence of large rhodoliths living on sandy bottoms along much of the eastern shelf of that island at depths of about 30–40 m from 3 to 6 km offshore. Shore deposits with smaller rhodoliths (2–3 cm in diameter), also with a lumpy morphology,

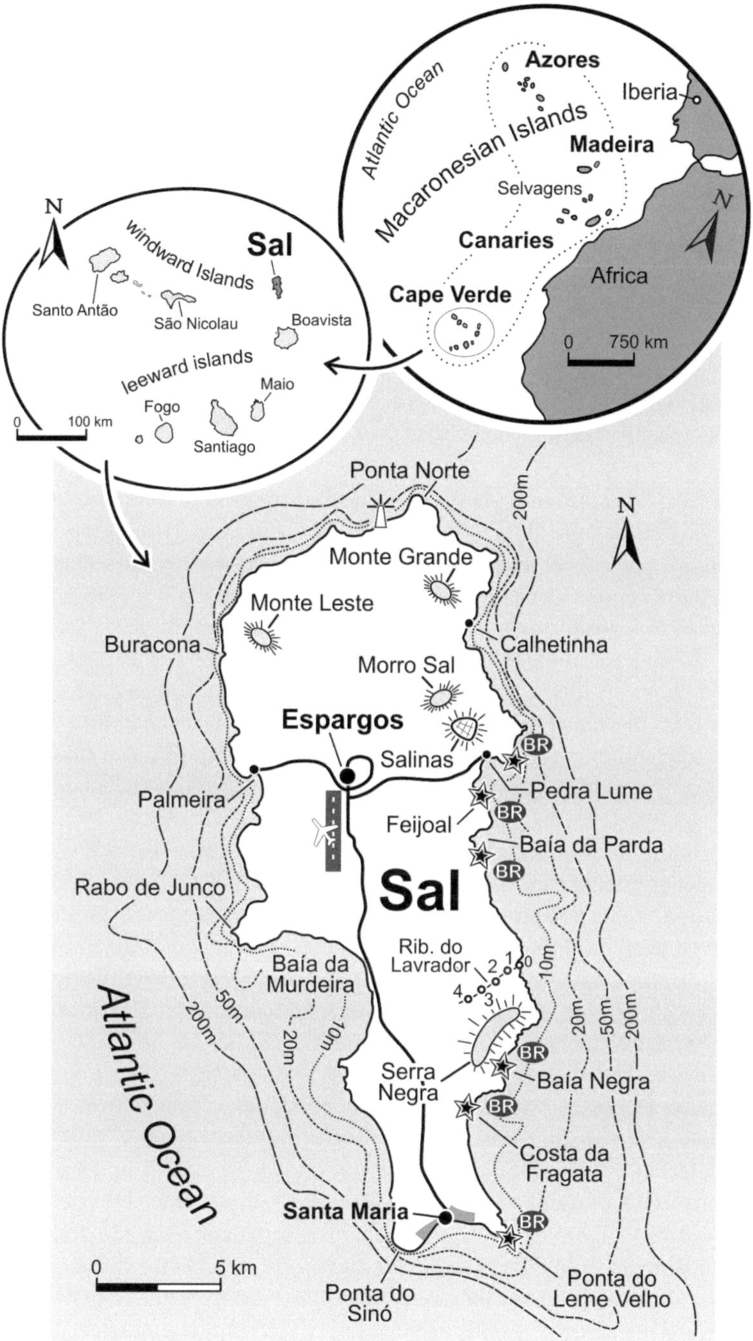

Fig. 9.3 Maps for Sal in the Cape Verde archipelago, showing its location in the North Atlantic Ocean with respect to the other islands in the Cape Verdes, as well as the island's shelf bathymetry. Stars labeled BR mark localities with extensive beach rock with composition dominated by rhodoliths

Fig. 9.4 Supra-tidal deposits with Recent rhodoliths on Sal in the Cape Verde archipelago. (**a**) Large, lumpy rhodoliths at Baía Negra on the east coast. (**b**) Small, lumpy rhodoliths from the north shore of Baía da Murdeira. (**c**) Detail of rhodoliths and other debris from the same locality

occur mixed with coral and shell debris in the supra-tidal zone above the beach around Baía da Murdeira (Fig. 9.4b, c). The material is entirely modern, showing no indication of time averaging with elements of substantially different ages. For example, there is no evidence to suggest that pieces of shells from the same species are degraded to a greater or lesser extent by boring organisms, nor is there any evidence of differences in diagenesis. According to the criteria of Brandt (1989), the associated shells in this assemblage correspond to taphonomic grade C because breakage is in the range of 50–75 %, although many of the individual rhodoliths remain whole and only moderately abraded. Emplacement of these smaller rhodoliths can only be related to wave refraction from the north along the island's western shelf.

Quite another feature that attests to the power of the northeast trade winds relates to beach deflation and the entrainment of dune sand through valleys that intersect the east coast. Table 9.1 provides data on the relative richness of beach and dune sand derived from rhodoliths, mollusk shells, and other organics in contrast to lithic materials of a volcanic origin from a succession of sample localities through the narrow Ribeira do Lavrador (Fig. 9.3). Topographically, Sal is the lowest island in the Cape Verde archipelago and the trade winds sweep across the island unimpeded. Valley dunes in the Ribeira do Lavrador stretch inland from the beach for 2 km. The concentration of crushed rhodolith debris in these dunes ranges from 50 to 67 %. Concentration of volcanic lithics varies depending on the degree to which basalt ridges project from the sides of the valley (i.e. dune sample 3 in Table 9.1). The beach from which the dune sand is derived gives a sample with 62 % domination by rhodolith grains (Table 9.1). Assignment of a taphonomic grade following the scheme of Brandt (1989) is awkward, but must correspond to the lowest grade D on the basis of deconstruction of the source rhodoliths recognizable only by cell structures retained in carbonate grains of fine sand ($\leq$0.25 mm in diameter). Potentially, the dune sand could include the degraded material from coralline algal crusts. If so, the contribution from this source is minor, because such crusts are not observed living on the rock platforms that skirt the east coast of the island.

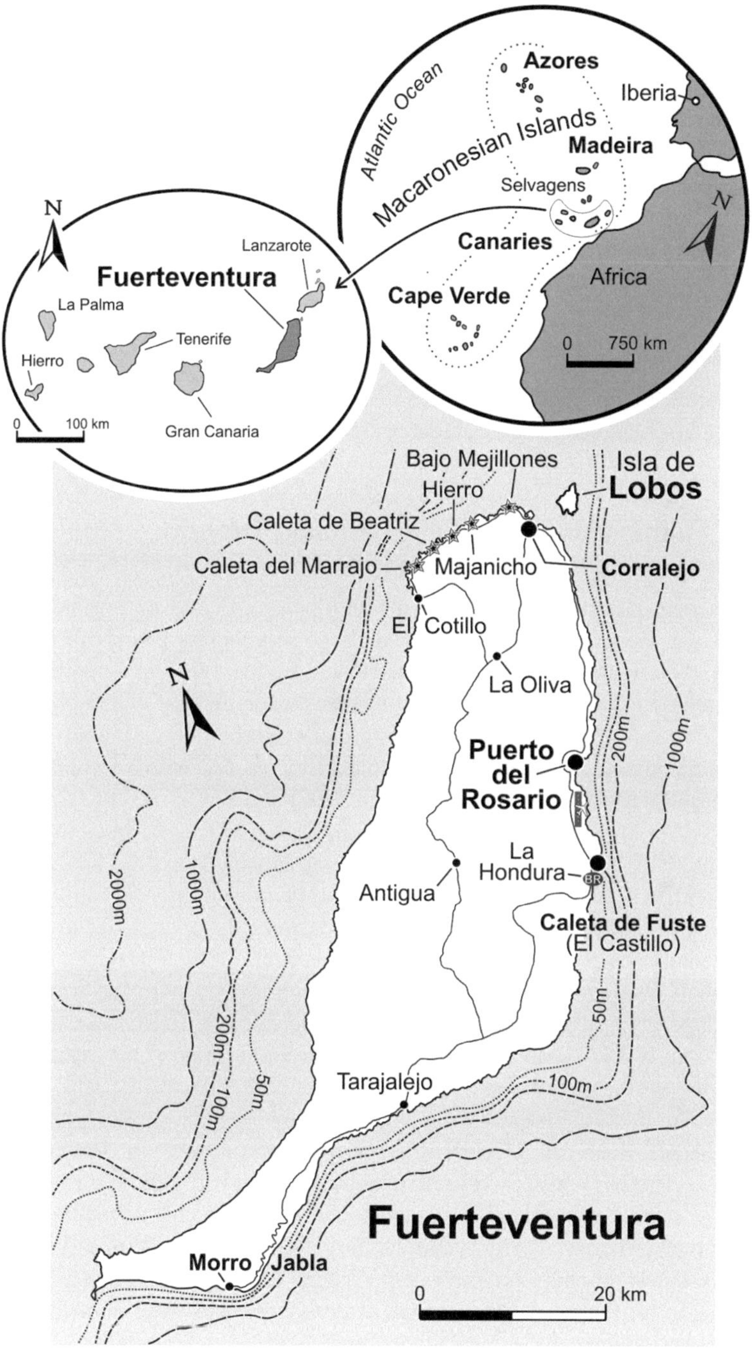

Fig. 9.5 Maps for Fuerteventura and related Isla de Lobos in the Canary archipelago, showing their location in the North Atlantic Ocean with respect to other islands in the archipelago, as well as the island's bathymetry

Table 9.1 Modern beach and dune-sand composition based on point counts from petrographic thin sections using epoxy-bound sediment prepared from samples collected in the Ribeira do Lavrador and adjoining beach (see Fig. 9.3 for locations on Sal)

	Beach sample		Dune sample 1		Dune sample 2		Dune sample 3		Dune sample 4	
Clast category	No.	%	No.	%	No.	%	No.	%	No.	%
Red algae	246	61.5	333	66.6	200	50	242	62	281	66
Bivalves	104	26	128	25.6	117	29.3	122	31	117	27.5
Gastropods	4	1.0	4	0.8	13	3.3	2	0.5	1	0.2
Corals	2	0.5	3	0.6	3	0.8	2	0.5	1	0.2
Foraminifera	6	1.5	3	0.6	3	0.8	2	0.5	3	0.7
Echinoderms	10	2.5	5	1	5	1.3	2	0.5	4	0.9
Bryozoa	19	4.75	10	2	10	2.5	1	0.25	15	3.5
Basalt and lithics	9	2.25	14	2.8	48	12	17	4.4	3	0.7
Total	400	100	500	100	399	100	390	99.7	425	99.7

9.5.2 Fuerteventura and Lobos (Canary Islands)

Located on the eastern edge of the Canary archipelago, Fuerteventura and its associated islet of Lobos share an oceanic rise with the neighboring island of Lanzarote to the north. This feature is distinctly demarcated by the 1,000-m isobath (Fig. 9.5) surmounted by islands with a northeast-southwest orientation. Although it sits leeward of Lanzarote, Fuerteventura is strongly affected by the wind field. The island is the archipelago's oldest and most tectonically complex with an uplifted slab of sedimentary Jurassic seafloor (Steiner et al. 1998) and a volcanic core complex that dates to 31 Ma in age (Gutiérrez et al. 2006). With a coastal perimeter of 240 km, Fuerteventura also qualifies as one of the mid-sized islands in the Canary group.

The break in slope on marine shelves around Fuerteventura is asymmetrical with respect to the shoreline. To the east, it occurs between the 50 and 100-m isobaths at a distance between 1 and 4 km from shore. This configuration provides a relatively narrow shelf conducive to the formation of rhodolith beds. On the west side, the marine slope is more gradual, overall. There, the distance from shore to the 50-m isobath is rarely 1 km, and often extends to 7 or 9 km. Long-shore currents stimulated by the northeast trade winds are active on both sides of the island. Modern rhodolith beds are widespread around Fuerteventura based on studies by the Atlantic Environmental Research Center at La Laguna University in Tenerife (Óscar Monterroso Hoyos, personal communication, February 2012). The shallow shelf between Fuerteventura and neighboring Lanzarote is especially advantageous to the production of rhodoliths. This includes the islet of Lobos opposite the town of Corralejo. The semi-enclosed Playa de la Concha on the south side of Isla Lobos is a natural trap for rhodoliths from which extensive deposits were removed and processed for lime during the nineteenth century.

Coastal deposits highly enriched by rhodoliths are extensive along the entire north shore of Fuerteventura (Johnson et al. 2012). A colossal accumulation of

Fig. 9.6 Supra-tidal deposits with Recent rhodoliths on Fuerteventura in the Canary arachipelago. (**a**) Small, lumpy rhodoliths (popcorn-size) from a platform overwash deposit at Caleta del Bajo de Mejillones on Fuerteventura Island. (**b**) Detail of the deposit with an excavation showing underlying rhodolith sand (meter stick for scale). Note: the lab deposit with whole rhodoliths is 2–3 cm thick

contemporary rhodoliths occurs as a platform over-wash deposit at Caleta del Bajo de Mejillones (Fig. 9.6). This deposit forms a long supratidal beach that sits above the landward termination of an extensive shore platform exposed at low tide to a width of 130 m (Fig. 9.6a). The stranded rhodoliths that remain whole are uniformly lumpy in morphology and small, ranging from 1 to 3 cm in diameter. The shape of whole rhodoliths plot out as highly spherical, based on a large random sample (Johnson et al. 2012, see their fig. 9). Were these rhodoliths to be cemented together, they would constitute a packstone. Based on grid counts from the deposit surface (50 cm by 50 cm), 20–25 % of the rhodoliths are whole; 60–65 % are broken and worn; and 10–15 % of the deposit consists of basalt pebbles (1–3 cm in diameter). Following the scheme of Brandt (1989), a taphonomic grade C can be assigned. The composition is monospecific, most likely belonging to a species of *Lithothamnion* (Rafael Riosmena-Rodrígez, pers. commun. February 2012). They make a surface layer 5 cm thick on mixed carbonate and basaltic sand (Fig. 9.6b). A second layer of whole rhodoliths occurs below a layer of calcarenite sand.

Farther west near Majanicho (Fig. 9.5), rhodoliths are mounded high in another overwash deposit left as a berm that rises 1.4 m above the adjacent shore platform. As at Caleta del Bajo de Mejillones, the rhodoliths here exhibit a popcorn morpholgy, locally called confites. During our survey of the north shore (July 2011), strands of brown algae were found still affixed to some of the rhodoliths at both Bajo de Mejillones and Majanicho. In Brazil, this phenomenon is known as the *arribadas* (Dias and Villaça 2012) and it is related to the seasonal arrival on beaches of rhodoliths with attached plumes of foliose algae that catch in the wind. This observation shows that brown algae and coralline red algae grew together in shallow, euphotic waters. In the case of the Fuerteventura beaches, tides and ocean swells empowered by the trade winds were instrumental in the mass transfer of rhodoliths across broad rock platforms.

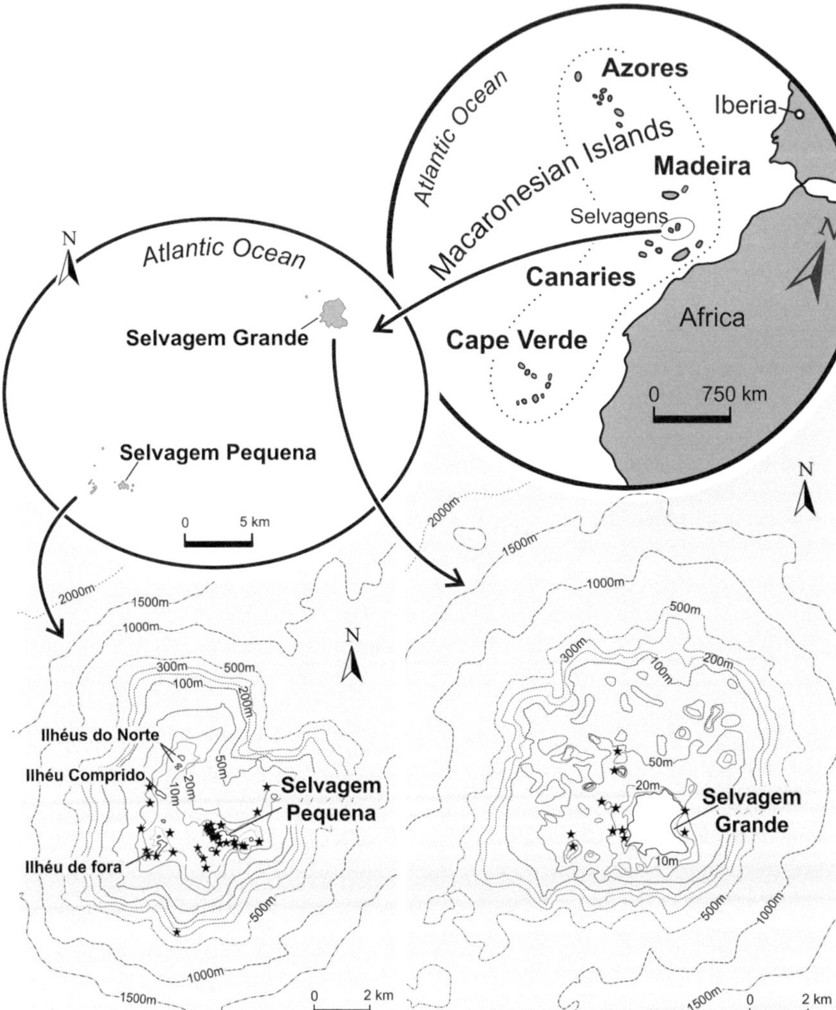

Fig. 9.7 Maps for Selvagem Pequena and Grande (the Savage Islands) showing the bathymetry around what were twin volcanic peaks. Stars mark locations where rhodoliths have been recovered (Courtesy of EMEPC, Portugal)

Wrack lines consisting of brown algae mixed with rhodoliths of assorted sizes appear in the inner cove at Majanicho. In bays farther west at Caleta de Beatriz and Caleta del Marrajo (Fig. 9.5), the wrack cover is even more extensive. At Playa Hierro, the beach is composed almost entirely of very coarse carbonate sand (1–2 mm in diameter) composed of rhodolith fragments. Many of the larger fragments correspond to the distal tips of closely branched rhodoliths. A dune field covering 40,000 m^2 occurs inland immediately south of Caleta del Marrajo (Fig. 9.5), composed of coarse to medium carbonate sand (0.25–1.0 mm in diameter) derived from finely crushed rhodoliths. In contrast to other dunes on Fuerteventura, the sand here is almost pure white. Asymmetrical dune ripples confirm the southerly traction of these rhodolith sands (Johnson et al. 2012). Based on sedimentology, the correlative taphonomic grades for these beach and dune sands exceed the maximum low grade of D by Brandt (1989).

9.5.3 Selvagens (Savage Islands)

The Selvagens (also known as the Savage Islands) constitute the eroded remains of two twin volcanic edifices that rise over 3,000 m above the surrounding Madeira Abyssal Plain (Figs. 9.1 and 9.7). These structures exhibit a long and complex emergence history with several episodes of submergence and re-emergence above sea level, starting around the late Oligocene (Ferreira et al. 1985; Geldmacher et al. 2001; Mata et al. 2013). This history resulted from an intermittent magmatic life separated by long periods of volcanic quiescence and erosion combined with vertical tectonic movements and the effects of glacio-eustatic oscillations. The islands were completely truncated by marine erosion during phases of volcanic inactivity. Today, the archipelago is composed of two main islands, Selvagem Grande and Selvagem Pequena with numerous adjacent islets. These constitute the residual reliefs of the largely razed summits of the NE and SW twin volcanic edifices. As defined by their respective 80-m isobaths, the islands are perched on extensive marine shelves much of which lies above the 20-m isobaths. The surviving land areas correspond to less than 8 % of the area above those particular isobaths. Coasts around Selvagem Grande are rocky and cliff-bound. Except for a few highly restricted pocket beaches dominated by gravel in more protected bays, the shores are stripped of fine sediments. Conversely, the shores of Selvagem Pequena are low-lying and bound by extensive shore platforms with some perched pocket beaches on the leeward side rich in very coarse bioclastic sand rich in rhodolith material (1–2 mm in diameter). Some of this sand is deflated by eolian processes and consequently Selvagem Pequena is extensively covered by Holocene eolianites.

Living rhodoliths (mostly *Lithothamnion corallioides* (P.Crouan & H.Crouan) P.Crouan & H.Crouan) are present on the shelves of both Selvagem Grande and Pequena. They are known wash up on an intertidal shore platform of the leeward side of Selvagem Grande. More importantly, numerous occurrences are recorded at depths ranging from 8 to 27 m, particularly on the leeward side of the islands (see Fig. 9.8). Surprisingly, living rhodoliths were found even at depths of 277 m during a ROV dive on the southern flank of Selvagem Pequena (Mónica Albuquerque, Sept. 2013, personal communication). Rhodolith debris is seldom found on the sediment-barren rocky shores of Selvagem Grande, but constitutes about 20 % of the bioclastic component of beach sediment on Selvagem Pequena, and about 25 % of the adjoining dune deposit at the island's main landing place. The preferential occurrence of rhodoliths on the leeward side of Selvagem Grande and Pequena is probably due to the very high-energy hydrodynamics on the windward side.

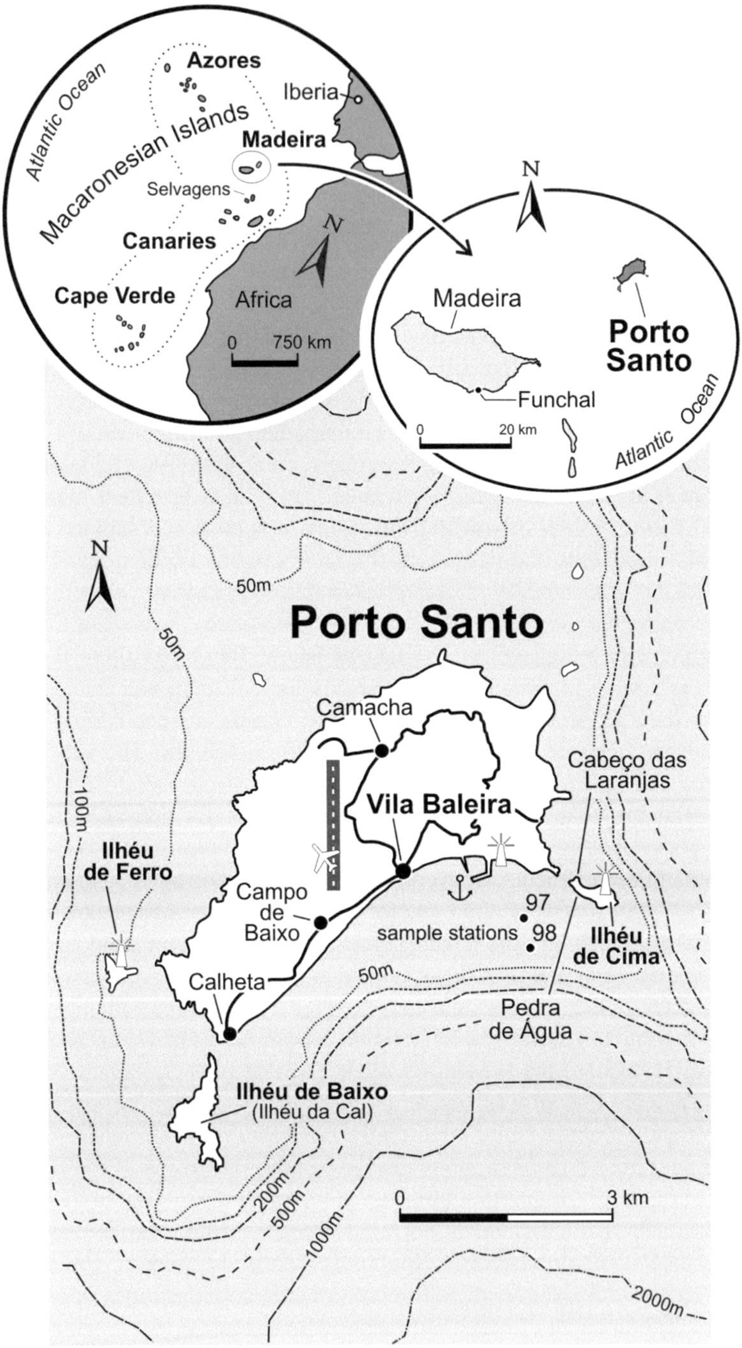

Fig. 9.8 Maps for Porto Santo and related islets in the Madeira archipelago, showing the surrounding bathymetry with sample stations 97 and 98, where contemporary rhodoliths have been dredged, as well as key locations where Miocene rhodoliths occur on Ilhéu de Cima and Ilhéu de Baixo

It is noteworthy that Miocene marine calcarenites occur at Selvagem Grande, intercalated between the two main volcanic units (Geldmacher et al. 2001; Mata et al. 2013). These bioclastic-rich sediments crop out along a continuous band 75–90 m above present sea level all around the island's perimeter and directly above a wave-cut erosional unconformity. The stratigraphic interval attests to a period during which the edifice was effectively a shallow-water sandy bank, and it is a prime target for future investigation with respect to sand derived from fossil coralline algae.

9.5.4 Porto Santo (Madeira Islands)

Porto Santo is an outlying island in the Madeira archipelago situated 50 km northeast of the principal island of Madeira (Fig. 9.8). Two phases of volcanic development include a submarine trachytic to basaltic core complex with ages ranging from 18.8 to 13.5 Ma and a subaerial basaltic to hawaiitic complex dated between 14 and 10.2 Ma (Ferreira 1985). Based on the island's surrounding bathymetry, the present land surface of 41 km^2 that includes several associated islets suggests a loss of about two-thirds the former maximum size. With an island shape that is elongated and oriented northeast to southwest, the impact of ocean swells is greatest against the northern shores. An extensive rock platform is a prominent fixture on the north-facing windward side of the island, while the south-facing leeward shore shelters a 7-km long beach. The golden sands of the Porto Santo beach account for 20 % of the island's circumference, whereas steep rocky sea cliffs comprise the other 80 %.

Ilhéu de Cima off the southeast end of Porto Santo and Ilhéu de Baixo off the southwest end, form major physical extensions that define a broad, south-facing bay from which modern rhodoliths (*Sporolithon* sp.) have been recovered. A survey conducted by the Portuguese National Hydrographic Institute (IH 2008) detected the presence of maërl beds at two sample localities near Ilhéu de Cima (Fig. 9.8, sample stations 97 and 98) at depths of 25 m and 34 m, respectively. Reports from divers in this area had already revealed the occurrence of maërl beds, but there is no evidence of rhodoliths or rhodolith debris washed onto the present-day beaches. Implications for the modern-day scarcity of rhodoliths compared to a middle Miocene surfeit of rhodoliths in the same area are treated in the discussion section (below).

9.5.5 Vila Franca do Campo (Azores Islands)

In the Azores, the Ilhéu de Vila Franca do Campo (better known as Vila Franca) is an islet located 500 m off the south coast of São Miguel (Figs. 9.9 and 9.10a). It remains as an eroded tuff cone built by hydromagmatic (surtseyan) activity of likely

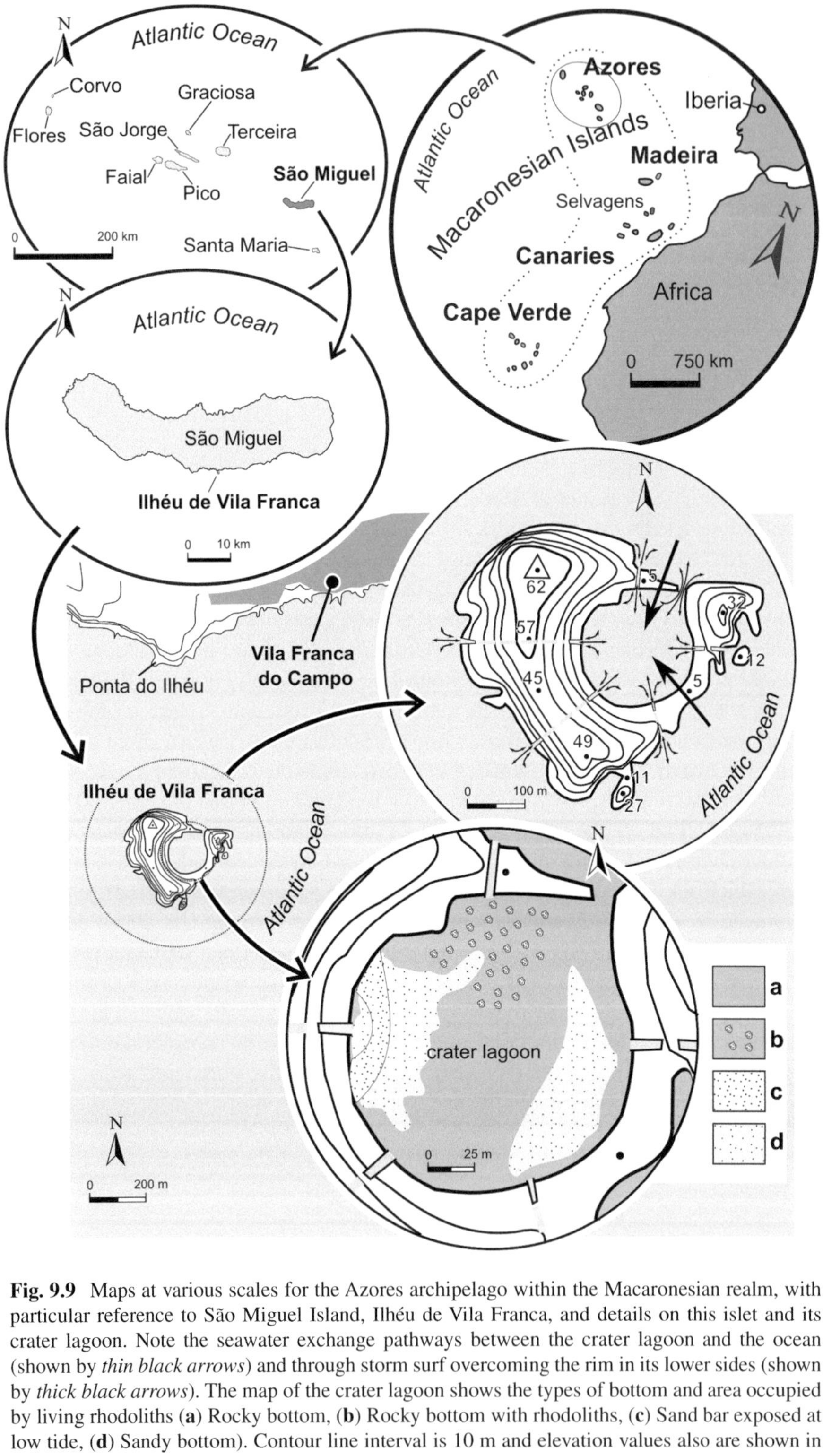

Fig. 9.9 Maps at various scales for the Azores archipelago within the Macaronesian realm, with particular reference to São Miguel Island, Ilhéu de Vila Franca, and details on this islet and its crater lagoon. Note the seawater exchange pathways between the crater lagoon and the ocean (shown by *thin black arrows*) and through storm surf overcoming the rim in its lower sides (shown by *thick black arrows*). The map of the crater lagoon shows the types of bottom and area occupied by living rhodoliths (**a**) Rocky bottom, (**b**) Rocky bottom with rhodoliths, (**c**) Sand bar exposed at low tide, (**d**) Sandy bottom). Contour line interval is 10 m and elevation values also are shown in meters

Fig. 9.10 Crater islet of Vila Franca do Campo located 1 km off the southern coast of São Miguel in the Azores. (**a**) View to the south from the outside looking into the breached wall of the crater. (**b**) View to the north from the inside looking through the crater's breached wall towards neighboring São Miguel

Holocene age (Abreu et al. 2002). The cone's outer slopes were significantly truncated by marine erosion, particularly on the southeast and north sides. The islet exhibits a nearly perfect circular crater 150 m in diameter with a shallow-water lagoon. Because the crater floor is below sea level, seawater flows freely through several extensional cracks across the crater rim (locally called "golas"). A channel on the northern side of the crater is breached by marine erosion and this channel was enlarged during historical times. Where marine erosion resulted in greater coastal retreat on the southeast side, the crater rim is low enough to allow storm surf to surge into the lagoon. On the southern and western sides, however, the crater rim rises more than 50 m, making an efficient barrier against the dominant westerly waves and particularly the southwesterly storms with strong winter surf (Quartau et al. 2010). Due to this peculiar setting, the Vila Franca lagoon constitutes a uniquely calm, shallow, and clear body of water during most of the year (Fig. 9.10b). This setting is disturbed only by strong storms, when surf surges through the crater rim and channel. Even during fair-weather, however, the lagoon benefits from the constant exchange of nutrient- and oxygen-rich water with the surrounding ocean.

The Vila Franca lagoon exhibits an average depth of 1–2 m and its lowest point is at a depth of about 5 m. The bottom is mostly rocky, except on southeast and west sides of the crater where it is sandy (Fig. 9.9, facies a and d, respectively, from the map inset). On the west side, a small crescent-shaped sand bar exists in front of one of the 'golas' (to which is clearly related) being exposed at low tide (Fig. 9.9, facies c from the map inset). A community of rhodoliths covers an area of 525 m^2 on the

Fig. 9.11 Holocene beachrock on basalt platform at Costa da Fragata on the east shore of Sal in the Cape Verde islands. (**a**) Beachrock surface approximately 5 m^2 in area with abundant rhodoliths and scattered basalt boulders (figures for scale). (**b**) Close-up view of eroded surface cut through rhodoliths up to 12 cm in diameter

northwest floor of the lagoon over a rocky bottom (Fig. 9.9, facies b from the map inset). The following species are represented: *Neogoniolithon brassica-florida* (Harvey) Setchell and Mason (1943) sensu Penrose (1996) and *Lithophyllum crouanii* Foslie (1898) sensu Chamberlain et al. (1988) (Rosas-Alquicira et al. 2009). *N. brassica-florida* has a lumpy growth form and a size that ranges from 3 to 8 cm in diameter, whereas *L. crouanii* has an encrusting growth form and averages 2.8 cm in diameter. All rhodoliths are spheroidal in shape, monospecific and nucleated, with nuclei smaller than 50 % of the volume of each rhodolith (Rosas-Alquicira et al. 2009). This accumulation represents an in situ, well protected rhodolith assemblage that if preserved intact would represent a high taphonomic grade A in the scheme of Brandt (1989), albeit one clearly lacking any trace of a mudstone.

9.6 Taphofacies Registered by Fossil Rhodoliths

Outcomes in rhodolith taphofacies and related sedimentological characteristics of the Macaronesian realm (Fig. 9.2) fall into a dozen scenarios divided between processes that operate over the long term at lower energy levels as every-day events and those that act under bursts of higher energy as short-term events. Most scenarios are well illustrated by examples of rhodolith deposits preserved in the rock record. Basic to the broad range of taphofacies achieved by rhodoliths is the function of their spherical to subspherical shape that permits the algal constructions to roll away from places where they thrive under ideal growing conditions. Larger rhodoliths with a more discoidal shape are more difficult to shift and spherical rhodoliths with a large lithic nucleus may also be resistant to movement away from their life zone due to greater mass. Dense concentrations of rhodoliths are found living from the low intertidal zone to depths of 150 m, limited by access to sufficient sunlight to promote growth and enough water motion to prevent burial by sediment (Foster 2001). Under conditions of exceptional water clarity, rhodoliths may continue to grow very slowly at depths of 200 m.

Delivery to bathyal depths between 200 and 500 m (Fig. 9.2) by the traction of currents sweeping rhodoliths over the break in an insular shelf cannot be verified on the basis of actual geological deposits. Referring to algal nodules near São Vicente in the Cape Verde archipelago, Van Der Land (1993, p. 42) reported the following: “Sand with numerous calcareous nodules may occur down to several hundred meters, apparently because of downward transport; at depths of over 100 m the algae are of course mostly dead.” Conditions clearly exist today and have existed in the past to accommodate such deposits on marine slopes around islands, but subaerial exposure on any of the Macaronesian islands is precluded by limitations on island uplift and/or fall in eustatic sea level.

9.6.1 Long-Term Lower Energy Deposits

Beach and beachrock deposits, as well as berm and dune deposits are the kind of accumulations that accrue in intertidal to supratidal zones extending to more elevated areas close to shore on a regular basis over the long term. Where rhodoliths live on the sea floor relatively near the coast on volcanic islands, they have the potential to be rapidly buried and preserved in life position by tephra or other volcanic products.

9.6.1.1 Rhodoliths Preserved in Life Settings

Preservation of rhodoliths under normal growth conditions is a rarity in the panoply of taphonomic modes in the Macaronesian realm (Fig. 9.2). Fossil rhodoliths that show nucleation around a rock core give some measure of the relative distance from a paleoshore and the possible water depth at which the rhodoliths started to form and enlarge by concentric growth. Basalt cobbles and pebbles eroded from sea cliffs on volcanic islands provide rock cores for potential rhodoliths. Smaller islands in the Madeira, Canary, and Cape Verde archipelagos are extremely arid and the seaward transfer of coarse siliciclastics due to storm-related runoff is virtually nonexistent, whereas those factors must be taken into account for larger islands such as the main island of Madeira and the Gran Canaria in the Canary Islands. Well rounded to subrounded clasts that undergo frequent turning in wave-agitated waters make ideal substrates for coralline red algae. The dynamics of wind-driven waves insure that energy is expended in the intertidal zone, where the erosion of sea cliffs and subsequent rounding of loosened clasts is most intense. Larger clasts remain close to shore, whereas lesser clasts in the form of smaller pebbles and granules are readily carried offshore. This scenario entails normal, every-day activity at relatively low-energy levels over the long term.

The study by Santos et al. (2012) from a middle Miocene paleoshore at Pedra de Água on Ilhéu de Cima off Porto Santo (Fig. 9.8) provides one of the best examples of fossil rhodoliths preserved in the setting where they originally lived and grew in the immediate study area. Cemented in a coarse, sandy matrix exposed on a bedding plane over an area of 12 m^2, a sample of 100 rhodoliths was evaluated for overall size and the extent to which nucleation occurred around a rock core. Belonging to the genera *Lithophyllum*, *Sporolithon* and *Hydrolithon*, these rhodoliths make a thin deposit (essentially a single nodule deep) strewn on a level surface between and among basalt mounds rising above about a half-meter. The surface density is roughly 15 rhodoliths per 100 cm^2 and the rhodoliths are well preserved. The basalt mounds are extensively encrusted by hermatypic corals dominated by *Tarbellastrea* and *Solenastrea*. The rhodoliths are large with an average diameter of 6.4 cm. Nearly all are nucleated around a rock core. The average diameter of rock cores in the sample was 4.7 cm. Hence, the average thickness of algal crust built up around the rock cores remained relatively thin with little variation between smaller and larger rhodoliths. Preservation of both the coral community elevated on the basalt mounds and the associated rhodolith community was achieved through catastrophic burial of a small embayment along a rocky shore by volcanic lapilli and tephra. The scheme of Brandt (1989) would assign a taphonomic grade of A, although the associated matrix consists of carbonate sand with no trace of mudstone.

Another example of a deposit formed by rhodoliths preserved while still close to living conditions comes from a Pleistocene rocky shore in the Cape Verde archipelago on Santiago at Portinho da Mulher Branca (Johnson et al. 2012, see their fig. 8). The present-day sea cliff at Portinho exposes a cross section through a Pleistocene basalt cliff with abutting limestone consisting of recurrent fining-upward cycles from 10 to 20 cm in thickness that feature bands of small, whole rhodoliths. The limestone from this exposure is more than 3 m thick containing ten or more couplets. Sediments were washed into a sheltered rocky cove by storm waves that arrived from the southeast or by wave refraction from the northeast. The rhodolith layers are stacked level to one another and the individual rhodolith beds at the bottom of each couplet are rarely more than one or two rhodoliths deep. Fall-out of fine calcarenite sand from the water column buried the rhodoliths in each succeeding couplet.

Rhodoliths preserved under growth conditions also are found at Lagoinhas on the north coast of Santa Maria in the Azores. This Pleistocene outcrop sits 7.4 m above sea level on a former shore platform attributed to Marine Isotope Stage 5e (Ávila et al. 2002). The base of the sequence marks a contact with eroded basalts and dykes. Directly overlying the unconformity is a basal conglomerate that fills depressions nearly 80 cm deep. A calcareous algal mat (crustal biostrome) with a maximum thickness of 30 cm overlies the conglomerate. The mat's upper surface exhibits fractures due to subaerial exposure and desiccation. The sequence continues with a 70-cm thick layer of poorly consolidated bioclastic white-yellowish sands, capped by colluviual breccias (Ávila et al. 2002). The conglomerate filling the basal pools consists of large boulders embedded in bioclastic sediments and well-rounded volcanoclastic pebbles. Some of these pebbles are encrusted by coralline algae as rhodoliths. These clast-encrusting rhodoliths have a maximum diameter of 5 cm, but the algal thickness around the volcanic nucleus is quite thin. The calcareous algal mat emplaced directly over the pools suggests that the pre-existing rhodoliths grew in place. The coralline algal mat clearly acted as a binder for the sediment below.

9.6.1.2 Beachrock Deposits

Beachrock forms within the intertidal zone on shores in tropical or subtropical regions worldwide where seawater is supersaturated with respect to calcium carbonate. Precipitation of carbonate cement provides the bond that holds together a conglomeratic mix of coarse-grained clastic and non-clastic sediments and the detritus of worn shells, corals, or other organic materials including whole rhodoliths (Fig. 9.2). The east shore of Sal in the Cape Verde archipelago provides outstanding examples of Holocene or Recent beachrock filled with whole rhodoliths and often the scattered shells of the large gastropod *Strombus latus*. Exposed by the tides, the beachrock forms narrow bands 2–3 m wide that extend along the shore for tens of meters. At least six locations with extensive beachrock are found between Ponta do Leme Velho in the south and Pedra Lume as intermittent deposits stretching over a distance of 22 km (Fig. 9.3). Rhodoliths in the beachrock at Costa da Fragata (Fig. 9.11a) are packed in close proximity, generally no more than two or three nodules deep. The rhodoliths are lumpy in morphology and show extensive wear. The surface layer is truncated by erosion (Fig. 9.11b), showing that essentially none of the rhodoliths are nucleated around a rock core. Basalt clasts of various sizes also are incorporated into the deposit, which sits directly on a basalt platform. Because the rock platform is both wide and laterally coextensive and the surf along this coast is strong, it suggests that the rhodoliths were introduced from farther offshore together with *Strombus* shells that normally occupy a sandy bottom in life.

Small patches of Holocene beachrock occur near the lighthouse on the north side of Isla Lobos (Fig. 9.5) close to Fuerteventura in the Canary archipelago. Likewise, extensive beachrock occurs on west side of Ponta Cais and nearby shores along Praia Real at the north side of Ilha do Maio in the Cape Verde archipelago (Fig. 9.12).

Beach sands composed of medium to fine-grained calcarenite and even calcilutite derived from coralline red algae may be converted into beachrock (Fig. 9.2). North from Caleta de Fuste on the east shore of Fuerteventura (Fig. 9.5), a long stretch of the beach exhibits a thin cover of modern sand that drapes a beachrock deposit with the same composition. Dipping gently seaward, these deposits consist of equal mixtures of white rhodolith calcarenite and black basaltic sand. This is the La Hondura locality, listed by Zazo et al. (2002, their table 2) as beachrock dating from 2.1 to 2.6 Ka.

9.6.1.3 Berm Deposits

Berms develop in the supratidal zone due to a combination of storms that coincide with high tides (Fig. 9.2). They may exhibit an over-steepened front related to subsequent erosion, and also a wedge shape in cross section depending on a covered rise in elevation. A well formed berm deposit of possible Holocene age that consists entirely of rhodoliths is exposed in cross section at Baía Negra on the east shore of Sal in the Cape Verde archipelago (Fig. 9.13a). This feature stands approximately 4 m above mean sea level, where it is isolated from modern shore processes. The deposit is in the incipient stages of burial by sand dunes.

9.6.1.4 Dune Deposits

A small subset of global dune systems consists of those from coastal areas where the chief component of sand-size grains comes from an organic source. Technically, the term eolianite applies to this special case for carbonate deposits generally formed between the latitudes of 20° and 40° in both hemispheres (Brooke 2001). All of the Macaronesian island groups fall within this latitudinal range, expect for the Cape Verde archipelago (Fig. 9.1). That rhodoliths and potentially coralline algal crusts contribute a significant volume of carbonate sand to a coastal dune is a novel concept (Fig. 9.2), documented for Pleistocene eolianites from Maio and São Nicolau in the Cape Verde archipelago (Johnson et al. 2013) but also known earlier from a study on eolianites from Porto Santo in the Madeira group (Soares 1973). The Porto Santo study by Soares (1973) found extensive Pleistocene dunes composed between 50 and 70 % by debris derived from coralline red algae.

Pleistocene sand dunes originally were mapped by Serralheiro (1970) on Maio, usually regarded as one of the leeward islands in the Cape Verdes. The large transverse dune at Lomba Greija near the east coast of Maio (Fig. 9.12) was restudied in order to relate cross-bed orientations to the trade winds and to analyze the composition of sand grains using petrographic thin sections. Although the Lomba Greija eolianite suffers from dissolution that produced many voids around ghost grains, those bioclasts that could be identified are dominated 74 % by coralline red algae (Johnson et al. 2013). Grains from this Pleistocene dune deposit typically fall into the category of a medium sand (0.2–0.4 mm in diameter). In contrast, Pleistocene transverse dunes of similar size on the west coast of Maio (Serralheiro 1970) lack any trace of rhodolith material. Today, detritus from mollusk shells is the principal contributor to the modern beach and dune sands on the west shore of Maio between Morro and Calheta. Deflation of Pleistocene beaches with much the same composition dominated by mollusk debris probably fed the fossil dunes now directly inland from Morro. Hence, there is a strong bimodality in terms of composition between associated beach and dune systems on the opposing east and west sides of the island.

Overlooked by Serralheiro (1970), a small Pleistocene dune is exposed in a quarry on the southeast coast of Maio at the Ribeira da Lagoa (Fig. 9.13b). The entire quarry is cut into the more steeply dipping leeward beds of a former dune.

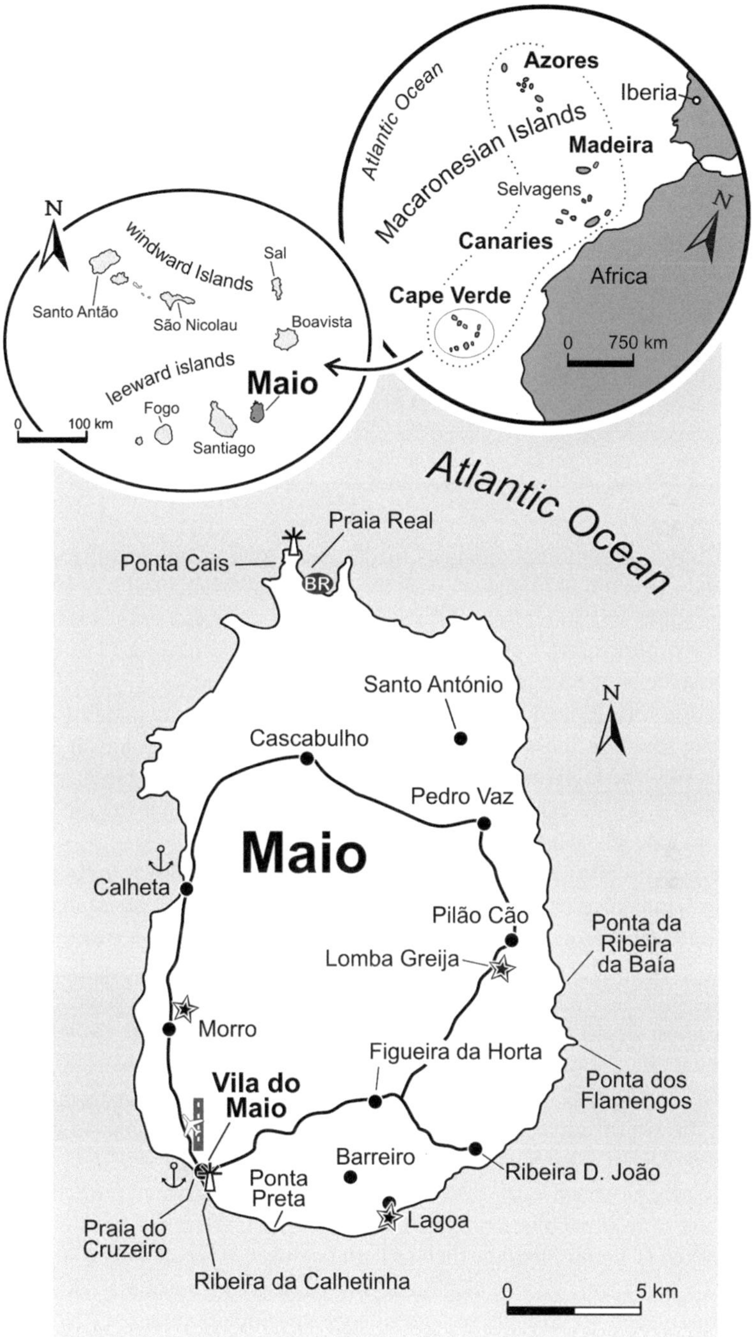

Fig. 9.12 Maps showing location of Maio in the Cape Verde archipelago and Atlantic Ocean with features that include fossil dunes (*star* symbol) and beachrock (= *BR*)

Fig. 9.13 Berm and dune deposits from Sal and Maio in the Cape Verde archipelago. (**a**) Holocene or younger berm composed entirely of large rhodoliths at section at Baía Negra on Sal (see Fig. 9.4). (**b**) Quarry exposure showing dune rocks near the Ribeira da Logoa on the southeast shores of Maio with inset from a petrographic thin section showing large bioclasts derived from rhodolth material, as well as equally rounded clasts eroded from basalt

Results from a petrographic analysis of the dune rock at this locality reveal very coarse sand (1–3 mm in diameter) with the highest enrichment of bioclasts coming from rhodoliths (60 %), followed by mollusks (10 %), and only minor amounts derived from byrozoans, echinoderms, corals, and foraminifera (4 % combined). Characteristics of the Lagoa dune are shown in thin section (Fig. 9.13b, inset). Black sand grains derived from basalt account for 26 % at this locality, five times greater than found in the more inland dune at Lomba Greija (Johnson et al. 2013, their table 1a). The higher specific gravity of basalt compared to bioclasts is sure to have an effect on the capacity of wind to transport sediments for long distances inland from a beach or an exposed valley wall.

Differences in island size, topography, and shape in regard to orientation with prevailing winds also exert a profound effect on dune development and composition. In the Cape Verdes, for example, São Nicolau is about 1.3 times as large as Maio with a maximum elevation nearly three times as great. São Nicolau is unusual in that the island is stretched out east to west in shape, and thus oriented to receive the full impact of the trade winds against a long northern shore. By comparison, Maio is elongated north to south in shape (Fig. 9.14) with the result that the trade winds have a more direct impact on its eastern shore and shelf. Pleistocene shoulder dunes on the south side of São Nicolau at Covoadinha de Chacina were found to contain a surprisingly high yield of rhodolith bioclasts comprising 96 % of all identified bioclasts (Johnson et al. 2013). Exceptionally strong wave refraction around the east end of São Nicolau combined with the baffle effect caused by high topographic relief is a combination that apparently nurtured a rhodolith bank off the island's southeast end and stimulated the deflection of local onshore winds from the southeast as clearly recorded by the stoss-slope on shoulder dunes at Covoadinha de Chacina.

Fig. 9.14 Examples of rhodoliths trapped in tidal pools. (**a**) Modern rhodoliths from the inner basalt platform at Majanicho on the north shore of Fuerteventura in the Canary islands. (**b**) Large rhodoliths (up to 7 cm in diameter) in a Holocene tidal pool exposed on the basalt platform at Costa da Fragata on the east shore of Sal in the Cape Verde islands

9.6.2 Short-Term Higher Energy Deposits

Major storms, hurricanes, and tsunamis occur with a much higher input of energy than under fair-weather conditions and such energy quickly dissipates over the short term measured in hours and days. The surface winds or shock waves that accompany these events generate ocean swells that radically decrease in wavelength and increase in wave height crossing insular shelves. Considerable energy is expended against the shore, but also takes a toll in back-surge across the marine shelf. On a geological time scale, the deposits from such events have a way of amalgamating to leave a distinctive presence in the rock record. Dispersal of rhodoliths by high-energy events from the banks where they naturally grow but remain susceptible to movement at depths below normal wave base has a profound outcome in rhodolith taphonomy. Inclusion of abundant rhodolith materials in storm beds, hurricane deposits, and tsunami deposits is a distinctive aspect of geological formations on some islands in the Macaronesian realm.

9.6.2.1 Tide Pool and Platform Over-Wash Deposits

In a geological sense, supra-tidal deposits can be related to massive over-wash deposits formed by a huge volume of rhodoliths swept landward from subtidal banks during major storms (Fig. 9.2). Many older Macaronesian islands, such as Maio, Sal, and Fuerteventura, exhibit basalt platforms that commonly exceed 100 m

Fig. 9.15 Platform over-wash deposits from the Pleistocene of Fuerteventura in the Canary Islands. (**a**) Quarry exposure at Aljibe de la Cueva near El Cotillo showing a rhodolith bed on a relatively flat unconformity surface formed by basalt. (**b**) Close-up view of the basal limestone layer consisting of rhodoliths 3–5 cm in diameter

in width on the windward shores. The rosy tint retained by modern rhodoliths sometimes trapped in distal tide pools indicates that coralline red algae are still alive. In contrast, tide pools adjacent to the shore or in a supratidal position characteristically include crystallized salt deposits together with bleached rhodoliths. Modern tidal-pool deposits filled with small, loose rhodoliths from the north shore of Fuerteventura (Fig. 9.14a) are compared in terms of rhodoliths size and tide-pool size with a cemented Holocene deposit with large rhodoliths from Costa da Fragata on the east shore of Sal (Fig. 9.14b). Other examples of tide-pool deposits with fossil rhodoliths are found in the classic Pleistocene outcrops explored by Darwin in 1832 on "Quail Island" in the Praia harbor area on Santiago in the Cape Verde archipelago (see Johnson et al. 2012, their profile 3).

Near the town of El Cotillo on the west side of Fuerteventura (Fig. 9.5), the quarry at Aljibe de la Cueva exposes a 1.4-m thick layer of rhodoliths that sits for the most part directly on basalt. The underlying unconformity surface is uneven with depressions that in places capture a basal conglomerate loaded with basalt clasts (Fig. 9.15a, left-center). The quarry floor exposes basalt that extends in the seaward direction for a considerable distance, suggesting the existence of a former rock platform. The massive layer in the overlying Pleistocene formation is completely dominated by lumpy rhodoliths from 3 to 5 cm in diameter (Fig. 9.15b) that are more than 50 % broken and otherwise worn. As the whole and broken rhodoliths are in direct contact with one another, the deposit constitutes a packstone. In terms of the scheme formulated by Brandt (1989), the taphonomic grade C is represented. It is argued that all the rhodoliths in this layer arrived during a single over-wash event. Interestingly, a dune deposit is the next layer above the rhodoliths in the quarry exposure (see Sect. 7.1.4, above). At the very least, a short temporal hiatus separated the rhodolith bed from the overlying dune but a longer break cannot be ruled out. A beach formed by crushed rhodoliths developed sooner or later after the storm event and occupied the adjacent rock platform, where wind deflation acted to feed the adjacent dune. Now undergoing burial by a modern sand dune, the Holocene berm deposit at Baía Negra on Sal in the Cape Verde archipelago (Fig. 9.13a), represents a similar scenario.

Much of the inspiration for the concept of over-wash deposits capable of transporting rhodoliths across rock platforms up to 95 m wide in the Cape Verde and Canary islands comes from the study by Ledesma-Vázquez et al. (2007) in the Gulf of California. There on Isla Coronados, layers of crushed rhodoliths drape over the top of a barrier formed by rocky islets positioned across the front of a Pleistocene lagoon. Major storms that swept up the Gulf of California from the south are the only possible agency with sufficient strength to transport rhodolith sand over the top of the solid barrier and into the paleolagoon. In this case, layers of rhodolith sand dip away from the inner face of the rock barrier and prograde into the lagoon. Each layer probably represents a separate storm event.

9.6.2.2 Tempestite Deposits

Fossil accumulations of whole and crushed rhodoliths become entrained in subtidal tempestite deposits as a result of storm waves (Fig. 9.2). During a storm, the period and height of the waves will change, reaching a much deeper zone at a depth known as the storm wave base. There are several effects associated with such storms. Among them are significant increases in: (1) local sea level due to storm surge which may be abetted by the tidal cycle; (2) the reach of waves into the supratidal zone and beyond; and (3) the capacity of wave action to disturb deeper parts of the marine shelf. As a consequence of storm surge and waves breaking above the fair-weather location of a beach, there is added potential for coastal erosion shown by the removal of sediments from the shore and upper shore face to be transported away into deeper areas. Crushed rhodoliths may be incorporated into more offshore deposits derived from living rhodolith beds during storm events. The extreme agitation of rhodoliths affected by storm waves accounts for extensive fragmentation of rhodoliths that normally aggregate in living banks only a few decimeters thick.

In order to interpret a fossil rhodolith deposit as a tempestite, an array of structures must be present in those beds consisting of sand-size, crushed rhodolith matrix. These include hummocky cross stratification (HCS), as well as swaley cross stratification (SCS) that are considered genetically connected to storms. Dumas and Arnott (2006) suggested that HCS optimally forms above (but near) storm wave base where aggradation rates during storms are high enough to preserve hummocks. According to the same authors, SCS is also hypothesized to be deposited by an aggrading bed between fair-weather and storm wave base, but in shallower water where aggradation rates are low enough to cause preferential preservation of swales. The landward position of SCS versus HCS is supported by observations from the rock record (Dott and Bourgeois 1982; Dumas and Arnott 2006), where SCS generally occurs above HCS in ancient lower shoreface and shallow shelf facies. Both

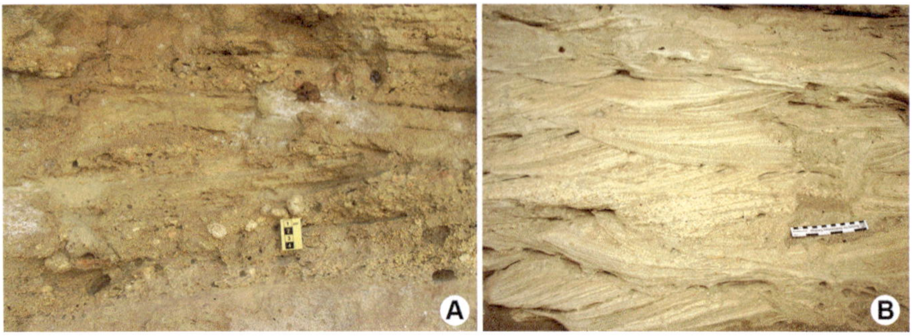

Fig. 9.16 Tempestite deposits from a Pleistocene outcrop exposed in the Ribeira da Calhetinha on the south shore of Maio in the Cape Verde archipelago. (**a**) Whole rhodoliths and coarse rhodolith debris captured in a layer that is truncated at an angle by an overlying bed. (**b**) Swaley units up to 45 cm wide that represent an amalgamation of separate storm events (scale = 15 cm)

structures represent the oscillation associated with wave action on the sediments deposited, once the peak of the storm has waned below normal wave base but remains above storm wave base.

Outstanding examples that demonstrate SCS and lag deposits associated with rhodolith debris captured by tempestites come from Pleistocene outcrops at Praia do Cruzeiro, Ribeira da Calhetinha, and Ponta Preta on the south coast of Maio (Fig. 9.12) in the Cape Verde archipelago. An exposure high in the 12-m thick sequence from the Ribeira da Calhetinha on the east side of Vila do Maio clearly shows some whole rhodoliths within a non-horizontal bed that is truncated at an angle by the succeeding bed (Fig. 9.16a). This is a normal feature for SCS, because these layers were deposited during the storm while the water column was still moving under oscillatory conditions below normal-wave depth. Also from the Ribeira da Calhetinha, the middle part of the canyon exposure exhibits multiple events with SCS (Fig. 9.16b) in what is interpreted as reactivation surfaces for an amalgamated sequence. Each in the succession of truncated swales is related to a different storm event.

Pleistocene sections with good evidence for HCS are recorded from Santiago at Darwin's Quail Island (Ilhéu de Santa Maria) and for SCS at nearby Portinho da Mulher Branca (Johnson et al. 2012, from the top of sections 5 and 9, their figs. 4 and 7).

Santa Maria in the Azores archipelago is the source for more tempestite deposits that incorporate fossil rhodoliths. This island is prolific with exposures of upper Miocene to lower Pliocene sedimentary outcrops, many of which exhibit evidence of storm-induced mass transport. An outstanding example is found on the island's southern coast at Malbusca. This sedimentary deposit is interbedded within effusive submarine volcanic sequences, now exposed approximately 20 m above sea level and covering a lateral distance of 250 m. At Malbusca, underlying pillow lavas formed a structural relief filled by conglomeratic sediments composed of rhodoliths in rock-forming quantities. Rhodoliths are mostly between 2 and 6 cm in diameter and show a wide variety of growth forms from encrusting to warty, fruticose, and lumpy modes. The individual nuclei of these rhodoliths were provided either by bioclasts or volcanic pebbles. Based on a study by Rebelo et al. (2014), the rhodoliths from this locality are taxonomically diverse including at least six taxa (*Lithophyllum prototypum* (Foslie) Foslie, *Lithophyllum* sp., *Spongites* sp, *Hydrolithon* sp., *Phymatolithon calcareum* (Pallas) Adey & McKibbin and cf. *Phymatolithon* sp.). The rhodolith-bearing bed is overlain by a 10-m thick calcarenite that exhibits low-amplitude, high wavelength HCS. In this case, deposition is interpreted to have taken place under high-energy levels over the seafloor at considerable depths, based on the fact that the deposit is capped by submarine lava flows with a thickness of 60–80 m.

9.6.2.3 Hurricane Deposits

As described by Williams and Delinger (2013), hurricanes create a combination of very low atmospheric pressure and strong onshore winds that generate substantial storm surge. Storm surges that rise several meters or more above normal tide level may easily overtop coastal barriers that include wide rock platforms and low, rocky shores. Potentially, sediments eroded by hurricanes from shallow marine, tidal flat, beach, and fore-dune areas, may be transported far inland by the storm surge and redeposited in terrestrial environments. In the northern hemisphere, storm surge from hurricanes makes landfall in the right-front quadrant of the storm.

To date, only a single locality has been identified in the Macaronesian realm that qualifies as a storm deposit on such an enormous scale that it is best interpreted as the result of a super storm of near-hurricane strength. The site entails the middle Miocene formations at the Hill of Oranges on Ilhéu de Cima off Porto Santo (Fig. 9.8). The full succession of limestone strata is banked against a former rocky shore with at least 6 m of paleorelief (Johnson et al. 2011, their fig. 3). The first of four successive rhodolith beds is the most massive, reaching 2.6 m in thickness. Fossil rhodoliths in this deposit include coralline red algae representing at least three genera, including *Sporolithon*, *Lithothamnion*, and *Neogoniolithon*. Some exhibit rock cores, but most do not. The rhodoliths are overwhelmingly lumpy in morphology and typically quite large (7–16 cm in diameter). Rhodolith shapes in this deposit are mostly spheroidal, but there is a trend toward some ellipsoidal rhodoliths (Johnson et al. 2011, their fig. 7). No discoidal forms are present in the 224 rhodoliths measured. The initial rhodolith layer in this deposit is partially exhumed as a level shelf covering 450 m^2 that buries substantial crevasses and sea stacks on a former rocky shore facing to the southeast. Based on grid counts, the density is 40–48 whole rhodoliths in 1 m^2, but with a matrix composed of basalt pebbles and extensive bioclastic debris derived from broken rhodoliths, corals, echinoids, and disarticulated bivalve shells. Applying the scheme of Brandt (1989), a taphonomic grade between B and C is warranted, although the deposit is a wackestone. The deposit's location fits perfectly with the expected brunt of a tropical cyclone sending wave surge from its right-front quadrant. Succeeding rhodolith beds represent separate storm events that were less severe with intervening layers of volcaniclastic sediments that include SCS structures (Johnson et al. 2011, their fig. 10).

9.6.2.4 Tsunami Deposits

According to Puga-Bernabéu et al. (2007), tsunamis are among the most powerful events on the surface of the earth, and globally they have a relatively high recurrence interval. These "harbor waves" may be induced by earthquakes, volcanic slides, or bolide impacts, and they are responsible for the erosion of huge amounts of sediment from shallow marine and coastal settings that become redeposited in either marine or continental environments well above sea level (Fig. 9.2). Tsunami-related deposits are called tsunamites. Especially in nearshore settings, tsunami waves produce major effects on the seafloor, although the resulting deposits (scour-and-fill structures) commonly are eroded due to subsequent reworking by currents and waves. Materials in siliciclastic systems generally remain unlithified on the sea floor, where particles of different sizes (silt to large boulders) are easily mobilized by a tsunamis wave. In contrast, sea-floor cementation is high in warm-water carbonate systems, where tropical tsunamites consist of large blocks swept up by tsunami waves (Puga-Bernabéu et al. 2007).

The only known example of a tsunamite with rhodolith debris preserved above present sea level in the Macaronesia realm is from a study by Paris et al. (2011) based on a calcretised and fossiliferous conglomerate composed of volcanic rocks and marine bioclasts at several locations near Tarrafal in the northwest part of Santiago in the Cape Verde archipelago. Marine fossils incorporated into the tsunamite include corals and calcareous algae (i.e. rhodoliths) attributed to *Lithothamnium* and *Lithophyllum* (Paris et al. 2011, p. 134). As no rhodolith material is illustrated, we are unable to comment on the size, shape, or level of breakage involved. However, in terms of taphonomy, the deposit clearly represents significant transport under a regime of very high energy. Dating of a coral sample from the Tarrafal tsunamite yielded a corrected U-Th age of 123 ± 3.9 ka. The embayment at Tarrafal is situated 55 km northeast from the still active volcano on Fogo. It is speculated that the extensive Upper Pleistocene tsunamite deposits around Tarrafal were generated by massive flank failure on the volcano at Fogo (Fig. 9.1). This kind of event represents a relatively low-frequency, but very high-magnitude hazard. Given the past and present abundance of rhodoliths around the older Cape Verde islands, their inclusion in ancient and potential future tsunamites is not surprising.

9.7 Fossil Evidence for Rhodoliths and Biodiversity

Bioconstruction and bioerosion are important driving processes and modifiers in modern marine environments. Comparable processes are known from fossil examples to have operated in former marine settings. Organisms that undertake bioconstruction include many marine invertebrates such as corals, bivalves, and bryozoans among many others. Those responsible for boring activities include epifaunal or infaunal elements as indicated by superficial scratchings to deep borings that bioeroded the host substrate represented by lithified seabed, skeletons of living and dead organisms, and rock clasts (Taylor and Wilson 2003). Encrusting organisms and borers found on fossil rhodoliths have received relatively little attention concerning their potential as ecosystem engineers or calcium-carbonate factories (Checconi and Monaco 2008; Checconi et al. 2010; Nebelsick and Bassi 2000; Bassi et al. 2011, 2012; Baarli et al. 2012). However, studies on the distribution, diversity, and preservation of such assemblages associated with coralline red algae provide invaluable data not only for understanding hard substrate biotas, but also for the dynamics of carbonate sedimentary successions.

9.7.1 Epifaunal Encrustations

The extent of epifaunal encrustation on rhodoliths depends on the benthic setting, rate of sedimentation, frequency of overturning, growth rates of encrusting fauna, and density of bioeroders sharing the same rhodolith. Many individual components attached to rhodoliths may form macroids on their own account, as for example coraliths, bryoliths, balanuliths, ostreoliths and vermetuliths (Baarli et al. 2012). These forms remain to be identified in Macaronesia. Other common types of epifauna on rhodoliths include serpulids and foraminfera. All of the epifaunal taxa cited above are commonly present on rhodoliths in Macaronesia, although different sorts tend to dominate in different deposits.

Studies on the epifauna that encrust fossil rhodoliths are few in Macaronesia. Grid counts on the extensive bedding surfaces at the Hill of Oranges were tallied for the kinds and numbers of encrusters on rhodoliths (Johnson et al. 2011). Two types include those that originally lived on the rhodoliths when the rhodoliths were still alive, and those that arrived after the stranding event when the rhodoliths expired. The difference can be discerned by preferential growth limited to one side of a rhodolith, which is most likely to occur after the rhodolith was cemented in place and no longer free to roll. Serpulids are the most common type of epifauna and they occur on all sides of the rhodolits. These were observed on approximately 1 % of the rhodoliths examined. Bryozoans were even less common. Both were likely to have been undercounted, because the survey was conducted as a field study. Only a few cross-sections through rhodoliths were slabed and checked for encrustations at different stages of rhodolith growth. However, the results suggest that the rhodoliths were lightly encrusted under living conditions. In contrast, large corals (*Tarbellastrea* sp., *Isophyllastrea* sp., and *Solenastrea* sp.) were found to grow on top of the rhodoliths (Fig. 9.17a) in the central, most distal part of the deposit. All the encrusting corals are preserved right-side up in growth position and must have encrusted the rhodoliths after the stranding event. Less commonly, balanid barnacles and spondylid bivalves also colonized these large rhodoliths.

Another detailed study of encrustations and borings on rodoliths is ongoing with respect to an Upper Pleistocene deposit from Praia on Santiago in the Cape Verde archipelago. Here the rhodoliths are relatively small and have a complex history that entails transport into a shallow estuary where they banked below a cliff line. The extent of epifaunal encrustation appears to be light, but is high in terms of the

Fig. 9.17 Rhodoliths from the middle Miocene deposit at Orange Hill (Ilhéu de Cima, Porto Santo). (**a**) Rhodolith encrusted by a large coral skeleton (*Tarbellastrea* sp.). (**b**) Cross-section through a rhodolith nucleated around a rock core showing borings (*G Gastrochaenolites*, *Tp Trypanites*, *E Entobia*)

numbers of organisms and colonies. The colonies inhabiting these rhodoliths are tiny and include a host of encrusters. The only large organism is the bivalve *Chama gryphoides*. One to ten small specimens of this oyster typically encrust a single rhodolith (Fig. 9.18a). Bryozoans are quite diverse with eight taxa present. Most inhabit cryptic spaces (e.g. borings or the inside of *Chama* shells), although one bryozoan, *Trematooecia* sp., seems to prefer a more exposed setting on the smooth exterior of the coralline red algae (Fig. 9.18b). Vermetids, serpulids and spirobids (*Spirorbis* sp.), and foraminifera are also present but not in great numbers. The epifauna from this locality lived among fairly dense infestations of large *Trypanites* borings on any single rhodolith.

Marine shelves on which rhodoliths thrive around the Macaronesian islands are subject to strong wave action under persistent winds (Brand 2011). The highly spheroidal shape of living and fossil rhodoliths sampled from this particular realm (Johnson et al. 2011, 2012, 2014) is consistent with susceptibility to rolling and frequent overturning. Such conditions are unfavourable for many kinds of relatively long-lived epifauna such as corals. The example from the Miocene Hill of Oranges is typical of rhodoliths relocated by major storms, where corals are interpreted to have encrusted the rhodoliths after they became locked in position among shallow subtidal to intertidal rocks. Rhodoliths from the Praia location might be expected to have hosted an epifauna consisting of both hard and soft-bodied organisms due to the more sheltered setting in which they lived. However, it is possible these rhodoliths experienced either a short residence time in their final resting environment or the makers of the *Trypanites* borings affected the settling of epibionts in a negative way.

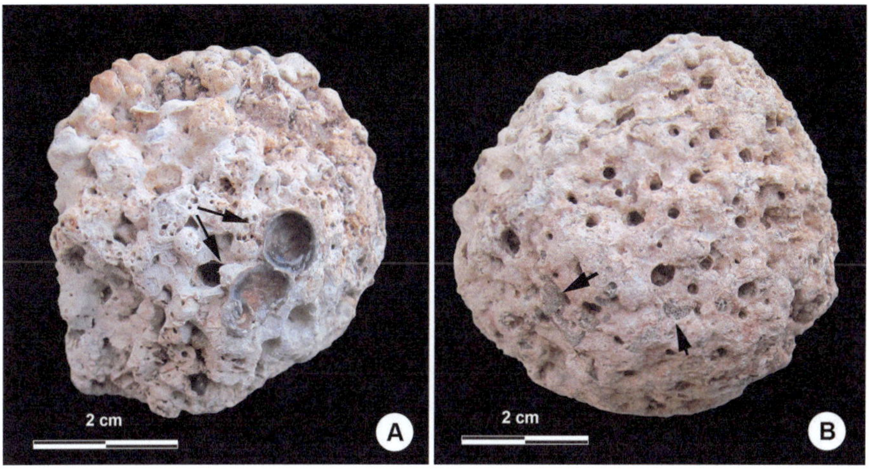

Fig. 9.18 Rhodoliths from the Upper Plesitocene deposit at Travessa da Cruz Vermelha in Praia (Santiago). (**a**) Rhodolith with encrusting oysters (*arrows*). (**b**) Rhodolith with encrusting bryozoans (*arrows*); Note: extensive openings at the surface belong to *Trypanites* isp

9.7.2 Bioeroding Organisms

Bioerosion is a persistent feature found both on the surface and in cross sections on all rhodolith samples studied to date from islands in the Macaronesian realm. Nevertheless, the diversity of trace fossils associated with rhodoliths is not very high. The most typical macrobioerosion structures are those produced by endolithic bivalves (*Gastrochaenolites*), sipunculids (*Trypanites*), and polychaete annelids (*Caulostrepsis*, *Maeandropolydora*). Shallow borings such as the structures produced by sponges (*Entobia*) may be missing due to some degree of reworking and erosion, although many rhodolith samples are unabraded.

Rhodoliths from the mid-Miocene Hill of Oranges on Ilhéu de Cima at Porto Santo (see above under Sect. 10.6.2.3) mainly represent robust growth forms that conform to a lumpy morphology (Johnson et al. 2011). The gradual growth of these large rhodoliths allowed time for colonization by sipunculid annelids and endolithic bivalves. The most abundant ichnofossils include *Trypanites* isp. (55 %) and *Gastrochaenolites* isp. (45 %), ubiquitous to all rhodoliths surveyed in the study. The density of borings can be seen in cross section on the example of an Orange Hill rhodolith from the nucleus to the last algal growth stage (Fig. 9.17b). As seen in cross section, the maximum number recorded on any single rhodolith was 22 borings. Other polychaete borings (*Caulostrepsis* isp.) are evident but much less significant. *Entobia* isp. also is present, but was detected only on a few rock cores (labelled in Fig. 9.17b). Overall, the high number of borings/rhodolith signifies an extended residency time on the seafloor prior to the major storm event that swept such a massive deposit of rhodoliths against a rocky shore at the Hill of Oranges. In the event that boring activity had continued after the rhodoliths expired and the deposit was stabilized or otherwise cemented in place, it can be expected that the addition of later borings would be more numerous on the exposed side. The later borings also might be expected to cross through older borings. No evidence of an imbalance in the location of borings or any overlap was observed.

A useful comparison comes from rhodoliths at Pedra de Água (Santos et al. 2012), on the opposite side of Ihéu de Cima from the Hill of Oranges. At Pedra de Água, the rhodoliths also are lumpy in morphology and their size (see Sect. 10.6.1.1) is comparably large. However, the majority at this locality are nucleated around large pebbles and cobbles. Cross-sections clearly show a pattern of symmetrical accretion around the core with few empty voids within the accretion, both factors indicating high energy with frequent overturning. In this case, the rhodoliths show reduced levels of bioerosion with fewer numbers of small *Gastrochaenolites* present. Comparing rhodoliths at Pedra de Água on the southwest side of Ihéu de Cima with those at the Hill of Oranges on the northeast side of the islet, they are of comparable size, but the thickness of the deposits and their packing density differs profoundly (Johnson et al. 2011; Santos et al. 2012). The rhodolith assemblages from the Hill of Oranges were transported to the site from deeper, offshore bank, whereas those at Pedra de Água were part of a near-shore community (Santos et al. 2012), but with limited bioerosion activity by bioeroders.

In another study still in progress, rhodoliths were collected from an Upper Pleistocene outcrop in Praia (Santiago, in the Cape Verde archipelago). Some preliminary results from this study are available for comparison with the two localities from Porto Santo previously described. The rhodoliths range from 3 to 6.6 cm in diameter. From a sample of 47 specimens, 24 % reveal a rock core or void where growth commenced around an unknown object. In the remaining 76 %, the growth of coralline alga is continuous from the center without any trace of nucleation. Some 82 % of rhodoliths in the study sample are extensively bored. Analysis of bioerosion structures reveals that *Trypanites* isp. is the most common and abundant trace from the rhodolith-boring assemblage and it is fairly evenly distributed both at the surface (Fig. 9.18b) and internally on all nodules. Bivalve borings represented by *Gastrochaenolites* isp. are less relatively frequent on rhodoliths from this site, with only 33 % affected. Some remains of the producer may still be found in situ inside of the related borings. The polychaete annelid borings referred to as *Caulostrepsis* isp. are uncommon and poorly preserved. No sponge borings (*Entobia* isp.) were found.

Water energy is a critical factor to keep in mind, because it is directly related to the frequency with which rhodoliths are overturned, as well as conditions of local sedimentation and the rate of larvae dispersal, all factors responsible for bioerosion rates. According to Hutchings et al. (1992), some variations in colonization among sites are explained by the fact that most organisms that create borings are recruited as pelagic larvae, and that local wind-generated currents are responsible for their dispersal. Such local-scale patterns of water movement may contribute to the differences in colonization between sites with rhodoliths. Although such ichnogenera have invariably been reported as proxies for paleobathymetry, in connection with rhodolith assemblages they appear to be appropriate indicators of residency time and water energy, as in agreement with Bassi et al. (2012).

9.8 Discussion

9.8.1 Taphonmic Grades Derived from Rhodolith Data

The method of determining taphonomic grades based on post-mortem breakage and corrosion in marine invertebrates advocated by Brandt (1989) as an elaboration on the system of Brett and Baird (1986) is definitive in setting categories that conform to strict definitions using percentages, but in some cases those divisions may under value the extent of relocation experienced by rhodoliths. This liability is demonstrated by modern coastal settings in the Macaronesian islands where massive quantities of rhodoliths are transported across broad rock platforms during storms and left stranded in a supratidal environment physically remote from the subtidal setting where they formerly lived without suffering substantial physical decay. Bleaching and death ensues in a supratidal setting, where burial by adjacent dunes may occur. Application of sedimentological criteria, such as the amount of matrix and the sorting of grain sizes provides additional information, for example, where beach sands and related dune sands are dominated by carbonate clasts derived from severely abraded rhodoliths. Brandt (1989, her table 1) gives such sedimentologic criteria equal weight in the correlation of taphonomic grades.

The basic approach to taphonomic studies leads back to the seminal contribution by Johnson (1960), in which three theoretical models for marine assemblages were introduced on a graph plotting post-mortem transportation effects (x-axis) against exposure effects (y-axis). Essentially, Johnson (1960) outlined three possible taphonomic grades arrayed between sudden burial of a marine community more-or-less intact (minus perhaps only the associated soft-body organisms) and the amalgamation of shelly by-products after relatively long periods of transportation and exposure. The numerical system advocated by Brandt (1989) divides taphonomic grades into four outcomes: A, B, C, and D, although further considerations of sedimentology allow for the division into grades D_1 and D_2. Grade assignments based on percentages alone in rating rhodolith deposits are vulnerable to instances where the material undergoes significant transportation without suffering much mechanical breakdown.

Further complication has to do with characteristics of rhodolith banks under life conditions. Steller et al. (2009) show a core drawn from a rhodolith bank in the western Gulf of California, in which the uppermost one or two rhodoliths are alive but the layer below is composed of broken rhodolith debris mixed with dark, organic sediment. In the event such a bank becomes fossilized intact, application of the taphonomic grades defined by Brandt (1989) are problematic because the living bank undergoes its own mechanical breakdown sitting in place. In terms of geologic time, there is no distinction between the living rhodolliths above and the degraded rhodoliths that occur below absent a clear disconformity. That is, the ongoing mechanical decay in a living rhodolith bed is a continuous process that is gradational in effect.

The range of taphofacies advanced for modern and fossil rhodolith beds in this contribution (Fig. 9.2), is easily related to the scheme originally laid out in three relative grades by Johnson (1960). Many of the scenarios described in Sect. 10.6 (above) apply to taphofacies deposited due to transportation during relatively prolonged exposure times during which mechanical breakdown and amalgamation with materials of potentially somewhat different ages has taken place. Assignment to taphonomic grades A to D according to Brandt (1989) based on the amount of breakage and corrosion found in fossil rhodoliths can be calculated, but we consider grade assignments equated with sedimentologic criteria to be more transparent.

9.8.2 Taphofacies Related to Studies on Regional Climatic Change

Prevailing wind patterns and storm tracks are inconstant through geologic time. Conditions were different for the Macaronesian realm during middle Miocene time 15 Ma ago, when the North Atlantic Ocean was not as wide as today and the climatic imbalance between a glaciated Antarctic Continent and minimal ice cap at the North Pole pushed the Inter-Tropical Convergence Zone (ITCZ) out of the equatorial region into the Northern Hemisphere (John et al. 2003; Holbourn et al. 2010). With an ITCZ displaced far north of the Macaronesian realm during mid-Miocene time, the trade winds and hurricane tracks in that region would have been much altered compared to today. This situation appears to coincide with the Middle Miocene Climatic Optimum (MMCO), when rhodoliths were abundant in the shelf waters around islands in the Madeira archipelago. Regionally, the trade winds would have been reversed, blowing from the southeast to the northwest (as now in the tropical Southern Hemisphere), but also brushing parallel to the West African coast to stimulate sustained marine upwelling. A major disruption to the region might also result from a northward shift in the staging area and related storm tracks of mid-Miocene hurricanes.

Under a prolonged Pliocene-Pleistocene cool-down in the Northern Hemisphere (Ravelo et al. 2004), the ITCZ was rebalanced against circulation patterns in the Southern Hemisphere. Thus with regard to surviving fossil rhodolith deposits in the Macaronesian island groups, there is potential for significant differences between early and middle Miocene deposits as opposed to later Miocene, Pliocene, and Pleistocene deposits. This change may account for the relative scarcity of rhodoliths in Madeira waters, today. Middle Miocene taphofacies have yet to be checked in other Macaronesian island groups in order to test for possible differences in climatic regime.

9.9 Future Perspectives

Island groups in the Macaronesian realm demonstrate a rich geological history of limestone strata with abundant rhodoliths that range in age from the middle Miocene to the Holocene. Coastal deposits on certain islands also help to characterize the present-day fecundity of offshore rhodolith beds. In comparison with one another, these past and contemporary deposits illustrate a wide range of taphonomic and sedimentological associations that represent all possible life and post-mortem outcomes in the natural history of rhodoliths. The age of volcanic islands is the principal factor that influenced the distribution of former rhodolith banks. Age is directly correlated with the time available for the erosion of broad insular shelves capable of fostering growth of rhodoliths under favorable conditions. Such conditions are shaped by the waves and currents stimulated by atmospheric circulation patterns as now in place with the northeast trade winds.

The few islands such as Fogo in the Cape Verde archipelago and El Hierro in the Canary archipelago (Fig. 9.1) that remain volcanically active today, exhibit narrow marine shelves that rapidly slope into deeper water. The existing beaches around these islands are composed of black, volcanic sand. Rhodolith beach debris is rare to non-existent because the offshore banks that normally shelter rhodoliths have not had sufficient time to develop. In contrast, older islands such as Maio and Sal in the Cape Verde archipelago or Fuerteventura in the Canary archipelago preserve extensive coastal deposits rich in fossil and contemporary rhodoliths.

Among the early naturalists who visited the islands of Macaronesia, only Charles Darwin understood the potential significance of rhodoliths as a major contributor to limestone. He exclaimed this insight with considerable zeal during his earliest years as a geologist and naturalist soon after his travels on the H.M.S. *Beagle* (Darwin 1839, 1844). In particular, the young Darwin was much impressed by the limestone band defined below and above by co-extensive basalt layers along the southeast coast of Santiago in the Cape Verde archipelago.

The diversity of sedimentary facies filled by the rhodoliths exported from their natural habitat spans a range of depositional settings preserved as tempestites, platform over-wash deposits, as well as beachrock, beach, berm, hurricane, tsunami, and coastal dune deposits. Among these, the tempestites, platform over-wash deposits, beachrock deposits, and eolianites all record information on the direction and source of rhodolith materials transported into the deposits. This is valuable information that contributes to a better understanding of how patterns of atmospheric and marine circulation changed over the Macaronesian realm during the last 25 Ma or so, preceding and following the middle Miocene Climatic Optimum. The thesis that profound changes in prevailing winds and shifts in hurricane tracks occurred over the Macaronesian realm during the transition from middle Miocene to early Pliocene times requires more research.

Additional field studies should be undertaken on all island groups in Macaronesia. This survey points out, for example, how little is known regarding possible rhodolith deposits on one of the region's oldest and most inaccessible island groups, the

Selvagens. In the Cape Verde archipelago, future research on rhodolith deposits of all ages and conditions can be expected on Boa Vista and São Nicolau. Research on limestone bands from the northeast shores of Santiago will provide geographic balance to the classic studies on deposits from the Praia area in the southeast part of the island initiated by Darwin (1839, 1844) and enlarged by Johnson et al. (2012). Likewise, new research is needed on fossil deposits from the islands of Lanzarote, Gran Canaria, and Tenerife in the Canary archipelago. Ongoing research in the Azores also will provide more insight on why those islands are so different from others in Macaronesia with respect to rhodolith deposits in the geological record and the general lack of coastal rhodolith deposits today. In all cases, fresh efforts can be expected to yield information on changing environmental patterns in atmospheric and marine circulation as keyed to directional data preserved by deposits as different as tempestites and eolianites.

Acknowledgments Funding for fieldwork in the Madeira, Canary, and Cape Verde island groups from 2011 to 2013 was provided under grant CGL2010-15372-BTE from the Spanish Ministry of Science and Innovation to project leader Eduardo Mayoral (University of Huelva). E. Mayoral and A. Santos acknowledge the support given by the research group RNM-276. Esther Martín González and Carolina Castillo (La Laguna University, Canary Islands) are thanked for their help and expert guidance on Fuerteventura during the 2011 field season of research. We are grateful to the rangers at the nature preserve in the Savage Islands for providing beach and dune samples from the landing place on the southwest side of Selvagem Pequena. Thanks also go to Mónica Albuquerque at EMEPC for her assistance in providing dredge data from the Selvagems. We are grateful to Julio Aguirre Rodriguez (University of Granada) for engaging us in stimulating discussions that helped to clarify outlooks on taphonomy and to an anonymous reviewer for comments that helped to improve this paper.

References

Abreu MM, Ribeiro L, Arsénio P, Bulcão L (2002) Dinâmica Geomorfológica como Metodologia Básica na Avaliação da Sensibilidade da Paisagem. Aplicação ao Ilhéu de Vila Franca do Campo, S. Miguel, Açores, 49–58. In: Associação Portuguesa de Geomorfólogos (ed) Contribuições para a Dinâmica Geomorfológica, vol 1. Publicações da Associação Portuguesa de Geomorfólogos, Lisboa, pp 1–140

Ávila SP, Amen R, Azevedo JMN, Cachão M, García-Talavera F (2002) Checklist of the Pleistocene marine molluscs of Praínha and Lagoínhas (Santa Maria island, Azores). Açoreana 9(4):343–370

Baarli BG, Santos A, Silva CM, Ledesma-Vázquez J, Mayoral E, Cachão M, Johnson ME (2011) Diverse macroids and rhodoliths from the Upper Pleistocene of Baja California Sur, Mexico. J Coast Res 28:296–305

Baarli BG, Santos A, Mayoral EJ, Ledesma-Vázquez J, Johnson ME, da Silva CM, Cachão M (2013) What Darwin did not see: pleistocene fossil assemblages on a high-energy coast at Ponta das Bicudas, Santiago, Cape Verde Islands. Geol Mag 150:183–189

Baarli BG, Cachão M, da Silva CM, Johnson ME, Mayoral EJ, Santos A (2014) A middle miocene carbonate embankment on an active volcanic slope: Ilhéu de Baixo, Madeira Archipelago, Eastern Atlantic. Geol J 49:90–106

Bassi D, Humblet M, Iryu Y (2011) Recent ichnocoenosis in deep water macroids, Ryukyu Islands, Japan. Palaios 26:232–238

Bassi D, Iryu Y, Humblet M, Matsuda H, Machiyama H, Sasaki K, Matsuda S, Arai K, Inoue T (2012) Recent macroids on the Kikai-jima shelf, Central Ryukyu Islands, Japan. Sedimentology. doi:10.1111/j.1365-3091.2012.01333.x

Brand S (ed) (2011) African severe weather port guide from naval research laboratory in Monterey, California. http://www.nrlmry.navy.mil/ port_studies/africaports/Mindelo/index.html. Last modified Apr 2011

Brandt DS (1989) Taphonomic grades as a classification for fossiliferous assemblages and implications for paleoecology. Palaios 4:303–309

Brett CE, Baird GC (1986) Comparative taphonomy: a key to paleoenvironmental interpretation based on fossil preservation. Palaios 1:207–227

Brooke B (2001) The distribution of carbonate eolianite. Earth-Sci Rev 55:135–164

Chamberlain YM, Irvine LM, Walker RI (1988) A redescription of Lithophyllum crouanii (Rhodophyta, Corallinales) in the British Isles with an assessment of its relationship to L. orbiculatus. Br Phycol J 23:177–192

Checconi A, Monaco P (2008) Trace fossil assemblages in rhodoliths from the Middle Miocene of Mt. Camposauro (Longano Formation, Southern Apennines, Italy). Stud Trent Sci Nat Acta Geol 83:165–176

Checconi A, Bassi D, Carannante G, Monaco P (2010) Re-deposited rhodoliths in the Middle Miocene hemipelagic deposits of Vitulano (Southern Apennines, Italy): coralline assemblage characterization and related trace fossils. Sed Geol 225:50–66

Darwin C (1839) Journal and remarks, 1832–1836. In: FitzRoy R (ed) Narrative of the surveying voyages of his majesty's ships adventure and beagle between the years 1826 and 1836, Volume 3. Henry Colburn, London, pp 1–615

Darwin C (1844) Geological observations on the volcanic islands visited during the voyage of the H.M.S. Beagle. Smith, Elder & Co, London, pp 1–175

Dias GT, Villaça RC (2012) Coralline algae depositional environments on the Brazilian central southeastern shelf. J Coast Res 28:270–279

Dott RH Jr, Bourgeois J (1982) Hummocky stratification: significance of its variable bedding sequences. Bull Geol Soc Am 93:663–680

Dumas S, Arnott RWC (2006) Origin of hummocky and swaley cross-stratification—the controlling influence of unidirectional current strength and aggradation rate. Geology 34:1073–1076

Ferreira MP (1985) Evolução geocronológica e paleomagnética das ilhas do arquipélago da Madeira: uma síntese. Mem Not Mus Lab Miner Geol Univ Coimbra 99:213–218

Ferreira MP, Macedo CR, Ferreira JF (1988) K/Ar geochronology in the Selvagens, Porto Santo and Madeira islands (Eastern-Central Atlantic): a 30-m.y. spectrum of submarine and subaerial volcanism. Lunar Planets Inst 19:325–326

Foslie MH (1898) Some new or critical lithothamnia. Konglige norske Videnskapelige Selskap Skrifter 6:1–19

Foster MS (2001) Rhodoliths: between rock and soft places. J Phycol 37:659–667

Geldmacher J, Hoernle K, Bogaard P, Zankl G, Garbe-Schonberg D (2001) Earlier history of the ≥ 70 Ma-old Canary hotspot based on the temporal and geochemical evolution of the Selvagem Archipelago and neighboring seamounts in the eastern North Atlantic. J Volcanol Geotherm Res 111:55–88

Gutiérrez M, Casillas R, Fernández C, Balogh K, Ahijado A, Castillo C, Colmenero JR, García-Navarro E (2006) The submarine volcanic succession of the basal complex of Fuerteventura, Canary Islands: a model of submarine growth and emergence of tectonic volcanic islands. Bull Geol Soc Am 118:785–804

Herbert S (2005) Charles Darwin, geologist. Cornell University Press, Ithaca, pp 1–485

Holbourn A, Kuhnt W, Regenberg M, Schulz M, Mix A, Andersen N (2010) Does Antarctic glaciation force migration of the tropical rain belt? Geology 38:783–786

Hontoria F, Redón S, Maccari M, Varó I, Navarro JC, Ballell L, Amat F (2012) A revision of *Artemia* biodiversity in Macaronesia. Aquat Biosyst 8:25–31

Hutchings PA, Kiene WE, Cunningham RB, Donnelly C (1992) Spatial and temporal patterns of non-colonial boring organism (polychaetes, sipunculans and bivalve molluscs) in *Porites* at Lizard Island, Great Barrier Reef. Coral Reefs 11:23–31

Instituto Hidrográfico (2008) Caracterização dos depósitos sedimentares da Plataforma insular sul da Ilha do Porto Santo. Final technical report Project GM 52OP02, Hydrographic Institute Portuguese Navy. p 78

John CM, Mutti M, Adatte T (2003) Mixed carbonate-siliciclastic record on the North African margin (Malta) – coupling of weathering processes and mid Miocene climate. Bull Geol Soc Am 115:217–229

Johnson RG (1960) Models and methods for the analysis of the mode of formation of fossil assemblages. Geol Soc Am Bull 71:1075–1086

Johnson ME, da Silva CM, Santos A, Baarli BG, Cachão M, Mayoral E, Rebelo AC, Ledesma-Vázquez J (2011) Rhodolith transport and immobilization on a volcanically active rocky shore: middle Miocene at Cabeço das Laranjas on Ilhéu de Cima (Madeira Archipelago, Portugal). Palaeogeogr Palaeoclimatol Palaeoecol 300:113–127

Johnson ME, Baarli BG, Cachão M, da Silva CM, Ledesma-Vázquez J, Mayoral E, Ramalho R, Santos A (2012) Rhodoliths, uniformitarianism, and Darwin: pleistocene and recent carbonate deposits in the Cape Verde and Canary archipelagos. Palaeogeog Palaeoclim Palaeoecol 329–330:83–100

Johnson ME, Baarli BG, da Silva CM, Cachão M, Ramalho R, Ledesma-Vázquez J, Mayoral E, Santos A (2013) Coastal dunes with high content of rhodolith (coralline red algae) bioclasts: pleistocene formations on Maio and São Nicolau in the Cape Verde archipelago. Aeolian Res 8:1–9

Johnson ME, Ramalho RS, Baarli BG, Cachão M, da Silva CM, Mayoral EJ, Santos A (2014) Miocene-Pliocene rocky shores on São Nicolau (Cape Verde Islands): contrasting windward and leeward biofacies on a volcanically active oceanic island. Palaeogeogr Palaeoclimatol Palaeoecol 395:131–143

Ledesma-Vázquez J, Johnson ME, Backus DH, Mirabal-Dávila C (2007) Coastal evolution from transgressive barrier deposit to marine terrace on Isla Coronados, Baja Califonria Sur, Mexico. Cien Mar 33:335–351

Mata J, Fonseca PE, Prada S, Rodrigues D, Martins D, Ramalho R, Madeira J, Cachão M, Silva CM, Matias MJ (2013) O Arquipélago da Madeira. In: Dias R, Araújo A, Terrinha P, Kullberg JC (eds) Geologia de Portugal, vol. II. Geologia Meso-cenozóica, Escolar Editora, pp 691–746

Morri C, Bianchi CN (1996) Ecological niches of hermatypic corals at Ilha do Sal (Arquipélago de Cabo Verde). Bol Mus Munic Funchal 4:473–485

Nebelsick JH, Bassi D (2000) Diversity, growth forms and taphonomy: key factors controlling the fabric of coralline algae dominated shelf carbonates. In: Insalaco E, Skelton P, Palmer T (eds) Carbonate platform systems: components and interactions. Geol Soc London Spec Pubs 178:89–107

Paris R, Giachetti T, Chevalier J, Guillou H, Frank N (2011) Tsunami deposits in Santiago Island (Cape Verde archipelago) as possible evidence of a massive flank failure of Fogo volcano. Sediment Geol 239:129–145

Penrose DL (1996) Genus Neogoniolithon. In: The marine benthic flora of southern Ausgtralia. Rhodophyta. Part IIIB, Cracilariales, Rhododymeniales, Corallinales and Bonnemaisoniales. Australian Biological Resources Study, Canberra: 280–283

Puga-Bernabéu A, Martín JM, Braga JC (2007) Tsunami-related deposits in temperate carbonate ramps, Sorbas Basin, southern Spain. Sediment Geol 199:107–127

Quartau R, Trenhaile AS, Mitchell NC, Tempera F (2010) Development of volcanic insular shelves: insights from observations and modelling of Faial Island in the Azores Archipelago. Mar Geol 275:66–83

Ramalho R, Quartau R, Trenhaile AS, Mitchell NC, Woodroffe CD, Ávila SP (2013) Coastal evolution on oceanic islands: a complex interplay between volcanism, erosion, sedimentation, sea-level change and biogenic production. Earth-Sci Rev 127:140–170

Ravelo AC, Andreasen DH, Lyle M, Olivarez-Lyle A, Wara MW (2004) Regional climate shifts caused by gradual global cooling in the Pliocene epoch. Nature 429:263–267

Rebelo AC, Rasser MW, Riosmena-Rodríguez R, Neto AI, Ávila SP (2014) Rhodolith forming coralline algae in the Upper Miocene of Santa Maria Island (Azores, NE Atlantic): a critical evaluation. Phytotaxa 190:370–382

Rosas-Alquicira EF, Riosmena-Rodríguez R, Couto RP, Neto AI (2009) New additions to the Azorean algal flora, with ecological observations on rhodolith formations. Cah Biol 50:143–151

Rusu L, Soares CG (2012) Wave energy assessments in the Azores islands. Renew Energy 45:183–196

Santos A, Mayoral E, Johnson ME, Baarli BG, Silva CM, Cachão M, Ledesma-Vázquez J (2012) Basalt mounds and adjacent depressions attract contrasting biofacies on a volcanically active Middle Miocene shoreline (Porto Santo, Madeira Archipelago, Portugal). Facies 58:573–585

Serralheiro A (1970) Geologia da Ilha de Maio. Junta de Investigações do Ultramar, Lisbon, pp 1–103

Setchell WA, Mason LR (1943) Goniolithon and Neogoniolithon two genera of crustaceous coralline algae. Proc Natl Acad Sci 29(3–4):87–92

Soares AF (1973) A formação eolianítica da ilha de Porto Santo. Memórias e Notícias, Publ. Museu Lab Min. Geológico Univ. Coimbra Cent. Estud Geol 75:47–88

Steiner C, Hobson A, Favre P, Stampfli GM, Hernandez J (1998) Mesozoic sequence of Fuerteventura (Canary Islands): witness of early jurassic sea-floor spreading in the central Atlantic. Bull Geol Soc Am 110:1304–1317

Steller DL, Riosmena-Rodríguez R, Foster MS (2009) Libing rhodolith bed ecosystems in the Gulf of California. In: Johnson ME, Ledesma-Vázquez J (eds) Atlas of coastal ecosystems in the Western Gulf of California. University of Arizona Press, Tucson, pp 73–82

Taylor PD, Wilson MA (2003) Palaeoecology and evolution of marine hard substrates communities. Earth Sci Rev 62:1–103

Tores PC, Silva LC, Serralheiro A, Tassinari C, Munhá J (2002) Enquadramento geocronológico pelo metodo K/Ar das principais sequências volcano-estratigráficas da Ilha do Sal – Cabo Verde, vol 18. Garcia de Orta, Serie Geologia, Lisboa, pp 9–13

Van Der Land J (1993) Marine biota of the Cape Verde Islands. Cour Forsch Inst Senckenb 159:39–44

Williams H, Delinger E (2013) Contributions of Hurricane Ike storm surge sedimentation to long-term aggradation of southeastern Texas coastal marshes. J Coast Res 65:838–843, Sp Issue

Zazo C, Goy JL, Hillaire-Marcel C, Gillot PY, Soler V, González JH, Dabrio CJ, Ghaleb B (2002) Raised marine sequences of Lanzarote and Fuerteventura revisited – a reappraisal of relative sea-level changes and vertical movements in the eastern Canary Islands during the quaternary. Quat Sci Rev 21:2019–2046

Part III
Conservation Status of Rhodolith/Maërl at Major Ocean Basins

Chapter 10
North Atlantic Rhodolith Beds

Jazmin J. Hernandez-Kantun, Jason M. Hall-Spencer, Jacques Grall, Walter Adey, Fabio Rindi, Christine A. Maggs, Ignacio Bárbara, and Viviana Peña

10.1 Introduction

Aggregations of living unattached corallines, previously often referred to as nodules, were given the name rhodoliths by Bosselini and Ginsburg (1971). Adey and MacIntyre (1973) provided an early discussion of their genesis and distribution. Such aggregations have long been known as maerl in the North East Atlantic, a Breton term for unattached thalli that lack a shell or pebble core (Irvine and Chamberlain 1994). Here, we provide an overview of rhodolith/maerl occurrence in the colder/temperate waters of the North Atlantic and summarize the distribution, species composition, biodiversity and ecological importance of these habitats. We

J.J. Hernandez-Kantun (✉)
National Museum of Natural History, Smithsonian Institution, Washington, DC, USA

Irish Seaweed Research Group, Ryan Institute, National University of Ireland Galway, Galway, Ireland

Phycology Research Group, Ghent University, Ghent, Belgium
e-mail: jaz1083@gmail.com

J.M. Hall-Spencer
Marine Biology and Ecology Research Centre, Plymouth University, Plymouth PL4 8AA, UK

Shimonda Marine Research Centre, University of Tsukuba, Japan
e-mail: jason.hall-spencer@plymouth.ac.uk

J. Grall
Séries Faune-Flore, Observatoire UMS 3113, Institut universitaire Européen de la Mer, Plouzane 29280, France
e-mail: Jacques.grall@univ-brest.fr

W. Adey
Botany Department, National Museum of Natural History, Smithsonian Institution, Washington, DC, USA
e-mail: adeyw@si.edu

R. Riosmena-Rodríguez et al. (eds.), *Rhodolith/Maërl Beds: A Global Perspective*, Coastal Research Library 15, DOI 10.1007/978-3-319-29315-8_10

include the eastern coasts of Canada and the United States north of Cape Hatteras plus Greenland, Iceland, Europe and Macaronesia (Azores, Madeira and Canary Islands).

10.2 Distribution and Species Composition

Unattached coralline algal beds occur on open coasts, in tide-swept channels and in sheltered areas with weak currents, from the intertidal to 60 m depth with the deepest records found in Svalbard and Macaronesia (Cabioch 1974; Alfonso-Carrillo and Gil-Rodríguez 1982; Hall-Spencer et al. 2010; Teichert et al. 2012). These habitats are particularly abundant in Scotland, Ireland and Brittany (De Grave et al. 2000; Hall-Spencer et al. 2010) and in Newfoundland and Labrador (Adey and MacIntyre 1973). The more northern beds are less well known but existing literature indicates that these habitats may be important and extensive at high latitudes (Foslie 1895; Adey 1968; Sneli 1968; Gunnarsson 1977; Freiwald 1995; Teichert et al. 2012).

Many species of crustose coralline algae can continue to survive unattached in appropriate environmental conditions. However, fewer species commonly form rhodoliths or maerl. For example, *Lithothamnion lemoineae* Adey forms thick crusts in the North Atlantic subarctic that can become detached but rarely survive as rhodoliths. In contrast, *Clathromorphum compactum* (Kjellman) Foslie occasionally forms rhodolith beds in the Subarctic coating pebbles in narrow island passages with strong currents.

The North Atlantic harbours 13 rhodolith/maerl-forming species of coralline algae (Table 10.1). The four most widely recorded species are *Lithothamnion corallioides, L. glaciale, L. tophiforme* and *Phymatolithon calcareum* (Fig. 10.1). These

F. Rindi
Dipartimento di Scienze della Vita e dell'Ambiente, Università Politecnica delle Marche, Via Brecce Bianche, 60131 Ancona, Italy
e-mail: f.rindi@univpm.it

C.A. Maggs
School of Biological Sciences, The Queen's University of Belfast, 97 Lisburn Road, Belfast BT9 7BL, Northern, Ireland
e-mail: c.maggs@qub.ac.uk

I. Bárbara
Grupo de investigación BIOCOST, University of A Coruña, Coruña, Spain
e-mail: barbara@udc.es

V. Peña
Institut de Systématique, Evolution, Biodiversité, UMR 7205 ISYEB CNRS, MNHN, UPMC, EPHE, Muséum national d'Histoire naturelle (MNHN), Paris, France

Phycology Research Group, Ghent University, Ghent, Belgium
e-mail: vpena@udc.es

Table 10.1 Rhodolith/maerl-forming species of coralline algae in the Northern Atlantic. Information on classification, type locality and distribution are presented. Unverified records are not included

Order/family/species	Type locality	Distribution in the NE Atlantic
Corallinales		
Corallinaceae		
Lithophylloideae		
Lithophyllum crouaniorum Foslie	Berwick-upon Tweed, Northumberland, England	Azores (as *L. crouanii)*[1]
Lithophyllum hibernicum Foslie	Ballynakill Harbour, Galway, Ireland	Ireland[2,3], France[4], England[4],
Lithophyllum incrustans Philippi	Sicily, Italy	Ireland[4], France[4]
Mastophoroideae		
Neogoniolithon brassica-florida (Harvey) Setchell & L.R. Mason	Algoa Bay, Cape Province, South Africa	Azores[1]
Spongites fruticulosa Kützing	Mediterranean Sea	Madeira[5]
Hapalidiales		
Hapalidiaceae		
Melobesioideae		
Leptophytum foecundum (Kjellman) Adey	Kara Sea, Russia.	Newfoundland and Labrador Canada (Adey pers. comm.)
Lithothamnion corallioides (P.L. Crouan & H.M. Crouan) P.L. Crouan & H.M. Crouan	Rade de Brest, Finistère, France	France[2,5,6], Ireland[2,7,8], England[2], Spain[9], Madeira[5], Canary Islands[10]
Lithothamnion glaciale Kjellman	Spitzbergen, Svalbard, Norway.	North Atlantic Arctic and Subarctic[11,12,13,14], Svalvard[15], Iceland[16,17,18], Scandinavia[19], Norway[17,20], Britain[2,21], Ireland[2,21]
Lithothamnion tophiforme (Esper) Unger	Julianehaab, Greenland	Canada[11,12], Greenland[12], Iceland[17,18,22], Norway[12,17,20]
Mesophyllum sphaericum V. Peña, I. Bárbara, W.H. Adey, R. Riosmena-Rodríguez & H.G. Choi	Benencia Island, Ria de Arousa, Galicia, Spain	Spain[23,24]
Phymatolithon calcareum (Pallas) Adey & McKibbin	Falmouth, England	Sweden[25], Norway[17], British Isles[2], Ireland[26,27,28], France[29,30,31], Spain[9,32], Azores[4]

(continued)

Table 10.1 (continued)

Order/family/species	Type locality	Distribution in the NE Atlantic
Phymatolithon lusitanicum V. Peña	Con de Pego, Ría de Vigo, Galicia, Spain,	Spain, Portugal, Ireland[33]
Phymatolithon purpureum (P.L. Crouan & H.M. Crouan) Woelkerling & L.M. Irvine	Mingant, Brest, France	Norway[22], Britain and Ireland[8]

References: 1. Rosas-Alquicira et al. (2009); 2. Irvine and Chamberlain (1994); 3. Foslie (1899); 4. Hernández-Kantún et al. (2015a), 5. Cabioch (1974); 6. Cabioch (1969); 7. Bosence (1980); 8. Blunden et al. (1981); 9. Peña and Bárbara (2008b); 10. Alfonso-Carrillo and Gil-Rodríguez (1982); 11. Kulka et al. (2004); 12. Adey et al. (2005); 13. Gagnon et al. (2012); 14. Kamenos et al. (2012); 15. Teichert et al. (2012); 16. Foslie (1905); 17. Adey (1971); 18. Gunnarsson (1977); 19. Athanasiadis (1996); 20. Foslie (1895); 21. Strömfelt (1886); 22. Adey and Adey (1973); 23. Peña et al. (2011); 24. Peña et al. (2015a); 25. Suneson (1958); 26. Bosence (1976); 27. De Grave and Whitaker (1999); 28. De Grave et al. (2000); 29. Lemoine (1910); 30. Mendoza and Cabioch (1998); 31. Sauriau et al. (2012); 32. Adey and McKibbin (1970); 33. Peña et al. (2015b)

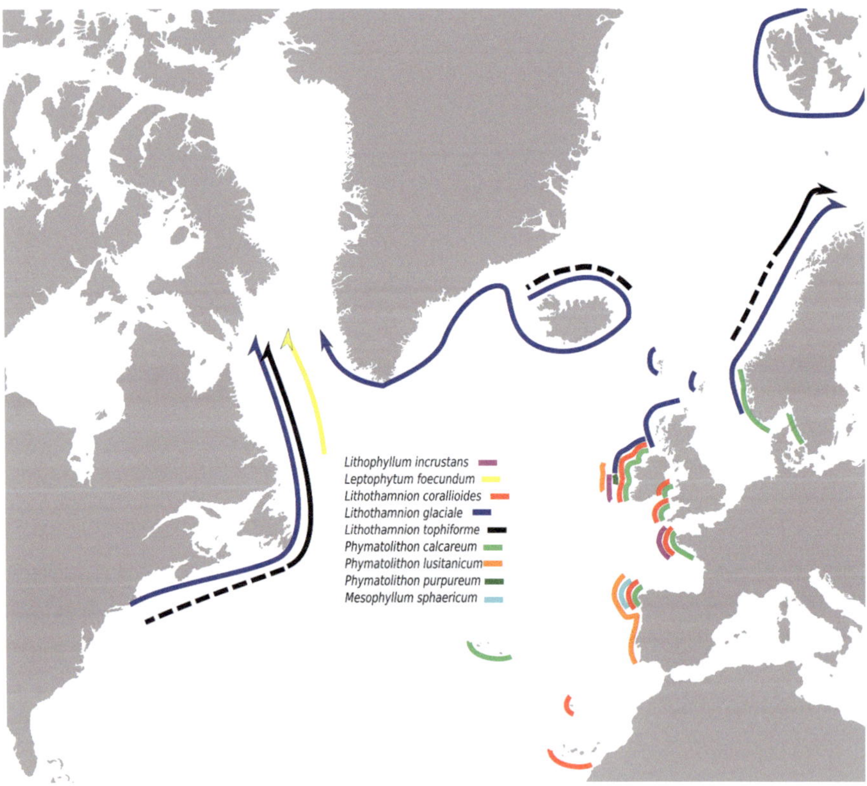

Fig. 10.1 Distribution map and regions of interaction for unattached species of coralline algae in the North Atlantic. Nine species are represented; *Lithothamnion glaciale* and *Phymatolithon calcareum* are the most widespread. Interrupted lines indicate that the abundance is lower than in other areas represented. *Arrows* indicate possible direction of distribution (information requires future verification)

four species have been exhaustively researched in the North Atlantic to find reliable identification features by Adey and McKibbin (1970), Irvine and Chamberlain (1994), Adey et al. (2005), and Teichert et al. (2012). In the northwestern Atlantic, *Lithothamnion glaciale* and *L. tophiforme* occupy different, overlapping ecological niches. *L. glaciale* is particularly widespread, ranging from the intertidal pools to moderately deep water and from highly exposed to protected shorelines in Subarctic and Boreal Regions (Adey 1966). In contrast, *L. tophiforme* is more abundant in Arctic and deeper waters (25–39 m, Adey et al. 2005). On the Labrador Coast, the two species typically occur together at 20–25 m. *Leptophytum foecundum*, in its rhodolith state, is colder Subarctic to Arctic in distribution and occurs primarily in shell and pebble beds in strong current zones. The cores of this species are usually pebbles or dead mollusc shells (Adey pers. comm.)

Currently, efforts are being made to understand the diversity of rhodolith-forming corallines in the North Atlantic area using molecular techniques (Pardo et al. 2014; Peña et al. 2014a, Peña et al. 2015a, b; Hernández-Kantún et al. 2014, 2015a, b). This approach has led to the recent description of two new species (*Phymatolithon lusitanicum* and *Mesophyllum sphaericum*) and the taxonomic re-assessment of *Lithophyllum incrustans* and *L. hibernicum* which form maerl beds in the NE Atlantic (Peña et al. 2011, 2015a, b; Hernández-Kantún et al. 2015a, b). The use of DNA sequences from type material will continue to improve the taxonomy of this group (Peña et al. 2014a; Hernández-Kantún et al. 2015a, b). Species of North Atlantic *Lithophyllum* will require considerable taxonomic work as there appear to be cryptic species, problems with the features used to separate species and the widespread use of incorrect names (Irvine and Chamberlain 1994). In the north Atlantic, *Lithophyllum* rarely contribute to maerl beds but at a few sites, such as Mannin Bay and Kingstown Bay (Ireland) and the Bay of Brest (France), *L. incrustans* (traditionally reported as *L. dentatum* and *L. fasciculatum*) can be abundant (Cabioch 1968; Hall-Spencer et al. 2010; Hernández-Kantún et al. 2015a, b).

The colder/temperate parts of the North Atlantic Ocean comprise four biogeographic regions: Arctic; Subarctic; Boreal (Celtic) and Lusitanian (Adey and Steneck 2001; Adey and Hayek 2011). The Subarctic Region transitions with the Boreal Region primarily in southern Iceland, the Norwegian Coast and to a lesser extent in northwestern Scotland. The North Atlantic Subarctic Region (NW Atlantic, S. Greenland and Spitzbergen and northern Iceland) as a Pleistocene, glacially-restricted outlier of the North Pacific Subarctic is relatively depauperate in seaweed species (Adey et al. 2008; Maggs et al. 2008), while the overlapping Boreal and Lusitanian Regions are relatively rich (Tittley 2002). Likewise the number of coralline algae species forming rhodolith/maerl beds is higher in European coastal Boreal/Lusitanian regions than the Subarctic Coasts. A recent study comprising maerl collections from the OSPAR maritime area observed that the majority of maerl-forming species have small distribution ranges, and revealed a latitudinal replacement of maerl species (Pardo et al. 2014) But also it has been detected taxa not assigned to any known species, and the existence of cold-water species described in the early twentieth century and never recorded again. Further studies, particularly in understudied regions of NE Atlantic, are necessary to disentangle the species diversity and to to fully clarify these relationships and to understand evolutionary patterns.

10.3 Biodiversity and Ecology

Unattached coralline algae produce banks of carbonate rich sediments that trap suspended particulates and infaunal faeces to build complex habitats that can be 1,000s of years old (Grall and Hall-Spencer 2003). They can also provide sediments for current transport to adjacent marine habitats (Farrow et al. 1978; Bosence 1979, 1980) and to coasts where carbonate sand dunes are important habitats for terrestrial organisms (Birkett et al. 1998a).

Rhodoliths and maerl are formed by coralline algae with relatively slow growth rates and a perennial life strategy; some thalli can live >100 years (Bosence 1983; Potin et al. 1990; Foster 2001; Blake and Maggs 2003; Bosence and Wilson 2003; Frantz et al. 2005). Many rhodoliths begin as crusts growing on rock; they break free from the substratum due to invertebrate boring and/or wave action and bioturbation, and continue growing unattached on surrounding or deeper sedimentary bottoms (Adey and MacIntyre 1973; Adey and Hayek 2011). Thallus fragmentation is the major mechanism of propagation of rhodolith beds (Bosence 1976; Johansen 1981). Coralline algae can accumulate in large beds with a three-dimensional matrix that provides a wide range of ecological niches (Bosence 1983; Birkett et al. 1998a; Hall-Spencer 1998; Barberá et al. 2003; Grall et al. 2006; Hinojosa-Arango et al. 2009).

These habitats have considerable ecological significance due to the high diversity of associated organisms (Fig. 10.2). There have been numerous studies of the associated flora and fauna of NE Atlantic beds (Cabioch 1969; Keegan 1974; Bosence 1979, 1980; Maggs 1983a, b; Hall-Spencer 1998, BIOMAERL Team 1999; Hall-Spencer and Atkinson 1999; Barbéra et al. (2004), Kamenos et al. 2004a, b, c; Hinojosa-Arango et al. 2009; Peña and Bárbara 2008a, b, 2010a, b, 2013; Peña 2010; Teichert et al. 2012; Scottish MPA Project 2013). In contrast, the number of studies available for the northwestern Atlantic is more limited: two studies in Canada include investigations on fish of rhodolith beds by Kulka et al. (2004) and the epi- and in-fauna by Gagnon et al. (2012).

A recent review of the associated flora of NE Atlantic rhodolith beds (Peña et al. 2014b) reported a total of 349 algal species, which is around 30 % of the total seaweed diversity in this region. The most diverse beds (150–257 species) occur in the temperate regions with 25–42 % of the total seaweed diversity recorded in the area. The flora of northern rhodolith beds (Iceland, Scandinavia) is less diverse but has not been studied in detail. The associated flora increases the heterogeneity and complexity of the habitat due to the presence of a wide range of algal morphologies, ranging from endolithic and crustose forms to large kelp. The associated flora of temperate NE Atlantic rhodolith beds is seasonal with a peak of diversity in the

Fig. 10.2 North Atlantic beds of maerl and rhodoliths. (**a**) Mannin Bay, Ireland, pale pink *Lithophyllum incrustans* (previously reported as *L. fasciculatum), vivid red/pink Phymatolithon calcareum.* (**b**) Iceland dredge sample, *Lithothamnion glaciale.* (**c**) Galicia, Spain, *Mesophyllum sphaericum.* (**d**) Dog Island, Northern Labrador, Canada, mix of *L. glaciale* and *L. tophiforme*

spring-summer period (Cabioch 1969; Maggs 1983a; Bárbara et al. 2004; Grall et al. 2006; Peña and Bárbara 2010a). Species with heteromorphic life-histories and crustose sporophyte phases are particularly abundant (Maggs and Guiry 1989; Peña and Bárbara 2010b, 2013). The presence of some algal species is largely confined to rhodolith beds, such as the Rhodophyta*Gelidiella calcicola*, *Cruoria cruoriaeformis*, *Gelidium maggsiae* and *Halymenia latifolia* and the Chlorophyta*Cladophora rhodolithicola* (Cabioch 1969; Maggs 1983a, b; Maggs and Guiry 1987, 1989; Rico and Guiry 1997; Bárbara et al. 2004; Grall et al. 2006; Leliaert et al. 2009; Peña and Bárbara 2010a, b, 2013).

The BIOMAERL project revealed 556 animal taxa associated with NE Atlantic maerl beds: 10 Porifera, 35 Cnidaria, 174 Annelida, 137 Crustacea, 138 Mollusca, 26 Echinodermata, 22 Bryozoa, 11 Ascidiacea, and 32 Pisces taxa (Barberá et al. 2003). A 10 year survey of 9 maerl beds in Brittany lists 1,204 fauna species, including 295 Annelida, 290 Arthropoda, 207 Mollusca, 35 Echinodermata, and 377 species of other phyla (Grall unpublished).

The three-dimensional matrix of coralline algal sediments enhances their biodiversity (Bosence 1979); this is evident when their diversity is compared with surrounding habitats; kelp forests harbour around 10 % fewer species of the NE Atlantic flora, seagrasses host around 5 % fewer species, and muddy and sand banks harbour 30–50 % fewer species than adjacent maerl beds (Whelan and Cullinane 1985; Shultze et al. 1990; Davison and Hughes 1998; Birkett et al. 1998b; Grall 2002; Otero-Schmitt and Pérez-Cirera 2002; Johnson et al. 2005). Maerl beds also provide brood strock areas for commercially important bivalves, such as scallops, as well as nursery areas for juvenile fish such as cod (Hall-Spencer et al. 2003; Kamenos and Moore 2004; Kamenos et al. 2004a, b, c). In the northwestern Atlantic, rhodolith beds of *Lithothamnion* typically support abundant scallops, (*Aequipecten irradians*) which are commercially harvested.

10.4 Conservation Status

In the North Atlantic, coralline algal sediments are dredged for use as a fertilizer on acidic soils, as a food additive for animals and humans, in water treatment filters and in cosmetics (Blunden et al. 1981; De Grave and Whitaker 1999; Foster 2001; Grall and Hall-Spencer 2003; Emerton and Choi 2008). This dredging of biodiverse nursery grounds is widely considered to be environmentally unsustainable (Barberá et al. 2003) and has ceased in the United Kingdom.

Although not as devastating as dredging, European maerl beds have also been impacted by aquaculture (Hall-Spencer et al. 2006; Hall-Spencer and Bamber 2007; Peña and Bárbara 2008a; Peña 2010), changes in current patterns associated with construction (Birkett et al. 1998a; Grall and Hall-Spencer 2003), dredge fisheries (Hall-Spencer 1998; Hall-Spencer and Moore 2000; Hall-Spencer et al. 2003; Hauton et al. 2003; Kamenos et al. 2003), as well as increased sedimentation and eutrophication (Hily et al. 1992; Grall and Glemarec 1997). There is also a growing realisation that these habitats may be especially vulnerable to ocean acidification since the high Mg-calcite skeletons of coralline algae dissolve easily as CO_2 levels rise (Nelson 2009; Büdenbender et al. 2011; Porzio et al. 2011; Diaz-Pulido et al. 2012; Noisette et al. 2013, see Chapter by Martin and Hall-Spencer).

Indirect effects to NE Atlantic maerl beds arise due to the spread of invasive species such as the gastropod *Crepidula fornicata,* the Rhodophyta*Antithamnionella spirographidis*, *A. ternifolia, Bonnemaisonia hamifera*, *Dasya sessilis, Grateloupia turuturu, Heterosiphonia japonica* and *Neosiphonia harveyi*, the Heterokontophyta*Colpomenia peregrina, Sargassum muticum* and *Undaria pinnatifida*, and the Chlorophyta *Codium fragile* (Grall and Hall-Spencer 2003; Peña et al. 2014b). Only through monitoring studies will we be able to detect ecological changes to these habitats (BIOMAERL Team 1999; Birkett et al. 1998a).

Research efforts have been made to understand the effects of physical factors on rhodolith-forming species in the North Atlantic, including the effect of temperature (Adey and McKibbin 1970; Blake and Maggs 2003; Wilson et al. 2004; Kamenos et al. 2008; Kamenos 2010; Kamenos and Law 2010); the influence of light (Adey

1970; Adey and McKibbin 1970; Adey and Macintyre 1973; Wilson et al. 2004; Rix et al. 2012; Teichert et al. 2013); the biological response of organic matter/burial (Wilson et al. 2004; Blake et al. 2007; Rix et al. 2012); the response to heavy metal contamination (Wilson et al. 2004) and the influence of ocean acidification (Hall-Spencer et al. 2006; Büdenbender et al. 2011; Burdett et al. 2012; Ragazzola et al. 2013; Noisette et al. 2013).

Many of these studies confirmed the negative effect of ocean acidification, rise of temperature and burial on the physiology of rhodolith-forming species, suggesting that a combination of physical stressors can affect coralline species and the flora and fauna assemblages associated with them. Also, the negative effects on the rhodolith bed structure and disturbance and loss of diversity of the associated flora and fauna has been documented, mostly due to anthropogenic activities such as bivalve fishing and aquaculture (BIOMAERL Team 1999; Barberá et al. 2003; Hall-Spencer et al. 2003, 2006; Hall-Spencer and Bamber 2007; Peña and Bárbara 2008a; Peña 2010). Apart from the decrease in the cover and thickness of the living maerl layer, a decrease in maerl size was observed in Galician beds impacted by mussel aquaculture (Peña 2010).

The EU Habitats Directive (92/43/EEC) includes maerl and lists *Lithothamnion corallioides* and *Phymatolithon calcareum* for protection. Maerl beds are also included in the list of threatened habitats within the NE Atlantic due to multiple threats from anthropogenic activities, high sensitivity and low recovery capacity (OSPAR 2006).

Many areas of maerl are currently included within areas that are listed as protected areas; in NW Spain for example 28 % of known maerl beds are located within protected areas with another 19 unprotected maerl beds proposed for future conservation actions (Peña and Bárbara 2006, 2009, 2010c).

There are still information gaps concerning the biological and ecological characteristics of maerl assemblages in Europe. Information on reproduction, size, distribution and diversity of coralline algae forming maerl beds is necessary to improve management of these habitats (OSPAR 2006). Barberá et al. (2003) suggested that maerl beds should be recognized as a specific habitat in the EU Habitats Directive and that the full range of maerl-forming species should be listed for conservation purposes otherwise the rarest types of maerl bed may not be given the high conservation status they require.

The development of newer methodologies for the characterization of maerl beds was also recommended by the Census of Marine Life project Natural Geography In Shore Areas (NaGISA, www.coml.org). A standardized methodology is required to monitor maerl beds in space and time for features such as reproductive effort, kind of reproduction, size, distribution, and diversity of coralline algae forming the habitat (Steller et al. 2007). Examination of these features can help to assess patterns typical of "pristine" vs "impacted" maerl beds in time (past vs present) and space (different beds located at distance of km, tens of km or hundreds of km) and can be used to evaluate the health of the habitats.

10.5 Conclusion

The four coralline algal species that most commonly grow unattached in the North Atlantic have been studied intensively in Europe but detailed information on the status and ecology of these habitats is lacking from the USA, Canada, Greenland and Iceland. There is a lack of long-term monitoring of these wonderful habitats throughout the North Atlantic which is a priority for management of these productive coastal systems. The activities that are most damaging for these habitats are now well known; the next steps are to raise awareness, enforce conservation legislation and promote sensible stewardship of these highly diverse habitats.

Acknowledgements JHK acknowledges support from CONACyt-Mexico, SEP-Mexico, Smithsonian Institution-USA and National University of Ireland, Galway. Studies on the Iberian Peninsula were supported by the research projects CGL2006-03576/BOS (Ministerio de Educacion y Ciencia and FEDER) and PGIDIT03PXIB10301PR (Xunta de Galicia). VP acknowledges support by Xunta de Galicia (Plan Galego de Investigación, Innovación e Crecemento, Plan I2C, 2011–2015), Universidade da Coruña and Programa Nacional de Movilidad de Recursos Humanos (Spain's Ministerio de Economía y Competitividad). JHS acknowledges support from the EU 'Mediterranean Sea Acidification under a changing climate' project (MedSeA; grant agreement 265103).

References

Adey WH (1966) Distribution of saxicolous crustose corallines in the northwestern North Atlantic. J Phycol 2:49–54

Adey WH (1968) The distribution of crustose corallines on the Icelandic coast. Sci Icel 1:16–25

Adey WH (1970) The effects of light and temperature on growth rates in boreal-subarctic crustose corallines. J Phycol 6:269–276

Adey WH (1971) The sublittoral distribution of crustose corallines on the Norwegian coast. Sarsia 46:41–58

Adey WH, Adey PJ (1973) Studies of the biosystematics and ecology of the epilithic crustose Corallinaceae of the British Isles. Brit Phycol J 8:343–407

Adey WH, Hayek L-AC (2011) Elucidating marine biogeography with macrophytes: quantitative analysis of the North Atlantic Supports the thermogeographic model and demonstrates a distinct subarctic region in the Northwestern Atlantic. Northeastern Naturalist 18: monograph 8: 1–128

Adey WH, MacIntyre IG (1973) Crustose coralline algae: a re-evaluation in the geological sciences. Geo Soc Am Bull 84:883–904

Adey WH, McKibbin DL (1970) Studies on the maërl species *Phymatolithon calcareum* (Pallas) nov. comb. and *Lithothamnium corallioides* Crouan in the Ria de Vigo. Bot Mar 13:100–106

Adey WH, Steneck, RS (2001). Thermogeography over time creates biogeographic regions: a temperature/space/time-integrated model and an abundance-weighted test for benthic marine al

Adey WH, Chamberlain YM, Irvine LM (2005) An SEM-based analysis of the morphology, anatomy and reproduction of *Lithothamnion tophiforme* (Esper) Unger (Corallinales, Rhodophyta), with a comparative study of associated North Atlantic Arctic/Subarctic Melobesioideae. J Phycol 41:1010–1024

Adey WH, Lindstorm SC, Hommersand MH, Muller KM (2008) The biogeographic origin of arctic endemic seaweeds: a thermogeographic view. J Phycol 44:1384–1394

Alfonso-Carrillo J, Gil-Rodríguez MC (1982) Sobre la presencia de un fondo de "Maerl" en las Islas Canarias. Collect Bot 13:703–708
Athanasiadis A (1996) Taxonomisk litteratur och biogeografi av Skandinaviska rödalger och brunalger. Algologia, Göteborg
Bárbara I, Cremades J, Veiga AJ (2004) Floristic study of a maërl and gravel subtidal bed in the Ría de Arousa (Galicia, Spain). Bot Complutensis 28:27–37
Barberá C, Bordehore C, Borg JA, Glémarec M, Grall J, Hall-Spencer JM, De La Huz C, Lanfranco E, Lastra M, Moore PG, Mora J, Pita ME, Ramos-Esplá AA, Rizzo R, Sánchez-Mata A, Seva A, Schembri PJ, Valle C (2003) Conservation and management of northeast Atlantic and Mediterranean maerl beds. Aquat Conserv 13:S65–S76
BIOMAERL Team (1999) Final Report, BIOMAERL project (Coordinator: Moore PG, University Marine Biological Station Millport, Scotland), EC Contract No. MAS3-CT95-0020, 1: 1–541, 2: 542–973 pp. + Appendix
Birkett DA, Maggs CA, Dring MJ (1998a) Maerl (Volume V). An overview of dynamics and sensitivity characteristics for conservation management of marine SACs. Acottish Association for Marine Science. (UK Marine SACs Project)
Birkett DA, Maggs CA, Dring MJ, Boaden PJS (1998b) Infralittoral reef biotopes with kelp species. An overview of dynamic and sensitivity characteristics for conservation management of marine SACs. Scott Assoc Mar Sci (UK Marine SACs Project)
Blake C, Maggs CA (2003) Comparative growth rates and internal banding periodicity of maerl species (Corallinales, Rhodophyta) from northern Europe. Phys Chem Chem Phys 42:606–612
Blake C, Maggs CA, Reimer P (2007) Use of radiocarbon dating to interpret past enviroments of maerl beds. Cienc Mar 33:385–397
Blunden G, Farnham WF, Jephson N, Barwell CJ, Fenn RH, Plunkett BA (1981) The composition of maërl beds of economic interest in northern Brittany, Cornwall and Ireland. In: Levring T (ed) Proceedings 10th international seaweed symposium. De Gruyter, Berlin
Bosence DWJ (1976) Ecological studies on two unattached coralline algae from western Ireland. Palaeontology 19:365–395
Bosence DWJ (1979) Live and dead faunas from coralline algal gravels, Co.Galway. Palaentology 22:449–478
Bosence DWJ (1980) Sedimentary facies, production rates and facies models for recent coralline algal gravels, Co. Galway, Ireland. Geophys J 15:91–111
Bosence DWJ (1983) Descriptions and classification of rhodoliths (rhodoids, rhodolites). In: Peryt TM (ed) Coated grains. Springer, Berlin
Bosence DWJ, Wilson J (2003) Maerl growth, carbonate production rates and accumulation rates in the northeast Atlantic. Aquat Conserv 13:S21–S31
Bosselini A, Ginsburg RN (1971) Form and intertidal structure of recent algal nodules (rhodolites) from Bermuda. J Geophys Res 79:669–682
Büdenbender J, Riebesell U, Form A (2011) Calcification of the Artic coralline red algae *Lithothamnion glaciale* in response to elevated CO_2. Mar Ecol Prog Ser 441:79–87
Burdett H, Aloisio E, Calosi P, Findlay HS, Widdicombe S, Hatton AD, Kamenos NA (2012) The effect of chronic and acute low pH on the intracellular DMSP production and epithelial cell morphology of red coralline algae. Mar Biol Res 8:756–763
Cabioch J (1968) Quelques particularités anatomiques du *Lithophyllum fasciculatum* (Lamarck) Foslie. Bulletin de la Société botanique de France 115:173–186
Cabioch J (1969) Les fonds de maërl de la Baie de Morlaix et leur peupplement vegetal. Cah Biol Mar 10:139–161
Cabioch J (1974) Un fond de maerl de l'Archipel de Madère et son peuplement végétal. Bulletin de la Société Phycologique de la France 19:74–82
Davison DM, Hughes DJ (1998) *Zostera* Biotopes (volume I). An overview of dynamics and sensitivity characteristics for conservation management of marine SACs. Scottish Association for Marine Science (UK Marine SACs Project)

De Grave S, Whitaker A (1999) A census of maërl beds in Irish waters. A quat Conserv 9:303–311
De Grave S, Fazakerley H, Kelly L, Guiry MD, Ryan M, Walshe J (2000) A study of selected maërl beds in Irish waters and their potential for sustainable extraction. Mar Resour Ser (Irel) 10:1–44
Diaz-Pulido G, Anthony KRN, Kline DI, Dove S, Hoegh-Guldberg O (2012) Interactions between ocean acidification and warming on the mortality and dissolution of coralline algae. J Phycol 48:32–39
Emerton V, Choi E (eds) (2008) Essential guide to food additives. Leatherhead Food International Ltd, London
Farrow GE, Cucci M, Scoffin TP (1978) Calcareous sediments on the nearshore continental shelf of western Scotland. P Roy Soc Edinb B 76:55–76
Foslie M (1895) The Norwegian forms of *Lithothamnion*. Kongelige Norske Videnskabers Selskabs Skrifter 1894:29–208
Foslie M (1899) Some new or critical lithothamnia. Det Kongelige Norske Videnskabers Selskabs Skrifter 1898:1–19
Foslie M (1905) Remarks on northern Lithothamnia. Det Kongelige Norske Videnskabers Selskabs Skrifter 1905:1–138
Foster MS (2001) Rhodoliths: between rocks and soft places. J Phycol 37:659–667
Frantz BR, Foster MS, Riosmena-Rodriguez R (2005) *Clathromorphum nereostratum* (Corallinales, Rhodophyta): the oldest alga? J Phycol 41:770–773
Freiwald A (1995) Sedimentological and biological aspects in the formation of branched rhodoliths in northern Norway. Beit Paläontol Österreichs 20:7–19
Gagnon P, Matheson K, Stapleton M (2012) Variation in rhodolith morphology and biogenic potential of newly discovered rhodolith beds in Newfoundland and Labrador (Canada). Bot Mar 55:85–99
Grall J (2002) Biodiversité spécifique et fonctionnelle du maerl : réponses a la variabilité de l'environnement côtier. PhD thesis, Université de Bretagne Occidentale, France, 302 pp
Grall J, Glemarec M (1997) Biodiversité des fonds de maerl en Bretagne: Approche fonctionelle et impacts anthropiques. Vie et Milieu 47:339–349
Grall J, Hall-Spencer JM (2003) Problems facing maerl conservation in Brittany. Aquat Conserv 13:S55–S64
Grall J, Le Loch F, Guyonnet B, Riera P (2006) Community structure and food web based on stable isotopes (δ15N and δ13C) analysis of a North Eastern Atlantic maerl bed. J Exp Mar Biol Ecol 338:1–15
Gunnarsson K (1977) Borugar a koralsetlogum i Amarfirdi. Hafransoknir 10:3–10
Hall-Spencer JM (1998) Conservation issues relating to maerl beds as habitats for molluscs. J Conchol Spec Publ 2:271–285
Hall-Spencer JM, Atkinson RJA (1999) *Upogebia deltaura* (Crustacea: Thalassinidea) in Clyde Sea maerl beds, Scotland. J Mar Biol Ass UK 79:871–880
Hall-Spencer JM, Bamber R (2007) Effects of salmon farming on benthic Crustacea. Cienc Mar 33:353–366
Hall-Spencer JM, Moore PG (2000) Scallop dredging has profound, long-term impacts on maerl habitats. ICES J Mar Sci 57:1407–1415
Hall-Spencer JM, Grall J, Moore PG, Atkinson RJA (2003) Bivalve fishing and maerl-bed conservation in France and the UK –retrospect and prospect. Aquat Conserv 13:S33–S41
Hall-Spencer JM, White N, Gillespie G, Gillham K, Foggo A (2006) Impact of fish farms on maerl beds in strongly tidal areas. Mar Ecol Prog Ser 326:1–9
Hall-Spencer JM, Kelly J, Maggs CA (2010) Background Document for maerl beds. OSPAR commission
Hauton C, Hall-Spencer JM, Moore PG (2003) An experimental study of the ecological impacts of hydraulic bivalve dredging on maerl. ICES J Mar Sci 60:381–392

Hernández-Kantún JJ, Riosmena-Rodríguez R, Adey WH, Rindi F (2014) Analysis of the *cox* 2–3 spacer region for population diversity and taxonomic implications in rhodolith-forming species (Rhodophyta, Corallinales). Phys Chem Chem Phys 190:331–354
Hernández-Kantún JJ, Rindi F, Adey HW, Le Gall L, Peña V, Gabrielson P (2015) Sequencing type material resolves the identity and distribution of the generitype *Lithophyllum incrustans*, and related European species *L. hibernicum* and *L. bathyporum* (Corallinales, Rhodophyta). J Phycol 51:791–807
Hernandez-Kantun JJ, Riosmena-Rodriguez R, Hall-Spencer J, Peña V, Maggs CA, Rindi F (2015) Phylogenetic analysis of rhodolith formation in the Corallinales (Rhodophyta). Eur J Phycol 50:46–61
Hily C, Potin P, Floch JY (1992) Structure of subtidal algal assemblages on soft-bottom sediments: fauna/flora interactions and role of disturbances in the Bay of Brest, France. Mar Ecol Prog Ser 85:115–130
Hinojosa-Arango G, Maggs C, Johnson M (2009) Like a rolling stone: the mobility of maërl (Corallinaceae) and the neutrality of the associated assemblages. Ecology 90:517–528
Irvine LM, Chamberlain YM (1994) Seaweeds of the British Isles. Volume 1 Rhodophyta Part 2B Corallinales, Hildenbrandiales. HMSO, London
Johansen HW (1981) Coralline algae, A first synthesis. CRC Publishing, Boca Raton
Johnson MP, Edwards M, Bunker F, Maggs C (2005) Algal epiphytes of *Zostera marina*: variation in assemblage structure from individual leaves to regional scale. Aquat Bot 82:12–26
Kamenos NA (2010) North Atlantic summers have warmed more than winters since 1353, and the response of marine zooplankton. Proc Natl Acad Sci U S A 107:22442–22447
Kamenos NA, Law A (2010) Temperature controls on coralline algal skeletal growth. J Phycol 46:331–335
Kamenos NA, Moore PG (2004) Maerl grounds provide both refuge and high growth potential for juvenile queen scallops (*Aequipecten opercularis* L). J Exp Mar Biol Ecol 313:241–254
Kamenos NA, Moore PG, Hall-Spencer JM (2003) Substratum heterogeneity of dredged vs undredged maerl grounds. J Mar Biol Assoc UK 83:411–413
Kamenos NA, Moore PG, Hall-Spencer JM (2004a) Small-scale distribution of juvenile gadoids in shallow inshore waters; what role does maerl play? ICES J Mar Sci 61:422–429
Kamenos NA, Moore PG, Hall-Spencer JM (2004b) Nursery-area function of maërl grounds for juvenile queen scallops *Aequipecten opercularis* and other invertebrates. Mar Ecol Prog Ser 274:183–189
Kamenos NA, Moore PG, Hall-Spencer JM (2004c) Attachment of the juvenile queen scallop (*Aequipecten opercularis* (L) to maerl in mesocosm conditions; juvenile hábitat selection. J Exp Mar Biol Ecol 306:139–155
Kamenos NA, Cusack M, Moore PG (2008) Coralline algae are global palaeothermometers with bi-weekly resolution. Geochim Cosmochim Acta 72:771–779
Kamenos NA, Hoey T, Nienow P, Fallick A, Claverie T (2012) Reconstructing Greenland ice sheet runoff using coralline algae. Geology G33405.1
Keegan BF (1974) The macrofauna of maërl substrates on the West Coast of Ireland. Cah Biol Mar 15:513–530
Kulka DW, Simpson MR, Hooper RG (2004) Changes in distribution and habitat associations of Wolffish (Anarhichidae) in the Grand Banks and Labrador Shelf. Research document for the Canadian Science Advisory Secretariat
Leliaert F, Boedeker C, Peña V, Bunker F, Verbruggen H, De Clerck O (2009) *Cladophora rhodolithicola* sp. nov. (Cladophorales, Chlorophyta), a diminutive species from European maerl beds. Eur J Phycol 44:155–169
Lemoine M (1910) Repartition et mode de vie du maërl (*Lithothamnium calcareum*) aux environs de Concarneau (Finistere). Ann L'Inst Oceanogr, Monaco 1(3):1–28
Maggs CA (1983a) Seasonal study of seaweed communities on subtidal maerl (unattached coralline algae) in Galway Bay, Ireland. Prog Underw Sci 9:27–40

Maggs CA (1983b) A phenological study of the epiflora of two maerl beds in Galway Bay. Unpublished Ph.D. thesis, Galway: University College, Galway, Ireland
Maggs CA, Guiry MD (1988 '1987') *Gelidiella calcicola* sp. nov. (Rhodophyta) from the British Isles and Northern France. Brit Phycol J 22(4):417–434
Maggs CA, Guiry MD (1989) A re-evaluation of the crustose red algal genus *Cruoria* and the family Cruoriaceae. Brit Phycol J 24:253–269
Maggs CA, Castilho R, Foltz D, Henzler C, Taimour J, Kelly J, Olsen J, Perez KE, Stam W, Vainola R, Viard F, Wares J (2008) Evaluating signatures of glacial refugia for North Atlantic benthic marine taxa. Ecology 89:S108–S122
Mendoza ML, Cabioch J (1998) Étude comparée de la reproduction de *Phymatolithon calcareum* (Pallas) Adey and McKibbin et *Lithothamnion corallioides* (P. and H. Crouan) P. and H. Crouan (Corallinales, Rhodophyta), et reconsidérations sur la définition des genres. Can J Bot 76:1433–1445
Nelson WA (2009) Calcified macroalgae, critical to coastal ecosystems and vulnerable to change: a review. Mar Freshw Res 60:787–801
Noisette F, Duong G, Six C, Davoult D, Martin S (2013) Effects of elevated p CO2 on the metabolism of a temperate rhodolith Lithothamnion corallioides grown under different temperatures. J Phycol 49:746–757
OSPAR (2006) Case reports for the initial list of threatened species and/or declining species and habitats in the OSPAR region. OSPAR Commission, 150 p
Otero-Schmitt J, Pérez-Cirera JL (2002) Infralittoral benthic biocenoses from northern Ría de Muros, Atlantic coast of northwest Spain. Bot Mar 45:93–122
Pardo C, López L, Peña V, Hernández-Kantún JJ, Le Gall L, Bárbara I, Barreiro L (2014) A multilocus species delimitation reveals a striking number of maërl species in the OSPAR region. PLoS One 9(8):e104073
Peña V (2010) Estudio ficológico de los fondos de maerl y cascajo en el noroeste de la Península Ibérica. Tesis doctoral, Universidade da Coruña
Peña V, Bárbara I (2006) Los fondos marinos de maërl del Parque Nacional de las Islas Atlánticas (Galicia, España): distribución, abundancia y flora asociada. NACC Nova Acta Cient Compostel Biol 15:7–25
Peña V, Bárbara I (2008a) Maërl community in the north-western Iberian peninsula: a review of floristic studies and long-term changes. Aquat Conserv 18:339–366
Peña V, Bárbara I (2008b) Biological importance of an Atlantic European maerl bed off Benencia Island (northwest Iberian Peninsula). Bot Mar 51:493–505
Peña V, Bárbara I (2009) Distribution of the Galician maerl beds and their shape classes (Atlantic Iberian Peninsula); proposal of areas in future conservation actions. Cah Biol Mar 50:353–368
Peña V, Bárbara I (2010a) Seasonal patterns in the maërl community of shallow European Atlantic beds and their use as a baseline for monitoring studies. Eur J Phycol 45:327–343
Peña V, Bárbara I (2010b) New records of crustose seaweeds associated with subtidal maërl beds and gravel bottoms in Galicia (NW Spain). Bot Mar 53:41–61
Peña V, Bárbara I (2010c) Plan integral de conservación de las algas formadoras de los fondos de maerl (*Lithothamnion corallioides* (P.L. Crouan and H.M. Crouan) y *Phymatolithon calcareum* (Pallas) Adey and McKibbin) en Galicia, 249 pp
Peña V, Bárbara I (2013) Non-coralline crustose algae associated with maerl beds in Portugal: a re-appraisal of their diversity in the Atlantic Iberian beds. Bot Mar 56(5–6):481–493
Peña V, Adey WH, Riosmena-Rodriguez R, Yung M-Y, Choi H-G, Afonso-Carrillo J, Bárbara I (2011) *Mesophyllum sphaericum* sp. nov. (Corallinales, Rhodophyta): a new maërl –forming species from the northeast Atlantic. J Phycol 47:911–927
Peña V, Hernandez-Kantun JJ, Grall J, Pardo C, Lopez L, Barbara I, Le Gall L, Barreiro R (2014a) Detection of gametophytes in the maerl-forming species *Phymatolithon calcareum* (Melobesioideae, Corallinales) assessed by DNA barcoding. Cryptog Algol 35(1):15–25

Peña V, Bárbara I, Grall J, Maggs CA, Hall-Spencer JM (2014b) The diversity of seaweeds on maerl in the NE Atlantic. Mar Biodiv 44(4):533–551

Peña V, De Clerck O, Afonso-Carrillo J, Ballesteros E, Bárbara I, Barreiro R, Le Gall L (2015a) An integrative systematic approach to species diversity and distribution in the genus *Mesophyllum* (Corallinales, Rhodophyta) in Atlantic and Mediterranean Europe. Eur J Phycol 50(1):20–36

Peña V, Pardo C, López L, Carro B, Hernandez-Kantun JJ, Adey WH, Bárbara I, Barreiro R and Le Gall L (2015b) *Phymatolithon lusitanicum* sp. nov. (Hapalidiales, Rhodophyta): the third most abundant maerl-forming species in the Atlantic Iberian Peninsula. Cryptogamie, Algol. 36(4): 429–459

Porzio L, Buia MC, Hall-Spencer JM (2011) Effects of ocean acidification on macroalgal communities. J Exp Mar Biol Ecol 400:278–287

Potin P, Floch JY, Augris J, Cabioch J (1990) Annual growth rate of calcareous red alga *Lithothamnion corallioides* (Corallinales, Rhodophyta) in the Bay of Brest, France. Hydrobiologia 204(205):263–267

Ragazzola F, Foster LC, Form AU, Buscher J, Hansteen TH, Fietzke J (2013) Phenotypic plasticity of coralline algae in a High CO_2 world. Ecol Evol 3:3436–3446

Rico JM, Guiry MD (1997) Life history and reproduction of *Gelidium maggsiae* sp. nov. (Rhodophyta, Gelidiales) from Ireland. Eur J Phycol 32:267–277

Rix LN, Burdett HL, Kamenos NA (2012) Irradiance-mediated dimethylsulphoniopropionate (DMSP) responses of coralline red algae. Estuar Coast Shelf Sci 96:268–272

Rosas-Alquicira EF, Riosmena-Rodriguez R, Cuto RP, Neto AI (2009) New additions to the Azorean algal flora, with ecological observations on rhodolith formations. Cah Biol Mar 50:1–9

Sauriau PG, Curti C, Jourde J, Aubert F, Cajeri P, Lavesque N, Dubois S, Lepareur F, Gouesbier C, Sauriau F, Sauriau M, Latry L, Leguay D, Robert S, Pineau P, Geairon P (2012) Le maerl algues corallinacees marines dans les Pertuis Charentais. Annls Acad L Rochelle 10:281–300

Scottish MPA Project (2013) Assessment against the MPA Selection Guidelines. Loch Sween possible nature conservation MPA. http://www.snh.gov.uk/

Shultze K, Janke K, Krüb A, Weidemann W (1990) The macrofauna and macroflora associated with Laminaria digitata and L. hyperborea at the island of Helgoland (German Bight, North Sea). Helgoländer Meeresun 44:39–51

Sneli JA (1968) The *Lithothamnion* community in Nord-Möre, Norway with notes on the epifauna of *Desmarestia viridis* (Müller). Sarsia 31:69–74

Steller DL, Riosmena-Rodriguez R, Foster MS (2007) Sampling and monitoring rhodolith beds. In: Rigby R, Iken K, Shirayama Y (eds) Handbook for sampling coastal seagrasses and macroalgae community biodiversity. Kyoto University Press, Kyoto

Strömfelt HFG (1886) Om algvegetationen vid Islands Kuster, pp. 1–89. Göteborg: Akademisk Afhandling

Suneson S (1958) *Lithothamnion calcareum* vid svenska västkusten. Bot Notiser 111:197–199

Teichert S, Woelkerling W, Rüggeberg A, Wisshak M, Piepenburg D, Meyerhöfer M, Form A, Büdenbender J, Freiwald A (2012) Rhodolith beds (Corallinales, Rhodophyta) and their physical and biological environment at 80°31′N in Nordkappbukta (Nordaustlandet, Svalbard Archipelago, Norway). Phys Chem Chem Phys 51:371–390

Teichert S, Woelkerling W, Ruggeberg A, Wisshak M, Piepenburg D, Meyerhofer M, Form A, Freiwald A (2013) Artic rhodolith beds and their environmental controls (Spitsbergen, Norway). Facies. doi:10.1007/s10347-013-0372-2

Tittley I (2002) Seaweed diversity in the North Atlantic Ocean. Arquipélago. Life Mar Sci 19A:13–25

Whelan PM, Cullinane JP (1985) The algal flora of a subtidal *Zostera* bed in Ventry Bay, southwest Ireland. Aquat Bot 23:41–51

Wilson S, Blake C, Berges JA, Maggs CA (2004) Environmental tolerances of free-living coralline algae (maerl): implications for European marine conservation. Biol Cons 120:279–289

Chapter 11
Mediterranean Rhodolith Beds

Daniela Basso, Lorenza Babbini, Angel Alfonso Ramos-Esplá, and Maria Salomidi

Abstract The available references to rhodolith beds have been analyzed for a total of 125 locations in the Mediterranean Sea, equally distributed in the eastern and western sub-basins. Mediterranean rhodolith beds occur from 9 to 150 m of water depth, with a mean depth of about 55 m in both sub-basins. Most rhodolith beds lay within the depth range 30–75 m, while those extending deeper than 75 m are about 18 % of the total, and those shallower than about 25 m are uncommon. The deepest and the largest Mediterranean rhodolith beds are both located in the Balearic Sea. Water motion from bottom currents, waves, and tides is needed to keep rhodoliths unburied, within mesotrophic to oligotrophic water conditions. Rhodolith beds are commonly small (<0.01 km^2) and multispecific, showing growth-form mixing and a much higher coralline biodiversity than Atlantic beds. They are vulnerable to physical disturbance by fishing gears and smothering, and to water pollution by organic enrichment and sewage. The existing instruments of legal protection appear ill defined, since these are based on scientific literature mostly derived from northern Europe, where specific and different environmental settings, species composition, depth distribution, and anthropogenic pressures occur. The protection of a specific habitat type cannot be effectively gained without access to geospatial and compositional data and an increased research effort is needed to improve taxonomic inventories, habitat mapping, and monitoring activities on a basin scale.

D. Basso (✉)
Department of Earth and Environmental Sciences, Univ. of Milano-Bicocca,
Piazza della Scienza 4, 20126 Milan, Italy
e-mail: daniela.basso@unimib.it

L. Babbini
ISPRA – Italian National Institute for Environmental Protection and Research,
Via V. Brancati 60, 00144 Rome, Italy
e-mail: lorenza.babbini@isprambiente.it

A.A. Ramos-Esplá
Centro de Investigación Marina de Santa Pola, Universidad de Alicante,
03080 Alicante, Spain
e-mail: alfonso.ramos@ua.es

M. Salomidi
Hellenic Centre for Marine Research, Institute of Oceanography,
PO box 712, 19013 Anavyssos, Greece
e-mail: msal@hcmr.gr

R. Riosmena-Rodríguez et al. (eds.), *Rhodolith/Maërl Beds: A Global Perspective*, Coastal Research Library 15, DOI 10.1007/978-3-319-29315-8_11

11.1 Introduction

After the pioneer research in the gulfs of Marseille and Naples (Marion 1883; Walther 1885; Funk 1927), the first scientific investigation specifically devoted to Mediterranean rhodolith beds dates back to 1940, when Dieuzeïde described the maërl of an Algerian fishing site at Castiglione (presently Bou Ismaïl). Other important contributions by Vatova (1935), Lemoine (1940), Parenzan (1960) and by the Marseille research group (Pérès 1952; Huvé 1955; Jacquotte 1962; Pérès and Picard 1956, 1964; Picard 1965) traced the baseline of our knowledge of the Mediterranean beds. Most of these early contributions relate to the Western Mediterranean, reflecting the higher research effort due to the presence of historical marine research centers (Station marine d'Endoume in France, Station d'Aquiculture et Pêche de Castiglione in the former French colony of Algeria, Stazione Zoologica di Napoli and Istituto Sperimentale Talassografico di Taranto in Italy).

At that time maerl was already well known from the Eastern Atlantic beds, where it was exploited as soil amendment since 1724 (Jacquotte 1962). The name "maërl" comes from the Breton language and means "marga", referring to an area of calcareous land or deposits of calcified algae. Since the Atlantic maerl was the reference, the Breton term was used to identify in the Mediterranean the same calcareous gravel composed of living and dead coralline branched thalli (Huvé 1955; Jacquotte 1962), although the occurrence of nodular shapes was already observed, sometimes under unusual terms like "tuberi" (=tubers, Walther 1885).

The first use of the term rhodolith in literature is found informally in Barnes et al. (1970), but its formal definition is due to Bosellini and Ginsburg (1971) as rhodolite, later corrected in rhodolith (Ginsburg and Bosellini 1973). Since then, rhodoliths are intended as unattached nodules formed by calcareous red algae and their growths, as part of a continuum spectrum of forms, with size spanning 2–250 mm of mean diameter (Bosellini and Ginsburg 1971). Based on this definition, the term rhodolith beds includes also the maërl beds sensu stricto and the calcareous *Peyssonnelia* beds (Basso et al. 2015) (Fig. 11.1).

This contribution is aimed at providing a baseline for future investigation on Mediterranean rhodolith beds, by summarizing our present knowledge of their geographic and depth distribution, areal extent, and composition, based on scientific papers and all other grey literature or information that can be acquired from scientific institutions.

11.2 Methods

Scientific publications from 1885 to 2013, containing reference to maërl and rhodolith beds have been analyzed for the extraction of data on location, depth, areal extent and coralline species composition. Further publications have been collected from Institutional sites (Regional sites, Governmental sites, Ministries of Environment,

Fig. 11.1 *Peyssonnelia rosa-marina* and *Peyssonnelia rubra* seabed in Korinthiakos Gulf (~25–35 m depth), Livadostra bay (Salomidi M. unpublished data) (Photo courtesy of Yiannis Issaris, www.yissaris.com)

national agencies for environmental protection, UNEP, RAC/SPA, etc.) from the EU and extra EU countries facing the Mediterranean Sea. Many references lacked precise indication of the areal extent of rhodolith beds, but we obtained their order of magnitude by plotting the depth range on the nautical maps or habitat maps whenever available, with a largely conservative approach. The definition of rhodolith bed follows Steller et al. (2003) and Basso et al. (2015). Data have been discussed separately for the western and the eastern sub-basins of the Mediterranean Sea, having their boundary fixed at the Messina Strait and across the Sicily channel, from Cape Lilibeo (Sicily), through the Adventure Bank to Cape Bon (Tunisia) (Fig. 11.2).

11.3 Geographic and Depth Distribution of Mediterranean Rhodolith Beds

The Mediterranean Sea is a basin of about 2.5 million km^2 surrounded by three continental regions (Europe, Asia, Africa) and 22 countries (Table 11.1). The name Mediterranean is derived from the Latin *mediterraneus*, meaning "in the middle of the land", since the basin is enclosed by a complex coastline with a total length of about 46,869 km, giving a high ratio of coastal area to total surface.

Rhodolith and maërl beds are reported for most of the Mediterranean coastal sectors, although a significant absence of reports is evident along the coasts of Eastern Adriatic, Libya, Egypt, Syria, Lebanon and the Black Sea (with an exception for the Sea of Marmara) (Figs. 11.2, 11.3, 11.4, and 11.5). Despite the Mediterranean Sea

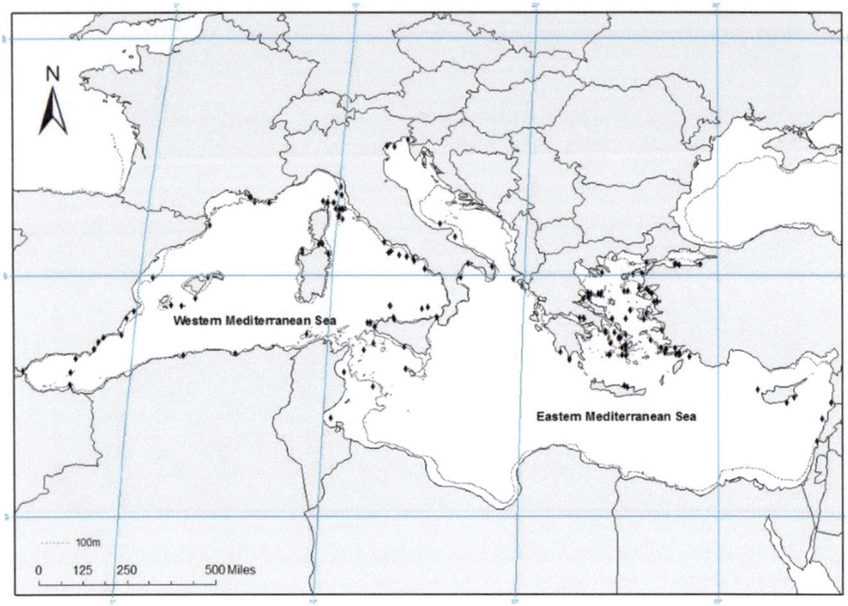

Fig. 11.2 The rhodolith beds of the Mediterranean Sea. The Eastern and Western Mediterranean sub-basins are conventionally separated at the sill of the Sicily channel (line from Cape Lilibeo to Cape Bon) and Messina Strait

is recognized since long time as a fundamental environmental and economic resource of the entire region, the available information is not equally distributed along its coastal area, due to the fragmentation of governmental authorities, and the large difference of knowledge and effort in marine research due to complex historical, political, and economic reasons. Thus, it is likely that the absence of records from the above-mentioned areas is at least partially a bias due to insufficient knowledge.

It is noteworthy that scattered rhodoliths occur in many infralittoral environments (among others: Trapani et al. 1999; García-March and Kersting 2006; Basso et al. 2008) and most of the coastal detritic bottoms (DC) of the Mediterranean (e.g. Astruch et al. 2012), potentially covering the seafloor from the lower limit of the *Posidonia oceanica* beds down to the disappearance of sciaphilous macroalge (= the circalittoral zone, Pérès and Picard 1964). According to Steller et al. (2003), only those soft-substrates with a >10 % cover of living corallines can be considered as rhodolith bed, but many old references provide only qualitative comments that could have been misinterpreted in more recent compilations. Another important source of possible error is the confusion of rhodolith beds with the coralligenous framework (C). The C "de plateau" can develop from rhodolith-rich soft substrates, and the two benthic associations can grade one into the other due to the progressive cementation by crustose corallines (Laborel 1961; Basso et al. 2007; Georgiadis

Table 11.1 Length of the coastline of the Mediterranean countries

Country	Coastline (km)
Albania	418
Algiers	1200
Bosnia Herzegovina	23
Croatia	5835
Cyprus	782
Egypt	955
France	1703
Greece	15,021
Israel	179
Italy	7375
Lebanon	225
Libya	1770
Malta	180
Monaco	4
Montenegro	394
Morocco	512
Slovenia	47
Spain	3562
Syria	183
Tunisia	1298
Turkey	5191
UK (Gibraltar)	12
Total	46,869

et al. 2009). Therefore, it is possible that some reports confused the rhodolith beds and the C, especially if the survey was ship-based on random indirect sampling.

After a critical review of the literature, we retained a total of 192 references relating to the 125 rhodolith bed locations (Figs. 11.2, 11.3, 11.4, and 11.5), with a considerable redundancy due to: (1) successive independent reports including different information; and/or (2) different bed types reported separately at adjacent depths in the same area, that cannot be resolved at the map scale (for example a *Peyssonnelia* bed adjacent to a maërl bed, as it occurs in the Naples area; Fig. 11.4). Ninety-nine references refer to the Western Mediterranean rhodolith beds, 93 to the Eastern (Fig. 11.2), pointing to an apparent equal coverage of both sub-basins. Actually only 31 % of the references contain sufficient information for the definition of the order of magnitude of the areal extent of the rhodolith bed, especially in the Eastern Mediterranean (Fig. 11.6). This bias is probably the reason for the apparent lack of small (<1 km^2) rhodolith beds in the Eastern Mediterranean, that actually occur there (Georgiadis et al. 2009) (Figs. 11.5 and 11.6).

The largest rhodolith beds so far identified in the Mediterranean, namely those exceeding a surface of 100 km^2, are located in the Balearic sea (about 300 km^2, Barberá et al. 2012, Fig. 11.3), followed by the northern tip of Corse (Bonacorsi 2012, Fig. 11.4) and the Maltese Islands (Micallef et al. 2012, Fig. 11.2).

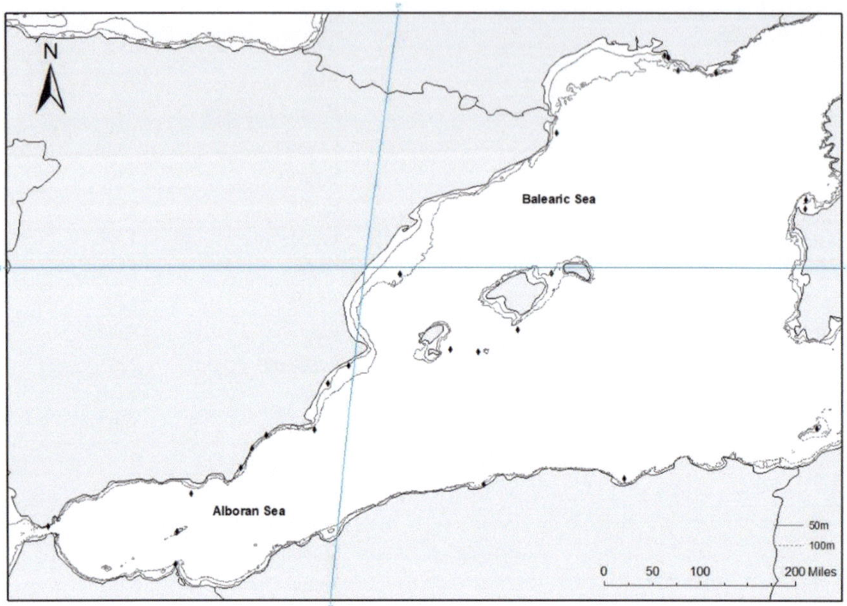

Fig. 11.3 Rhodolith beds of the westernmost Mediterranean Sea, including the coasts of Spain, Morocco, Algeria, and France (pro-parte)

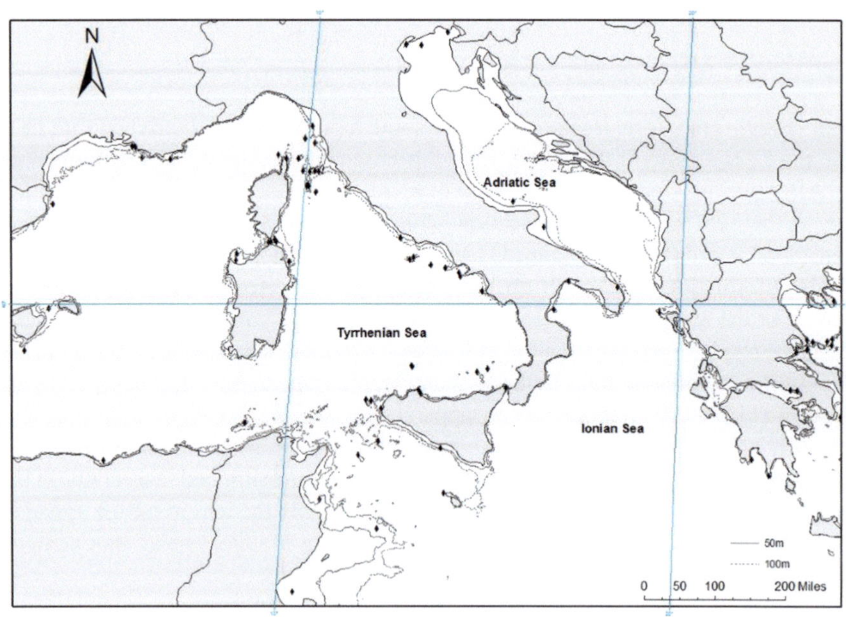

Fig. 11.4 Rhodolith beds along the coasts of France (pro-parte, including Corse), Italy, and Tunisia (pro-parte)

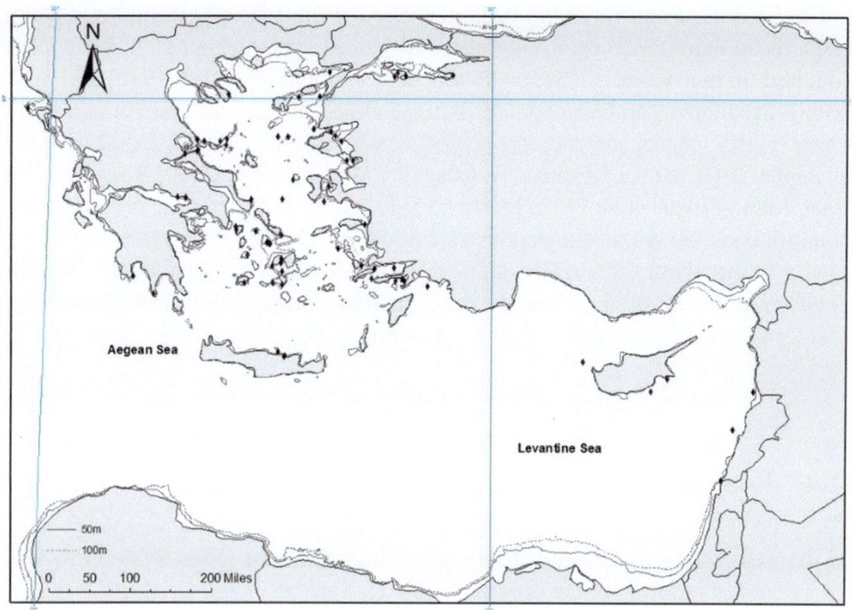

Fig. 11.5 Rhodolith beds of the Aegean Sea, Levantine, and Marmara Seas

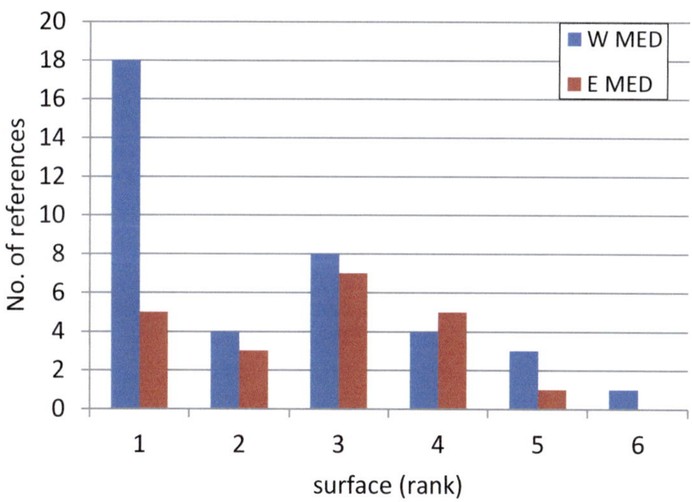

Fig. 11.6 Surface of Western and Eastern Mediterranean rhodolith beds compared. Surface is indicated as order of magnitude within six ranks: *1* = 0.001– 0.01; *2* = 0.02–0.1; *3* = 0.11–1; *4* = 1.01–10; *5* = 10.01–100; *6* > 100 Km^2

The Mediterranean rhodolith beds may occur from 9 to 150 m of water depth, with a mean depth of 55 m. Rhodolith beds shallower than about 25 m are uncommon, both in the Western (Tuscan Archipelago, Pignatti and Rizzi 1972) and in the Eastern Mediterranean (Amendolara Bank, Ionian Sea, Cecere and Perrone 1987; Trieste Gulf, Falace and Bressan 2003; Kuriat Islands, Ramos-Esplá and Ben Mustapha 2010; Gulf of Gabes, Tunisia, El Lakhrach et al. 2012; Rosh Haniqra, Israel, Ramos-Esplá et al. 2012; Marmara Sea, Müller 1985). On the contrary, most rhodolith beds lay within the depth range 30–75 m, and the beds extending deeper than 75 m are about 18 % of the total, more or less equally distributed in the two Mediterranean sub-basins. The deepest Mediterranean rhodolith bed has been recently discovered on the top of a seamount in the Balearic Sea (140–150 m) (Aguilar et al. 2009).

11.4 Ecology

On the basis of the observed distribution of rhodolith beds (Figs. 11.2, 11.3, 11.4, and 11.5), the variation of the oceanographic parameters characterizing the Eastern and the Western Mediterranean (higher salinity and temperature on the E, higher nutrient concentration on the W; Pinardi and Masetti 2000; Huertas et al. 2012) seem to play only a minor role in the coarse geographic and bathymetric pattern, leading to poor predictability of rhodolith bed occurrence (Martin et al. 2014). However, the geomorphology at sites of occurrence may provide some hints on their ecological requirements.

The Mediterranean rhodolith beds are mostly located around islands and capes (among others: Jacquotte 1962; García-Carrascosa 1987; Ballesteros et al. 1993; Atabey 1998; Schembri 1998; Bordehore et al. 2002a; Piazzi et al. 2002; Salomidi et al. 2009); on top of submarine plateaux, like the top of seamounts (Aktan 2012; Aguilar et al. 2009), marine terraces (Damiani et al. 1988; Savini et al. 2012), inter-island or island-mainland channels (Barberá et al. 2012; Bakir and Katağan 2005; Bracchi and Basso 2012; Sezgin et al. 2009; Georgiadis et al. 2009; Papakosta et al. 2012), and banks (Parenzan 1960; Cecere and Perrone 1987).

These habitats share some common features, such as a negligible sedimentary rate and the distance from significant nutrient influx. These environmental conditions can be primary characteristics of a specific geographic setting (as around small islands) or induced by hydrodynamics. Submarine reliefs are swept by currents that flow over their top by removing the finest particles (Savini et al. 2012). Water motion is enhanced through channels and around capes preventing fine sediment deposition. Rhodolith beds are very sensitive to burial (Bordehore et al. 2002b, 2003; BIOMAERL Team 2003; Riul et al. 2008; Bahia et al. 2010) and to increased organic carbon and nutrient load (Sanz-Lázaro et al. 2011), therefore water motion (waves, currents) is needed to maintain an unburied state within mesotrophic to oligotrophic conditions (Marrack 1999; Steller et al. 2003; Ramos-Esplá and Luque 2008). Indeed, rhodolith beds are frequently associated with ripple marks or submarine dunes, that are sedimentary structures indicating water motion over the seafloor (Babbini et al. 2006; Barberá et al. 2012; Micallef et al. 2012). The situation of the shallow-water rhodolith beds (<25 m) in the northern Adriatic Sea, Kuriat Islands and Gulf of Gabés (E Tunisian coast, Eastern Mediterranean) is intriguing, since these two large embayments are geomorphologically very different from submarine plateaux and inter-island channels. Interestingly, within the microtidal Mediterranean Sea, the maximum tidal range (and related water energy) is observed exactly in correspondence with the northern Adriatic (about 1 m), and in the Gulf of Gabés (about 1.5 m) (Lozano and Candela 1995; Tsimplis et al. 1995). This suggests that in addition to waves and bottom currents, also tidal currents may be in some cases a controlling factor for the distribution of these assemblages, as it happens for the NE Atlantic coasts.

11.5 Species and Growth-Form Composition

Mediterranean rhodolith beds include a suite of assemblages that have been referred to as "facies" of infralittoral and circalittoral biocoenoses (Pérès and Picard 1964; Picard 1965; Meinesz et al. 1983; Basso 1998; Basso et al. 2009; Relini and Giaccone 2009; Agnesi et al. 2011). Very few rhodolith beds have been investigated for a complete and exaustive floristic description, and facies frequently grade one into the other, hampering mapping and boundary definitions (Barberá et al. 2012).

Mediterranean rhodolith beds show a much higher coralline biodiversity compared with the NE Atlantic, both at the basin scale, and within each bed (Lanfranco et al. 1999; Bordehore et al. 2002a; Basso et al. 2014; Falace et al. 2014; Table 11.2), although DNA barcoding recently revealed an unexpected species diversity in the NE Atlantic (Pardo et al. 2014). Only 20 % of the reports refer specifically to dominant *L. corallioides* and *P. calcareum*, that are the major components of the NE Atlantic rhodolith beds. Monospecific or oligospecific beds are definitely not the rule, and species mixing is frequently accompanied by growth-forms mixing. Moreover, the superposition of different species on the same rhodolith is a rather common phenomenon, compared to other coastal areas worldwide.

Contrarily to oceans, the Mediterranean is a sill basin with a deficient hydrological budget compensated by the inflow of surface Atlantic waters though the Gibraltar strait (Pinardi and Masetti 2000; Huertas et al. 2012). On a global scale, the species composition of the maërl beds is temperature-related (Adey and Steneck 2001; Wilson et al. 2004; Pardo et al. 2014). Therefore, the Mediterranean is biogeographically at the crossroad of Indo-Pacific and Atlantic biota, with some rhodolith-forming corallines reflecting the present-day Atlantic tribute (for example *P. calcareum*, *L. corallioides*, *L. incrustans*), and other endemic species, derived from the complex geologic history and the Tethyan origin of this basin (Basso et al. 1996, 1997). Despite our long history of marine biological research, the coralline flora of the Mediterranean is probably still incompletely known (Bordehore et al. 2002a). Molecular systematics opened new frontiers to species delimitation and a growing

Table 11.2 List of calcareous red algal species composing the Mediterranean rhodolith beds

Family and subfamily	Species
Order Peyssonneliales	
Peyssonneliaceae	*Peyssonnelia polymorpha* (Zanardini) Schmitz
	Peyssonnelia rosa-marina Boudouresque et Denizot
	Peyssonnelia crispata Boudouresque et Denizot
Order Sporolithales	
Sporolithaceae	*Sporolithon ptychoides* Heydrich
Order Corallinales	
Corallinaceae	
Lithophylloideae	*Lithophyllum dentatum* (Kützing) Foslie
	Lithophyllum incrustans Philippi
	Lithophyllum racemus (Lamarck) Foslie
Neogoniolithoideae	*Neogoniolithon brassica-florida* (Harvey) Setchell et Mason
	Neogoniolithon mamillosum (Hauck) Setchell et Mason
	Spongites fruticulosus Kützing
Order Hapalidiales	
Hapalidiaceae	
Melobesioideae	*Lithothamnion corallioides* (Crouan et Crouan) Crouan et Crouan
	Lithothamnion crispatum Hauck
	Lithothamnion minervae Basso
	Lithothamnion valens Foslie
	Mesophyllum expansum (Philippi) Cabioch et Mendoza
	Mesophyllum philippii (Foslie) Adey
	Mesophyllum sphaericum Peña, Bárbara, Adey, Riosmena-Rodríguez et Choi
	Phymatolithon calcareum (Pallas) Adey et McKibbin

body of evidence is showing that an integrated approach is the most powerful tool to the study of biodiversity, especially in a group with strong morphological plasticity like coralline algae.

11.6 Rhodolith Beds as Hot Spots of Mediterranean Biodiversity

The rhodolith beds represent one of the hot points of the marine Mediterranean biodiversity, both in species and trophic guilds. In comparison with other habitats of the continental shelf, the biodiversity of rhodolith beds is very high, although the available information is still patchy and inhomogeneous (Picard 1965; Martin Sintes 1987; Ballesteros 1989; Glémarec 1997; Grall and Glémarec 1997; BIOMAERL 1998; Giménez-Casalduero et al. 2001; Covazzi Harriague et al. 2002; Ramos-Esplá and Luque 2004, 2008; Castriota et al. 2003, 2005; Gambi et al. 2009; Ordines and Massutí 2009; Barberá et al. 2012). The longevity of the algal thalli, together with their complex and open architecture, creates a permanent heterogeneous habitat in which abundant flora and fauna (generally small in size) live, seek refuge and/or feed (Soto 1990; Mannino et al. 2002; Templado et al. 2002, 2006; Castriota et al. 2003). About 1000 spp. of macro-organisms have been cited (300 macrophytes and 700 animals; Ramos-Esplá and Luque 2008), that represent, after the coralligenous and *Posidonia* beds, the most biodiverse community of the Mediterranean Sea (Ballesteros 2006). These beds can be considered both soft and hard substrata, depending on the point of view of the dweller (Jacquotte 1962; Keegan 1974; Hall-Spencer 1998), thus they allow for a great deal of interstitial life, and favour the settlement of a variety of sessile organisms and endobionts.

11.7 Conservation

Mediterranean rhodolith beds as biogenic calcareous formations are considered benthic habitats of high conservation interest for different reasons, such as their biogeographic uniqueness, and the three-dimensional structure hosting a great biodiversity (Basso et al. 2015). These assemblages are a matter of conservation concern because they represent a non-renewable resource due to the slow rate of growth and carbonate deposition (about 1 mm year^{-1}; Canals and Ballesteros 1997; Martin and Gattuso 2009; Basso 2012) and because they undergo various anthropogenic threats, ranging from the physical damage induced by human activities (i.e. impacts of towed fishing gear) to the degradation of the habitat and water quality (i.e. smothering effects due to the alteration of sedimentation rates such as derived from aquaculture activities; sewage, etc.) (Bordehore et al. 2002b, 2003; Grall and Hall-Spencer 2003; Barbera et al. 2003; Aguado-Giménez and Ruiz-Fernández 2012; Basso et al. 2015).

The vulnerability of these habitats has been recognized, over the course of the last decades, within various European and International frameworks through the adoption of a range of protection instruments such as Directives, Regulations or conventions. Within the European legislation two main tools have been put in place: the Habitat and Species Directive 92/43/EEC and the EC Council Regulation 1967/2006 EC. Particularly, the Habitat Directive in the Annex V includes *Phymatolithon calcareum* and *Lithothamnion corallioides* among those species subject to exploitation and for which Member States have to ensure appropriate management measures. The EC Council Regulation 1967/2006 EC concerning management measures for the sustainable exploitation of fishery resources in the Mediterranean Sea establishes the banning of fishing gears (i.e. trawls, seine nets, dredges) on coralligenous or maërl beds defined as "…a collective term for a biogenic structure due to several species of coralline red algae (Corallinaceae), which have hard calcium skeletons and grow as unattached free living branched, twig-like or nodule corallines algae on the seabed, forming accumulations within the ripples of mudflats or sandflats seabed. Maërl beds are usually composed of one or a variable combination of red algae, in particular, *Lithothamnion corallioides* and *Phymatolithon calcareum*". It is noteworthy that the overall rationale of these legal instruments for the protection of rhodolith beds is derived from the human use of this resource in the Atlantic, where they are currently actively exploited as a source of soil amendment, water filtration agent or as a remedy for osteoporosis (Salomidi et al. 2012 and references therein). On the contrary, corallines are not exploited in the Mediterranean Sea, since they are mostly not abundant and shallow enough for an economically convenient activity, rather they may undergo extraction for beach nourishment by dredging, often with significant environmental impact (Nicoletti et al. 2006; Gambi et al. 2009; Basso et al. 2015).

Due to the diverse and distinctive coralline algae composition of the Mediterranean rhodolith beds (Table 11.2), where the occurrence of *L. corallioides* and *P. calcareum* is not obvious, the need to grant a legal protection to the other characteristic species of the Mediterranean assemblages should be considered. Alternatively, given the heterogeneous composition and the difficulty of species identification, this habitat type as a whole should be considered for legal protection as a simpler and more efficient tool to prevent disturbances and/or destruction. In the Mediterranean Sea rhodolith beds are subject of a special plan for the protection within the framework of the United Nations Programme's Mediterranean Action Plan (UNEP-MAP-RAC/SPA 2008; Agnesi et al. 2009), which however is not mandatory for the national governments. This instrument is aimed at gathering coordinated actions to ensure special protection through marine protected areas, based on the presence of such assemblages and of initiatives for the improvement of their scientific knowledge (i.e. adequate taxonomic inventories, geographical distribution maps, monitoring activities). Nevertheless, the protection of a specific habitat type cannot be effectively gained without access to sound geospatial data and monitoring plans (Salomidi et al. 2012). To this purpose, modern advances in remote sensing and acoustic habitat mapping (Georgiadis et al. 2009; Savini et al. 2012; Barbera et al. 2012) can provide an effective tool to assessing the distribution, extent and state of the rhodolith beds, thus easing management and conservation actions.

Acknowledgments Part of the geographic data compilation for the Aegean Sea was performed within the MAREA "Mediterranean Sensitive Habitats" (MEDISEH) project (service contract SI2.600741) aiming to map and model the distribution of priority habitats along the Mediterranean coasts. The European Research Network COCARDE and the EU project MedSeA in the background of this work have created favorable conditions for exchange of data and ideas to grow.

References

Adey WH, Steneck RS (2001) Thermogeography over time creates biogeographic regions: a temperature/space/time-integrated model and an abundance-weighted test for benthic marine algae. J Phycol 37:677–698

Agnesi S, Annunziatellis A, Cassese ML, Di Nora T, La Mesa G, Mo G, Pergent-Martini C, Tunesi L (2009) Analysis on the coralligenous assemblages in the Mediterranean sea: a review of the current state of knowledge in support of future investigations. In: Pergent-Martini C, Brichet M (eds) UNEP – MAP – RAC/SPA Proc of the 1st Mediterranean symposium on the conservation of the coralligenous and other calcareous bio-concretions, Tabarka, Tunisia, 15–16 January 2009, pp 41–46

Agnesi S, Babbini L, Bressan G, Cassese ML, Mo G, Tunesi L (2011) Distribution of maërl facies and rhodolith associations in the Italian seas: current state of knowledge. Biol Mar Mediterr 18(1):50–51

Aguado-Giménez F, Ruiz-Fernández JM (2012) Influence of an experimental fish farm on the spatio-temporal dynamic of a Mediterranean maërl algae community. Mar Environ Res 74:47–55

Aguilar R, Pastor X, Torriente A, Garcia S (2009) Deep-sea coralligenous beds observed with ROV on four seamounts in the western Mediterranean. In: Pergent-Martini C, Brichet M (eds), UNEP – MAP – RAC/SPA Proc of the 1st Mediterranean symp on the conservation of the coralligenous and other calcareous bio-concretions, Tabarka, Tunisia, 15–16 January 2009, pp 148–150

Aktan Y (2012) On the occurrence of Coralligenous algae in the Johnston Bank (Aegean Sea). J Black Sea/Mediterr Environ 18(3):414–419

Astruch P, Boudouresque CF, Bonhomme D, Goujard A, Antonioli PA, Bonhomme P, Perez T, Ruitton S, De Saint-Martin T, Verlaque M (2012) Mapping and state of conservation of benthic marine habitats and assemblages of Port-Cros national Park (Provence, France, northwestern Mediterranean Sea). Sci Rep Port-Cros Natl Park 26:45–90

Atabey N (1998) Facies characteristics and geographic distribution of rhodoliths and maerls (red algae) in Southern shelf of the Sea of Marmara. Miner Res Expl Bull 120:55–61

Babbini L, Bressan G, Massa-Gallucci A, Buia MC, Gambi MC (2006) Segnalazione di una facies a maërl (Corallinales) lungo le coste dell'isola di Ischia. Biol Mar Mediterr 13(1):548–552

Bahia RG, Abrantes DP, Brasileiro PS, Pereira Filho GH, Amado Filho GM (2010) Rhodolith bed structure along a depth gradient on the northern coast of Bahia State, Brazil Braz. J Oceanogr 58(4):323–337

Bakir K, Katağan T (2005) Crustacean diversity of the coralligenous beds of Markiz Island (Aegean coast of Turkey). Crustaceana 78(7):873–883

Ballesteros E (1989) Composición y estructura de los fondos de maërl de Tossa de Mar (Gerona, España). Collectaea Bot 17(2):161–182

Ballesteros E (2006) Mediterranean coralligenous assemblages: a synthesis of present knowledge. Oceanogr Mar Biol Ann Rev 44:123–195

Ballesteros E, Zabala M, Uriz MJ, García-Rubiés A, Turón X (1993) El bentos: Les comunitats. En: Història natural de l'arxipèlag de Cabrera. J.A. Alcover, E. Balleteros, J.J. Fornós (eds). Monogr Soc Hist Nat Balears 2:687–730

Barbera C, Bordehore C, Borg JA, Glémarec M, Grall J, Hall-Spencer JM, De La Hoz CH, Lanfranco E, Lastra M, Moore PG, Mora J, Pita ME, Ramos-Espla AA, Rizzo M, Sanchez-Mata A, Seva A, Schembri PJ, Valle C (2003) Conservation and management of northeast Atlantic and Mediterranean maërl beds. Aquat Conserv Mar Freshw Ecosyst 13:S65–S76

Barberá C, Moranta J, Ordines F, Ramón M, de Mesa A, Diaz-Valdés M, Grau AM, Massutí E (2012) Biodiversity and habitat mapping of Menorca Channel (Western Mediterranean): implications for conservation. Biodivers Conserv 21:701–728. doi:10.1007/s10531-011-0210-1

Barnes J, Bellamy DJ, Jones DJ, Whitton BA, Drew E, Lythgoe J (1970) Sublittoral reef phenomena of Aldabra. Nature 225:268–269

Basso D (1998) Deep rhodolith distribution in the Pontian Islands, Italy: a model for the paleoecology of a temperate sea. Palaeogeogr Palaeocl 137:173–187

Basso D (2012) Carbonate production by calcareous red algae and global change. In: Basso D, Granier B (eds) Calcareous algae and global change: from identification to quantification. Geodiversitas 34:13–33

Basso D, Fravega P, Vannucci G (1996) Fossil and living corallinaceans related to the Mediterranean endemic species *Lithophyllum racemus* (Lamarck) Foslie. Facies 35:275–292

Basso D, Fravega P, Vannucci G (1997) The taxonomy of *Lithothamnion ramosissimum* Gümbel (non Reuss) Conti and *Lithothamnion operculatum* Conti (Rhodophyta, Corallinaceae). Facies 37:167–182

Basso D, Nalin R, Massari F (2007) Genesis and composition of the Pleistocene Coralligène de plateau of the Cutro Terrace (Calabria, southern Italy). N Jb Geol Paläont (Abh) 244/2:173–182

Basso D, Bernasconi MP, Robba E, Marozzo S (2008) Environmental evolution of the Marsala sound, Sicily, during the last 6000 years. J Coast Res 24(1):177–197

Basso D, Nalin R, Nelson CS (2009) Shallow-water *Sporolithon* rhodoliths from North Island (New Zealand). Palaios 24(2):92–103. doi:10.2110/palo.2008.p08-048r

Basso D, Rodondi G, Caragnano A (2014) Coralline species composition of Tyrrhenian maerl beds (Western Mediterranean). In: Bouafif C, Langar H, Ouerghi A (eds) UNEP/MAP – RAC/SPA Proceedings of the 2nd Mediterranean symp on the conservation of Coralligenous and other calcareous bio-concretions. Portorož, Slovenia, 29–30 October 2014, pp 197–198

Basso D, Babbini L, Kaleb S, Bracchi VA, Falace A (2015) Monitoring deep Mediterranean rhodolith beds. Aquat Conserv: Mar Freshw Ecosyst. Wiley Online Library doi:10.1002/aqc.2586

BIOMAERL Team (1998) Maerl grounds: habitats of high biodiversity in European waters. In: Barthel KG, Barth H, Bohle-Carbonell M, Fragakis C, Lipiatou E, Martin P, Ollier G, Weydert M (eds) Third European marine science and technology conference, (Lisbon, Portugal, 23–27 May 1998), vol 1, Marine systems. European Commission, Brussels, pp 170–178

BIOMAERL Team (2003) Conservation and management of northeast Atlantic and Mediterranean maerl beds. Aquat Conserv 13:S65–S76

Bonacorsi M (2012) Caractérisation des peuplements benthiques du Cap Corse (Méditerranée, France). Thèse Ecologie marine, Universite' de Corse, 159 pp

Bordehore C, Riosmena-Rodriguez R, Ramos-Esplá AA (2002a) Maerl-forming species in Alicante province (SE Spain): a taxonomic analysis. In: Proceedings of the 1st Mediterranean symposium on marine vegetation, Ajaccio, 3–4 October 2000. Mednature 1:101–104. RAC/SPA Tunis, Tunisia

Bordehore C, Borg JA, Lanfranco E, Ramos-Esplá AA, Rizzo M, Schembri PJ (2002b) Trawling as a major threat to Mediterranean maerl beds. Proc of the First Mediterranean Symposium on Marine Vegetation, Ajaccio, 3–4 October 2000. Mednature 1:105–109. RAC/SPA Tunis, Tunisia

Bordehore C, Ramos-Esplá AA, Riosmena-Rodriguez R (2003) Comparative study of two maerl beds with different otter trawling history, southeast Iberian Peninsula. Aquat Conserv 13:S43–S54

Bosellini A, Ginsburg RN (1971) Form and internal structure of Recent algal nodules (rhodolites) from Bermuda. J Geol 79:669–682

Bracchi VA and Basso D (2012) The contribution of calcareous algae to the biogenic carbonates of the continental shelf: Pontian Islands, Tyrrhenian Sea, Italy. In: Basso D, Granier B (eds) Calcareous algae and global change: from identification to quantification. Geodiversitas 34:61–76

Canals M, Ballesteros E (1997) Production of carbonate particles by phytobenthic communities on the Mallorca-Menorca shelf, northwestern Mediterranean Sea. Deep Sea Res II 44(3–4):611–629

Castriota L, Gambi MC, Zupo V, Sunseri G (2003) Structure and trophic ecology of a population of *Lysidice ninetta* (Polychaeta) associated to rhodoliths off the island of Ustica (Southern Tyrrhenian Sea). Biol Mar Mediterr 10(2):517–520

Castriota L, Agamennone F, Sunseri G (2005) The molluscan commuity associated with maërl beds of Ustica Island (Tyrrhenian Sea). Cah Biol Mar 46:289–297

Cecere E, Perrone C (1987) First contribution to the knowledge of macrobenthic flora of the Amendolara sea-mount (Ionian Sea). Oebalia 14:1–14

Covazzi Harriague A, Schiaparelli S, Panciroli H, Albertelli G (2002) Soft bottom mollusc communities of four south Tyrrhenian archipelagos and Ustica island (NW Mediterranean). P Italian Assoc Oceanol Limnol 15:63–74

Damiani V, Bianchi CN, Ferretti O, Bedulli D, Morri C, Viel M, Zurlini G (1988) Risultati di una ricerca ecologica sul sistema marino costiero pugliese. Thalass Salentina 18:153–169

El Lakhrach H, Hattour A, Jarboui O, Elhasni K, Ramos-Espla AA (2012) Spatial distribution and abundance of the megabenthic fauna community in Gabes gulf (Tunisia, eastern Mediterranean Sea). Mediterr Mar Sci 13(1):12–29

Falace A, Bressan G (2003) Changes of algal flora in the Gulf of Trieste (Northern Adriatic Sea). Bocconea 16:1033–1037

Falace A, Kaleb S, Agnesi S, Annunziatellis A, Salviati E, Tunesi L (2014) Macroalgal composition of rhodolith beds in a pilot area of the Tuscan Archipelago (Tyrrhenian Sea): primary elements to evaluate the degree of conservation of this habitat. In: Bouafif C, Langar H, Ouerghi A (eds), UNEP/MAP – RAC/SPA Proceedings of the 2nd Mediterranean symposium on the conservation of Coralligenous and other calcareous bio-concretions, Portorož, Slovenia, 29–30 October 2014, pp 213–214

Funk G (1927) Die algenvegetation des Golfs von Neapel. Pubbl. della Staz Zool di Napoli 7

Gambi MC, Buia MC, Massa-Gallucci A, Cigliano M, Lattanzi L, Patti FP (2009) The "pink mile": benthic assemblages of rhodolith and maërl beds (Corallinales) off the Island of Ischia (Thyrrhenian Sea). In: Pergent-Martini C, Brichet M (eds), UNEP – MAP – RAC/SPA Proceedings of the 1st Mediterranean symposium on the conservation of the coralligenous and other calcareous bio-concretions, Tabarka, Tunisia, 15–16 January 2009, pp 198–200

García-Carrascosa AM (1987) El bentos de los alrededores de las islas Columbretes. Elementos para su cartografía bionómica. En: *Islas Columbretes. Contribución al estudio de su medio natural.* LA Alonso Matilla, JL Carretero & AM García-Carrascosa (coord.). COPUT, Generalitat Valenciana, pp 363–390

García-March JR, Kersting DK (2006) Preliminary data on the distribution and density of *Pinna nobilis* and *Pinna rudis* in the Columbretes Islands Marine Reserve (Western Mediterranean, Spain). Poster presented at the International Congress on Bivalvia. Universitat Autònoma de Barcelona, Bellaterra, Catalunya, Spain, 22–27 July 2006

Georgiadis M, Papatheodorou G, Tzanatos E, Geraga M, Ramfos A, Koutsikopoulos C et al (2009) Coralligène formations in the eastern Mediterranean Sea: morphology, distribution, mapping and relation to fisheries in the southern Aegean Sea (Greece) based on high-resolution acoustics. J Exp Mar Biol Ecol 368:44–58

Giménez-Casalduero F, Rodriguez-Ruiz S, Vivas M, Ramos-Esplá AA (2001) Variaciones de las características estructurales de la comunidad de poliquetos asociada a dos fondos de maërl del litoral alicantino (sudeste de la península Ibérica). Bol Inst Esp Oceanog 17(1–2):191–201

Ginsburg RN, Bosellini A (1973) Form and internal structure of recent algal nodules (Rhodolites) from Bermuda: a reply. J Geol 81:239–241

Glémarec M (1997) La complexité architecturale des fonds de maérl en tant que source de biodiversité. In: Dauvin JE (ed) Les biocénoses marines et littorales françaises des côtes atlantique, Manche et Mer du Nord. Synthèse, menaces et perspectives. Muséum Nat Hist Nat:147–149

Grall J, Glémarec M (1997) Biodiversité des fonds de maërl en Bretagne: approche fonctionelle et impacts anthropiques. Vie Milieu 47(4):339–349

Grall J, Hall-Spencer JM (2003) Problems facing maërl conservation in Brittany. Aquat Conserv 13:55–64

Hall-Spencer JM (1998) Conservation issues relating to maërl beds as habitats for molluscs. J Conchol Spec Publ 2:271–286

Huertas IE, Ríos AF, García-Lafuente J, Navarro G, Makaoui A, Sánchez-Román A, Rodriguez-Galvez S, Orbi A, Ruíz J, Pérez FF (2012) Atlantic forcing of the Mediterranean oligotrophy. Global Biogeochem Cy 26:GB2022. doi:10.1029/2011GB004167

Huvé H (1955) Contribution à l'étude des fonds à *Lithothamnium* (?) *solutum* de la région de Marseille. Recl Trav St Mar Endoume 18:105–133

Jacquotte R (1962) Etude des fonds de maerl de Méditérranée. Recl Trav St Mar Endoume 26:141–235

Keegan BE (1974) The macrofauna of maërl substrates on the best coast of Ireland. Cah Biol Mar 15:513–530

Laborel J (1961) Le concrétionnement algal "coralligène" et son importance geomorphologique en Méditerranée. Recl Trav St Mar Endoume 23:37–60
Lanfranco E, Rizzo M, Hall-Spencer J, Borg JA, Schembri PJ (1999) Maerl-forming coralline algae and associated phytobenthos from the Maltese Islands. Cent Mediterr Nat 3(1):1–6
Lemoine M M.me. P. (1940) Les algues calcaires de la Zone Néritique. Mem Soc Biogéogr 7:75–138
Lozano CJ, Candela J (1995) The M2 tide in the Mediterranean Sea: dynamic analysis and data assimilation. Oceanol Acta 18(4):419–441
Mannino AM, Castriota L, Beltrano AM, Sunseri G (2002) The epiflora of a rhodolith bed from the island of Ustica (Southern Tyrrhenian Sea). Flora Mediterr 12:11–28
Marion AF (1883) Esquisse d'une topographie zoologique du golfe de Marseille. Ann Mus Hist Nat Marseille 1:1–108
Marrack E (1999) The relationship between water motion and living rhodolith beds in the south-Western Gulf of California, Mexico. Palaios 14:159–171
Martin Sintes D (1987) La comunidad de Anélidos Poliquetos de las concreciones de algas calcáreas del litoral catalán. Caracterización de las especies. P Dept Zool Barcelona 13:45–54
Martin S, Gattuso JP (2009) Response of Mediterranean coralline algae to ocean acidification and elevated temperature. Global Change Biol 15:2089–2100. doi:10.1111/j.1365-2486.2009.01874.x
Martin CS, Giannoulaki M, De Leo F, Scardi M, Salomidi M, Knitweiss L, Pace ML, Garofalo G, Gristina M, Ballesteros E, Bavestrello G, Belluscio A, Cebrian E, Gerakaris V, Pergent G, Pergent-Martini C, Schembri PJ, Terribile K, Rizzo L, Ben Souissi J, Bonacorsi M, Guarnieri G, Krzelj M, Macic V, Punzo E, Valavanis V, Fraschetti S (2014) Coralligenous and maërl habitats: predictive modelling to identify their spatial distributions across the Mediterranean Sea. Sci Rep 4:5073. doi:10.1038/srep05073
Meinesz A, Boudouresque CF, Falconetti C, Astier JM, Blanc JJ, Bourcier M, Cinelli F, Cirik S, Cristiani G, Di Geronimo I, Giaccone G, Harmelin JG, Laubier L, Lovric AZ, Molinier R, Soyer J, Vamvakas C (1983) Normalisation des symboles pour la représentation et la cartographie des biocénoses benthiques littorales de Méditerranée. Ann Inst Océanogr NS 59:155–172
Micallef A, Le Bas TP, Huvenne VAI, Blondel P, Huhnerbach V, Deidun A (2012) A multi-method approach for benthic habitat mapping of shallow coastal areas with high-resolution multibeam data. Cont Shelf Res 39–40:14–26. http://dx.doi.org/10.1016/j.csr.2012.03.008
Müller GJ (1985) The pre-coralligen community in the Marmara Sea. Rapp Comm Int Mer Medit 29:327–328
Nicoletti L, Paganelli D, Gabellini M (2006) Aspetti ambientali del dragaggio di sabbie relitte a fini di ripascimento: proposta di un protocollo di monitoraggio. Quaderno ICRAM 5:159 pp
Ordines F, Massutí E (2009) Relationships between macro-epibenthic communities and fish on the shelf grounds of the western Mediterranean. Aquat Conserv 19(4):370–383
Papakosta F, Hasiotis T, Paleokrassas A (2012) Biogenic formations in the Sikinos-Folegandros area. In: Proceedings of the 10th Hellenic symposium on oceanography and fisheries, 7–11 May, Athens, 41 p (in Greek)
Pardo C, Lopez L, Peña V, Hernández-Kantún J, Le Gall L et al (2014) A Multilocus species delimitation reveals a striking number of species of coralline algae forming maerl in the OSPAR Maritime Area. PLoS ONE 9(8):e104073. doi:10.1371/journal.pone.0104073
Parenzan P (1960) Aspetti biocenotici dei fondi ad alghe litoproduttrici del Mediterraneo. Rapp Comm Int Mer Méditerr 15:87–107
Pérès JM (1952) Notes sur les fonds de gravelle dans la région de Marseille comparés à ceux des côtes d'Algérie. Vie Milieu Suppl 2:208–215
Pérès JM, Picard J (1956) Recherches sur les peuplements benthiques de la Méditerranée Nord-Orientale. Résultats scientifiques des campagnes de la Calypso, Fascicule III, 214–291. Masson et Cie ed Paris
Pérès JM, Picard J (1964) Nouveau manuel de bionomie benthique de la Mer Méditérranée. Recl Trav St Mar Endoume 47:1–137

Piazzi L, Pardi G, Cinelli F (2002) Structure and temporal dynamics of a macroalgal assemblage associated with a rhodolith bed of the Tuscan archipelago (Tyrrhenian sea). Atti Soc Tosc Sci Nat Mem Ser B 109:5–10
Picard J (1965) Recherches qualitatives sur les biocoenoses marines des substrats meubles dragables de la région marseillaise. Recl Trav St Mar Endoume 36(52):1–160
Pignatti S, Rizzi L (1972) Raccolte di alghe bentoniche nelle acque dell'Arcipelago Toscano. Atti Ist Veneto Sci Lett ed Arti Cl Sci Mat e Nat 130:313–327
Pinardi N, Masetti E (2000) Variability of the large scale general circulation of the Mediterranean Sea from observations and modelling: a review. Palaeogeogr Palaeocl 158:153–173
Ramos-Esplá AA, Ben Mustapha K (2010) Etude des habitats marins et des principales espèces des îles Kuriat (Tunisie, 2008). Rapport de la mission. Contrat n° 16/CAR-ASP/2008, 91 pp
Ramos-Esplá AA, Luque AA (2004) Los fondos de "maerl". In: Luque AA, Templado J (Coords) Pradera y bosques marinos de Andalucía. Consejería de Medio Ambiente, Junta de Andalucía, Sevilla:221–236
Ramos-Esplá AA, Luque AA (2008) The seas of Spain. Maerl beds: a fragile oasis of marine life. Ministero del Medio Ambiente y medio rural y marino, pp 273–290
Ramos-Esplá AA, Riosmena-Rodriguez R, Galil B (2012) Contribution to the knowledge of maérl beds in the Levantine Basin. Eastern Mediterranean. IV International Rhodolith Workshop, Granada (Spain):63
Relini G, Giaccone G (2009) Gli Habitat prioritari del Protocollo SPA/BIO (Convenzione di Barcellona) presenti in Italia. Schede descrittive per l'identificazione. Biol Mar Mediterr 16(suppl1):1–266
Riul P, Targino CE, da Nobrega Farias J, Visscher PT, Horta PA (2008) Decrease in Lithothamnion sp. (Rhodophyta) primary production due to the deposition of a thin sediment layer. J Mar Biol Assoc UK 88(1):17–19
Salomidi M, Smith C, Katsanevakis S, Panayotidis P, Papathanassiou V (2009) Some Observations on the structure and distribution of several Gorgonian Assemblages in the Eastern Mediterranean Sea. In: Pergent-Martini C, Brichet M (eds) UNEP – MAP – RAC/SPA Proc of the 1st Mediterranean symp on the conservation of the coralligenous and other calcareous bio-concretions, Tabarka, Tunisia, 15–16 January 2009, pp 242–245
Salomidi M, Katsanevakis S, Borja A, Braeckman U, Damalas D, Galparsoro I, Mifsud R, Mirto S, Pascual M, Pipitone C, Rabaut M, Todorova V, Vassilopoulou V, Vega Fernandez T (2012) Assessment of goods and services, vulnerability, and conservation status of European seabed biotopes: a stepping stone towards ecosystem-based marine spatial management. Medit Mar Sci 13(1):49–88
Sanz-Lázaro C, Belando MD, Marín-Guirao L, Navarrete-Mier F, Marín A (2011) Relationship between sedimentation rates and benthic impact on Maërl beds derived from fish farming in the Mediterranean. Mar Environ Res 71(1):22–30
Savini A, Basso D, Bracchi VA, Corselli C, Pennetta M (2012) Maerl-bed mapping and carbonate quantification on submerged terraces offshore the Cilento peninsula (Tyrrhenian Sea, Italy). In: Basso D, Granier B (eds) Calcareous algae and global change: from identification to quantification. Geodiversitas 34:77–98
Schembri PJ (1998) Maerl ecosystems of the Maltese Islands. In: Dandria D (ed) Biology abstracts MSc, PhD 1998 and contributions to marine biology. Department of Biology, University of Malta, Msida, pp 35–37, iv+38pp
Sezgin M, Bakir K, Katağan T, Suat-Ateş A, Kirkim F (2009) Crustacean diversity of the coralligenous (maërl) beds of the Aegean sea coast of Turkey. In: Pergent-Martini C, Brichet M (Eds), UNEP – MAP – RAC/SPA Proc of the 1st Mediterranean symp on the conservation of the coralligenous and other calcareous bio-concretions, Tabarka, Tunisia, 15–16 January 2009, pp 253–255
Soto J (1990) Vegetación algal sobre sustrato móvil de la zona circalitoral del sureste de la Península Ibérica: una aproximación. Folia Bot Misc 7:43–49

Steller DL, Riosmena-Rodriguez R, Foster MS, Roberts CA (2003) Rhodolith bed diversity in the Gulf of California: the importance of rhodolith structure and consequences of disturbance. Aquat Conserv 13:S5–S20

Templado J, Calvo M, García-Carrascosa AM, Boisset F, Jiménez J (2002) Flora y fauna de la Reserva Marina de las Islas Columbretes. Secretaría Gral. Pesca Marítima. M° Agricultura, Pesca y Alimentación. Museo Nacional de Ciencias Naturales, CSIC, 263 pp

Templado J, Calvo M, Moreno D, Flores A, Conde F, Abad R, Rubio J, López-Fé CM, Ortiz M (2006). Flora y fauna de la Reserva Marina y Reserva de Pesca de la Isla de Alborán. Secretaría Gral. Pesca Marítima. M° Agricultura, Pesca y Alimentación. Museo Nacional de Ciencias Naturales, CSIC. 269 pp

Trapani F, Scotti G, Gianguzza P, Chemello R, Riggio S (1999) Struttura della malacofauna associata ai rodoliti dello Stagnone di Marsala (Sicilia occidentale). Biol Mar Mediterr 6(1):462–465

Tsimplis MN, Proctor R, Flather RA (1995) A two-dimensional tidal model for the Mediterranean Sea. J Geophys Res 100(C8):16223–16239

Vatova A (1935) Ricerche preliminari sulle biocenosi del Golfo di Rovigno. Thalassia 2:1–30

Walther J (1885) Le alghe calcarifere litoproduttrici del Golfo di Napoli e l'origine di certi calcarei compatti. Boll Regio Comitato Geol d'Italia 16:360–369

Wilson S, Blake C, Berges JA, Maggs CA (2004) Environmental tolerances of freeliving coralline algae (maerl): implications for European marine conservation. Biol Conserv 120:279–289

Chapter 12
South Atlantic Rhodolith Beds: Latitudinal Distribution, Species Composition, Structure and Ecosystem Functions, Threats and Conservation Status

Gilberto M. Amado-Filho, Ricardo G. Bahia, Guilherme H. Pereira-Filho, and Leila L. Longo

Abstract The largest continuous latitudinal distribution of rhodolith beds occur in the South Atlantic Ocean. Up to now rhodolith beds were referred exclusively to the western portion of the South Atlantic. Here we describe the recent advances in the South Atlantic taking into account latitudinal distribution, species composition, structure and ecosystem functions, threats and conservation status. Rhodolith beds have been mapped and ecologically described from extensive areas of the continental shelf (Abrolhos Bank), seamounts tops (Vitoria Trindade Chain), insular shelfs of oceanic islands (Fernando de Noronha Archipelago) and atolls (Rocas Atoll). Thirty three species of crustose coralline algae were recorded forming rhodoliths. Despite some initiatives, the richness of fauna associated with rhodoliths in SW Atlantic is still poorly known. Specific microbiome described associated with rhodoliths indicates important role in biomineralization process. The environmental services provided by the recently described rhodolith beds (Abrolhos Banks and Vitoria Trindade Seamounts) as calcium carbonate production, increase habitat complexity, benthic diversity and associated fish assemblages justify urgent actions to protect these ecosystems.

G.M. Amado-Filho (✉) • R.G. Bahia
Instituto de Pesquisas Jardim Botânico do Rio de Janeiro, Rio de Janeiro, Brazil
e-mail: gfilho@jbrj.gov.br

G.H. Pereira-Filho
Universidade Federal de São Paulo, Santos, Brazil

L.L. Longo
Universidade Federal do Espírito Santo, Vitória, Brazil

R. Riosmena-Rodríguez et al. (eds.), *Rhodolith/Maërl Beds: A Global Perspective*, Coastal Research Library 15, DOI 10.1007/978-3-319-29315-8_12

12.1 Introduction

The South Atlantic Ocean is the worldwide region where the largest continuous latitudinal distribution of rhodolith beds was seen. These beds are distributed at the western portion of the Ocean, specifically along the Brazilian coast. Up to now no references about rhodolith beds were registered for the South Atlantic coast of Africa nor for Antarctica.

As was indicated by Foster (2001) and Foster et al. (2013) the Brazilian coast supports the largest known rhodolith beds in the world, covering extensive areas of the north, northeastern, and southeastern Brazilian continental shelf (Kempf 1970; Milliman 1977; Amado-Filho et al. 2012a). However, the understanding of the latitudinal distribution, depth range occurrence, $CaCO_3$ production, ecological roles of these beds still have to be done. Most of the early studies were focused on the characterization of shelf sediments and prospects for potential commercial exploitation (Kempf 1970; Mabesoone et al. 1972; Milliman and Amaral 1974; Milliman 1977; Vicalvi and Milliman 1977). In the mid-1990s, some studies started to approach the rhodolith beds from a biological point of view, providing, for a limited area of the northeastern Brazilian coast, information on rhodolith mapping and bed structure (e.g., Testa 1997; Testa and Bosence 1999). Only at the 2000s an increase of studies related to different aspects of rhodolith beds was seen. These studies considered consistent information about meso and small scale distribution, bed structure, associated organisms, CCA species composition, $CaCO_3$ production and physiological aspects of rhodolith-forming CCA (e.g., Gherardi 2004; Riul et al. 2009; Villas-Boas et al. 2009; Amado-Filho et al. 2010; Amado-Filho et al. 2012a; Figueiredo et al. 2012; Moura et al. 2013). Despite the recent advances, most rhodolith beds in Brazil remain unmapped and the potential of their ecosystem services is unknown (Amado Filho and Pereira Filho 2012). This is largely because the beds occur mostly in mesophotic habitats (~30–120 m depth) that are difficult to adequately access. An example of gaps in knowledge of rhodolith beds is presented by Foster et al. (2013): "if hundreds of species were described from deeper areas, strongly suggesting that more species would be found as surveys continued. However, these collections were all made by dredging, and the few specimens obtained may not have been representative of local variation. Taxonomic analyses could be biased by describing new species based on a limited number of specimens". The recent advances in mixed-gas diving techniques, complemented by ROV observations and high resolution multibeam bathymetric mapping systems, have allowed much better determinations of the extent, structure, dynamics of rhodolith beds. In addition, we can collect samples with much more accuracy for taxonomic studies. The referred tools have been used to investigate extensive areas of the mesophotic zone of the continental shelf (e.g., Amazon River Mouth, Abrolhos Bank) (Amado Filho et al. 2012a; Moura et al. 2013, 2016), the tops of seamounts (Vitoria-Trindade Chain) (Pereira-Filho et al. 2012) and around oceanic islands (Pereira-Filho et al. 2011 and Amado-Filho et al. 2012b, 2016).

In this chapter, we will describe the recent advances in rhodolith studies in the South Atlantic taking into account the following aspects: distribution and mapping; species composition; structure and ecosystem functions of rhodolith beds; and threats, vulnerability and conservation status.

12.2 Distribution and Mapping

The highest concentration of rhodolith beds in the world were found in the Brazilian Economic Exclusive Zone and represents an estimated reservoir of 2×10^{11} tons of calcium carbonate (Milliman and Amaral 1974). The beds cover extensive areas of the continental shelf mainly at the mid and external portion of the shelf. With the actual information available, the range of the latitudinal distribution is between 1° N and 23° lat. S, with an additional isolated small specific bank southern localized at Arvoredo Island (27°17′12″S, 48°22′25″W) (Fig. 12.1a).

Recent samples collected on the shelf of the Amazon River delta indicate the presence of rhodoliths between 25 and 120 m depth. These dredge collected samples were composed of live CCA individuals (Moura et al. 2016) and have to be considered the north limit of occurrence for rhodolith bed distribution in the South Atlantic Ocean. Including this mentioned area, there is great information gaps in rhodolith beds distribution for extensive areas of the north and northeastern Brazilian shelf. The information available is restricted to sedimentary maps produced in the 1970s and re-interpretated by more recent reports, as Cavalcanti (2011), that normally mentioned rhodolith beds as bioclastic granules. Clearly, an effort can be made to produce basic information about extension, distribution depth range, composition and bed structure of rhodolith beds along 1400 km west-east extension range from the Amazon mouth (~01°00′ lat N, 47°50′ long W) to north of Rio Grande do Norte State (~05°05′ lat S, 35°25′ long W) (Fig. 12.1a).

For the northeastern Brazilian region there are some data available about rhodolith distribution. Testa and Bosence (1999) presented a detailed sedimentary characterization of shallow waters (0–20 m depth) at the northeast region of the Rio Grande do Norte State (5° S, 35° W) (Fig. 12.1c) and showed that rhodolith beds are an important feature in an area of extremely high hydrodynamic regime that result in a large scale marine bed forms within a shallow algal dominated carbonate zone. Riul et al. (2009) presented a characterization of rhodolith beds from 10 to 20 m depth between 6°58′ and 7°04′ lat S, approximately 5 km from the coast of João Pessoa city (Paraiba State). From this point to 13° lat S near Salvador city (Bahia State) there is a great gap about rhodolith beds distribution. Certainly, in this 750 km continental shelf extension there are large rhodolith beds to be mapped as was indicated by the sedimentary works of the 1970s (synthesis in Cavalcanti 2011). For a very specific point of the continental shelf of surroundings of Salvador city (north of Bahia State), Bahia et al. (2010) presented a characterization of rhodolith beds from 5 to 25 m depth (Fig. 12.1d). The continental shelf off the coast extending from 10 to 16°S, is unusually narrow (~20 km) and the shelf break is located

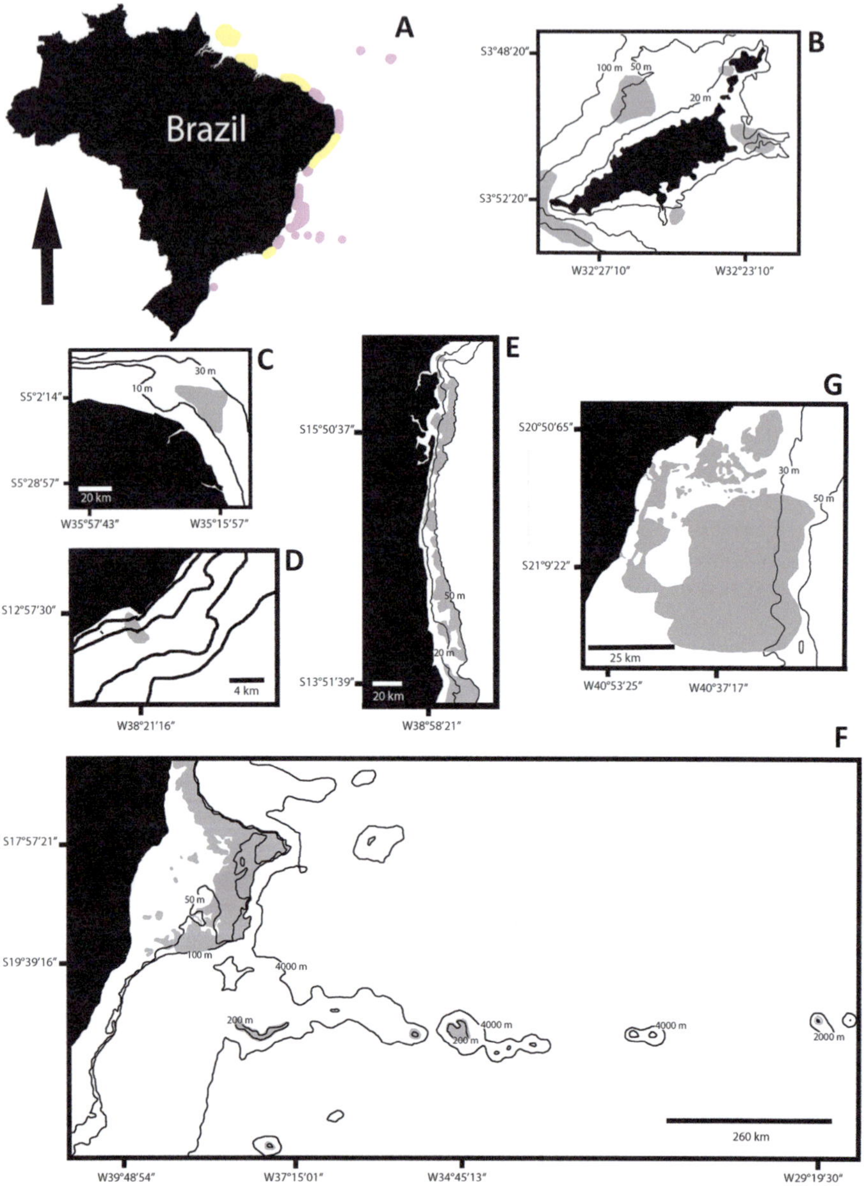

Fig. 12.1 Distribution of rhodolith beds in SW Atlantic Ocean. (**a**) Rhodolith beds covered areas, in pink confirmed mapped areas and in yellow suggested areas occupied by rhodolith beds; (**b–g**) Mapped areas, (**b**) Fernando de Noronha Archipelago (Amado Filho et al. 2012b), (**c**) Rio Grande do Norte State (Testa and Bosence 1999), (**d**) Salvador city area (Bahia et al. 2010), (**e**) Central Bahia State area (Dominguez et al. 2013), (**f**) Abrolhos Bank and Vitoria Trindade Chain (Amado Filho et al. 2013, Pereira Filho et al. 2012), (**g**) South of Espírito Santo State (Amado Filho et al. 2007, 2010; Dias and Villaça 2012; Villas-Boas et al. 2014)

between the 45 and 50 m isobaths and coincides with major hinge-lines of the marginal basins. Dominguez et al. (2013) evaluated the sedimentology of the continental shelf from 13 to 16° S (Fig. 12.1e) and showed that from 20 m depth to the limit of the shelf, sandy sedimentation is essentially bioclastic with major constituents represented by encrusting coralline algae. Gravelly sediments exclusively bioclastic constituted by rhodoliths are restricted to the outer shelf in waters usually deeper than 30 m, especially close to the shelf edge.

It is important to mention the information gap about rhodoliths distribution and mapping in mesophotic zones for the most part of the north and northeastern region of the SW Atlantic. In these regions, information about mesophotic zone is only available for insular shelves of Fernando de Noronha Archipelago and Rocas Atoll (Fig. 12.1a, b), both of them situated offshore of the Northeastern Brazilian coast (Amado-Filho et al. 2012b, 2016). By using Side Scan Sonar mapping and/or ROV imaging systems, the rhodolith beds were described occurring between the lowest limit of the shallow reef (Rocas Atoll) and rocky shores (Fernando de Noronha) where the seabed became flattened (10–20 m depth) down to the edge limit of the shelf ranging from 60 to 100 m depth.

The Abrolhos Bank with an area of 46,000 km^2, is well known to encompass the most important coralline reefs in the South Atlantic, with high levels of endemism. Moura et al. (2013) mapped the benthic megahabitats of Abrolhos Bank shelf at an unprecedented spatial scale and resolution in the SW Atlantic Ocean. This study presents a classification of the main benthic megahabitats and their spatial distribution based on sidescan sonar surveys coupled with groundtruthing with remotely operated vehicles (ROV) and mixed-gas diving. Three megahabitats were described: unconsolidated sediments, coralline reefs and rhodolith beds. The rhodolith megahabitat encompasses the largest continuous rhodolith bed in the world (20,904 km^2), covering an area comparable to that of coral reefs in the Caribbean (21,600 km^2) (Amado-Filho et al. 2012a). The rhodolith bed megahabitat is widely distributed, ranging from the mid shelf to the shelf break across most of the latitudinal range of the Bank (Fig. 12.1f). On the inner shelf, rhodolith beds intermingle with unconsolidated sediments and can be partially buried by sand or terrigenous mud (Moura et al. 2013).

Adjacent to the Abrolhos Bank is localized the Vitoria Trindade Chain (VTC; 20°S–21°S, 29–38°W), that is a 1150 km west-east chain of nine seamounts that includes two small and highly isolated islands at its eastern end (Pereira Filho et al. 2012). Previous biological sampling on the VTC was largely restricted to dredging, data from commercial fisheries, and a few scientific diving operations in the shallow waters surrounding Trindade Island. The insular platforms of Trindade Island, as well as three tops of seamounts within the VTC (Jaseur, Davis, and Vitoria seamounts), and from the Almirante Saldanha Seamount (22°30′00″ S – 37°S30′00″ W) were surveyed by using the combination tools, sidescan sonars, ROVs and mixed-gas technical dives (Pereira Filho et al. 2012). All sample sites presented predominantly flattened tops lying at average depths around 60 m (10–110 m). These flattened summits, with horizontal extensions of up to 50 km, apparently result from alternate growth and erosion of carbonate deposits over volcanic pedestals (Skolotnev et al. 2010). On the basis of bathymetry and extension of occurrence, the estimated area covered by rhodolith beds in the studied area is 1511 km^2 (850, 590, 40, 15, and 16 km^2, for Davis, Vitória, Jaseur and Almte, Saldanha Seamounts and Trindade shelf, respectively (Pereira Filho et al. 2012) (Fig. 12.1f).

From the south of Abrolhos Bank to the north of Rio de Janeiro State, including all the extension of Espírito Santo State (18.35° to 21.30°S), rhodolith have been mentioned as important feature for the continental shelf. This region is understood to be the transition between the tropical and subtropical Brazilian coast. Possessing high marine macroalgal diversity, the region comprises the richest marine flora in Brazil (Guimarães 2003). Its high diversity has been associated, in part, with the presence of extensive rhodolith beds, extending from the intertidal zone down to 100 m depth (Amado-Filho et al. 2007, 2010; Villas-Boas et al. 2014). As was pointed by Dias (2000), extensive areas at the edge of the shelf are composed by coralline algae growing in crusts superimposed that should correspond to rhodolith beds in the part. More intensive studies were made at the south of Espírito Santo State by Dias and Villaça (2012), Amado-Filho et al. (2007, 2010) and Villas-Boas et al. (2014) that showed high extension of rhodolith beds from shallow water through the mesophotic zone at the edge of the shelf (Fig. 12.1g).

The area known as Campos Basin (~100,000 km^2), localized at the north of Rio de Janeiro State, the most important offshore field of oil and gas production area in Brazil, presented in its mid and external shelves extensive rhodolith beds from 45 to 150 m depth (Giustina 2006). Although very scarce scientific published data is available about rhodolith distribution, potentially the area occupied by rhodoliths in Campos Basin (Fig. 12.1a) can correspond to that recently described by Amado-Filho et al. (2012a) for Abrolhos Bank. As the edge of the shelf of Campos Basin is very gently sloped, rhodolith beds with live CCA in its surface can be found down to 200 m depth (Giustina 2006). The need for improvement of basic information about rhodolith distribution and community structure in rhodolith beds at Campos Basin continental shelf have been recognized by the Brazilian Environmental Agency (IBAMA), specially for a better definition of sites for installation of new oil and gas productions platforms, ducts and structures.

Up to now the southern limit for the occurrence of extensive rhodolith beds in the southwestern Atlantic Ocean is the region of Cabo Frio localized at the south limit of Campos Basin. Figueiredo et al. (2012) evaluated photosynthetic and growth rates of two rhodolith forming CCA species from 90 to 100 m depth at Peregrino oil field, Cabo Frio. Cabo Frio is an upwelling region where carbonate deposits physiognomy shift from mostly corals and coralline algae to bryozoans in the central continental shelf to mollusks towards the south.

At the Marine Biological Reserve of Arvoredo Island, distant 800 km southwest from Cabo Frio region it is found a well studied and protected small rhodolith bed (area of 100 km^2) that occurs between 7 and 20 m depth at the northwestern portion of the Island (Gherardi 2004; Pascelli et al. 2013). Up to now, the Arvoredo Island is referred as the southern limit of occurrence of rhodolith beds in the SW Atlantic Ocean.

12.3 Rhodoliths Species Composition

The SW Atlantic has not only the largest area covered by rhodolith beds but also concentrates the highest number of CCA species associated to them in the world. According to the checklist provided by Bahia (2014) in addition to recent discoveries in Henriques et al. (2014a, b) and Bahia et al. (2015), there are 33 CCA species recorded for the rhodolith beds of the continental shelf, seamounts and oceanic islands of Brazil, most of which occurring in tropical waters (Table 12.1). Among them, 26 were identified based on morfo/anatomical features according to the modern understanding of CCA taxonomy (see Harvey and Woelkerling 2007 for a review). The other seven records still need further examination according to the modern taxonomy criteria.

The species richness of rhodolith-forming CCA in this area is more than twice than what is recorded for any other part of the world where rhodoliths are found (Table 12.2). Most rhodolith beds around the world are composed of two or three species (Konar et al. 2006) and each individual rhodolith is usually built by only one species. The SW Atlantic rhodolith-beds correspond to an exception to this pattern. Six or more species can be found within a relatively small area (Amado-Filho et al. 2010, 2012b) and is not uncommon to find up to four CCA species growing on only an individual rhodolith surface. The most commonly recorded species with the widest latitudinal range in this area are *Harveylithon rupestre* (Foslie) A. Rösler, *Lithothamnion crispatum* Hauck, *Melyvonnea erubescens* (Foslie) Athanasiadis & D.L. Ballantine, and *Sporolithon ptychoides* Heydrich (Bahia 2014). According to Pascelli et al. (2013) *Lithothamnion* appears to be the most important genus of rhodolith-forming genus in the SW Atlantic since *L. crispatum* is dominant species at the southern limit of this habitat and also dominates the largest rhodolith bed at Abrolhos Bank (Amado-Filho et al. 2012a).

Recent efforts to document the rhodolith-forming species in the southwestern Atlantic have promoted the discovery of several new occurrences (e.g. Amado-Filho et al. 2010; Bahia et al. 2011, 2014a; Henriques et al. 2012, 2014b) and the description of four new species (Villas-Boas et al. 2009; Bahia et al. 2014b, 2015; Henriques et al. 2014a). However, there is still much to be done in terms of rhodolith-forming species inventory, taxonomy and distribution. Most part of the enormous area of northeastern coast of Brazil covered by rhodolith beds has never been inventoried. This means that inevitably there are species as yet unrecorded and undescribed in this region. Furthermore, a name is just a label: knowing it does not imply that we really know about the named species. Once a species is named and its phenotype described, we should know about its variability and its life cycle, define its ecological niche, understand its role within communities and ecosystems (Boero 2010). These other tasks are far from being completed and they also deserve attention in further investigations.

The taxonomy of rhodoliths has been regarded as difficult because of their morphological variability and because they are reported to be frequently unfertile, and thus critical distinguishing characters are unable to be evaluated (Nelson et al. 2012). Molecular characters provide an appealing alternative that could widen the

Table 12.1 Rhodolith-forming CCA species recorded for the southwestern Atlantic (Brazilian coast)

Species	Distribution along the Brazilian coast
(Corallinales)	
Harveylithon rupestre (Foslie) A. Rösler, Perfectti, V. Peña & J.C. Braga	SE and NE (tropical)
Hydrolithon breviclavium (Foslie) Foslie	SE (tropical)
Lithophyllum corallinae (PL. Crouan & HL. Crouan) Heydrich	SE (tropical)
Lithophyllum depressum Villas-Boas, Figueiredo & Riosmena-Rodríguez	SE (tropical)
Lithophyllum johansenii Woelkerling & Campbell	SE (tropical)
Lithophyllum margaritae (Hariot) Heydrich	S (subtropical)
Lithophyllum rugosum (Foslie) Lemoine[a]	S (temperate)
Lithophyllum stictaeforme (Arechoug) Hauck	S, SE, and NE (temperate and tropical)
Neogoniolithon brassica-florida (Harvey) Setchell & L.R. Mason	SE and NE (tropical)
Neogoniolithon fosliei (Heydrich) Setchell & L.R. Mason	SE and NE (subtropical and tropical)
Pneophyllum fragile Kützing	S, SE, and NE (temperate and tropical)
Spongites fruticulosus Kützing	SE and NE (tropical)
Spongites yendoi (Foslie) Y.M. Chamberlain	SE (tropical)
Titanoderma prototypum (Foslie) Woelkerling et al.	SE and NE (tropical)
Titanoderma pustulatum (J.V. Lamouroux) Nägeli	SE and NE (subtropical and tropical)
(Hapalidiales)	
Lithothamnion brasiliense Foslie	SE and NE (tropical)
Lithothamnion crispatum Hauck	S, SE, and NE (temperate, subtropical and tropical)
Lithothamnion glaciale Kjellman	SE (tropical)
Lithothamnion muelleri Lenormand ex Rosanoff	SE (tropical)
Mesophyllum engelhartii (Foslie) Adey	S, SE and NE (temperate and tropical)
Melyvonnea erubescens (Foslie) Athanasiadis & D.L. Ballantine	S, SE and NE (temperate and tropical)
Phymatolithon calcareum (Pallas) W.H. Adey & D.L. McKibbin[a]	S and SE (temperate and subtropical)
Phymatolithon masonianum Wilks & Woelkerling[a]	S (temperate)
(Sporolithales)	
Sporolithon australasicum (Foslie) N. Yamaguishi-Tomita ex M.J. Wynne[a]	SE (tropical)
Sporolithon episporum (M.A. Howe) E.Y. Dawson	NE (tropical)
Sporolithon durum (Foslie) Townsend & Woelkerling[a]	SE and NE (tropical)
Sporolithon elevatum M.C. Henriques & R. Riosmena-Rodríguez	SE (tropical)
Sporolithon erythraeum (Rothpletz) Kylin[a]	NE (tropical)
Sporolithon molle (Heydrich) Heydrich	SE (tropical)

(continued)

Table 12.1 (continued)

Species	Distribution along the Brazilian coast
Sporolithon pacificum E.Y. Dawson[a]	SE and NE (tropical)
Sporolithon ptychoides Heydrich	SE, NE and N (tropical)
Sporolithon tenue Bahia et al.	NE (tropical)
Sporolithon yoneshigueae Bahia et al.	SE and NE (tropical)

Data based on Bahia (2014), Henriques et al. (2014a, b) and Bahia et al. (2015)
[a]Species await further examination in a modern context of CCA taxonomy, as a consequence, species names recorded in the literature must be treated with caution

Table 12.2 Species richness of rhodolith-forming CCA in different regions of the world where their flora was inventoried

Region	Species richness	References
Southwestern Atlantic (Brazil)	33	Bahia (2014); Henriques et al. (2014a, b); Bahia et al. (2015)
Europe and Azores	12	Peña and Barbara (2008) and Peña et al. (2011)
Australia and New Zeland	7	Harvey and Bird (2008) and Nelson et al. (2012)
Gulf of California	4	Avila and Riosmena-Rodríguez (2010)
Alaska	1	Konar et al. (2006)
Canada	1	Gagnon et al. (2012)

range of identifiable specimens to include sterile and abraded material (Broom et al. 2008), in addition to enable more robust inferences of evolutionary relationships. However, until now, published DNA sequence data on rhodolith-forming species in SW Atlantic are limited to four species (Bahia et al. 2014b, 2015, Sissini et al. 2014). There are ongoing research employing DNA sequence data in this area and they will help in some important taxonomic aspects of CCA species, such as the identification of reliable morphoanatomical characters for taxonomy, distinction of cryptic species, and identification of sterile material thereby overcoming a significant barrier to work on corallines.

12.4 Structure and Ecosystem Functions of Rhodolith Beds

The controversy if a rhodolith bed is a mineral deposit or a habitat that supports high diversity stayed in the past. Today it is known that a rhodolith bed typically forms as a living layer on top of an otherwise sedimentary bottom and harbors a diverse and often unique assemblage of associated species (Foster et al. 2013).

An effort has been made in the last 10 years to improve the knowledge about the structure of rhodolith beds in Brazil. In a study that evaluated the marine biota in the Brazilian exclusive economic zone (the marine area extending 320 km offshore), Lavrado (2006) has shown a Shannon-Wiener diversity index of 4.0 and 5.1 bits ind^{-1} for specific areas down to 250 m depth and associate this increase in diversity of epibenthic communities to the presence of rhodolith beds. Information about density, dimension, shape, percentage of live surface of CCA, associated fauna, growing rates and calcium carbonate mineralization were produced for specific areas of the Brazilian Economic Exclusive Zone. Bahia et al. (2010) concluded that the structure of rhodolith beds is clearly influenced by depth; comparing rhodolith dimension and density patterns in three areas of the Brazilian continental shelf with different slopes and width, it was shown that in areas with extensive width and gentle slope rhodoliths increase in size, decrease in density and change shape from spherical to discoid. On the other hand, with narrow shelves and steep slopes rhodoliths decrease in size (see Fig. 10 in Bahia et al. 2010) and increase in density.

The Abrolhos Shelf of Brazil boasts the world's largest expanse of rhodoliths (News of the week, *Science*, 4 May 2012). In this area, Amado-Filho et al. (2012a) besides mapping rhodolith distribution and extent, determined bed structure, growing rates and estimated annual $CaCO_3$ production. The mean relative cover of rhodoliths was 69.1 + 1.7 % while mean density was 211 + 20 nodules m^2. The mean rhodolith diameter was 5.9 + 0.2 cm and the mean percentage of live surface area was 57 + 5 %, indicating active CCA growth. While rhodolith diameter seemed to increase with depth, neither rhodolith diameter nor percent live surface area was significantly correlated with depth. They estimated that the mean rate of $CaCO_3$ production by these organisms is about 0.025 Gt/year, in line with the world's largest biogenic $CaCO_3$ deposits, and that the rhodolith beds of the Abrolhos Shelf alone contain approximately 5 % of the $CaCO_3$ inventory of all the world's carbonate banks. It is important to note that the rhodolith surface of most studied beds were composed by live CCA, indicating a dynamic biomineralization process even in areas with very low PAR conditions.

Pereira Filho et al. (2012) and Amado-Filho et al. (2012b) showed that mesophotic zone of seamounts tops and insular shelves of oceanic islands presented amazing benthic communities dominated by rhodoliths. The peculiar abundance of mounds of rhodoliths built by the tilefish *Malacanthus plumieri* is an example of dynamic biological interaction that occurs in a rhodolith bed (Fig. 12.2). As the tilefish transported rhodoliths to mounds located up to 40 m from the deepest boundary of the beds it acts as an important mechanism to increase the rhodolith bed area and favored the connectivity among shallow and mesophotic beds (Amado-Filho et al. 2012b; Pereira-Filho et al. 2015). Other interesting aspect of mesophotic communities is the fusion of rhodoliths (Fig. 12.3). Minireefs composed by aggregations of several rhodoliths were seen in depths between 50 and 100 m depth in areas of very clean water and with very low hydrodynamic pressure as observed at Fernando de Noronha Island, Rocas Atol and Vitoria Trindade Chain (Pereira-Filho et al. 2014; Amado-Filho et al. 2016). The increase of substrate stability with a reduction in rhodolith turnover can explain the fusion between them and the conse-

Fig. 12.2 Mound build by the tilefish Malacanthus plumieri *Malacanthus plumieri* at the deepest boundary of the rhodoliths at Fernando de Noronha Archipelago. Note that most of rhodoliths are reddish indicating active photosynthesis and present a similar size (Photo: Z Matheus)

Fig. 12.3 A group of fused rhodoliths from mesophotic insular zone of Rocas Atoll. Note the presence of sponges and ascidian (in *red* and *yellow* color) connecting rhodoliths to each other (Photo: GM Amado Filho)

Fig. 12.4 A typical mesophotic reef found at Vitoria Trindade Jaseur Semounts top (77 m depth). The increase of substrate stability with a reduction in rhodolith turnover can explain the presence of rhodolith associated long-living organisms such as corals (*Montastrea cavernosa* – *arrowhead*) and sponges (*Xestospongea* sp. – *arrow*)

quent succession from a typical rhodolith bed to a mesophotic complex reef (Fig. 12.4). We can relate the increase of the substrate stability observed at the actual mesophotic reefs with the sea level rise that occurred during the Holocene. The actual mesophic reefs were exposed to intense water movement in shallow water at the beginning of Holocene (11–8 kyrs before present). On one hand, the decrease of water movement reduced the rhodolith movement and on other hand the availability of light maintained CCA growth in deeper areas allowing the fusion between rhodoliths and formation of consolidated mesophotic reefs.

Despite their extensive area, the richness of fauna associated with rhodoliths in the SW Atlantic is poorly known, although the importance of rhodolith beds in supporting high diversity and abundance of marine animals and algae in comparison with surrounding habitats is recognized (Nelson 2009).

Among the initiatives to evaluate the fauna associated with rhodoliths we highlight the most important ones. Metri (2006) studied the associated fauna of a small subtropical rhodolith bed (100 m^2) localized at Arvoredo Island and reported 168 invertebrate taxa, being 136 solitary forms and 32 colonial ones. Among the solitary forms Polychaeta, Crustacea and Mollusca were the most important groups, while among the colonial forms the Porifera *Pachataxa* sp. and the Ascidiacea *Didemnum* sp. were registered on 90% of the samples. In the same area, Rocha et al. (2005) found 36 Ascidian species and Bouzon and Freire (2007) identified 31 decapod species. Figueiredo et al. (2007) showed that Polychaeta was the infaunal dominant group taxon associated with a shallow rhodolith bed at Abrolhos Archipelago. Berlandi et al. (2012) comparing the Polychaeta composition between different rhodolith forms (small rhodoliths with long branches and large rhodoliths with short branches) found 26 families, 4 exclusive to large rhodoliths and 9 exclusive to the small ones. Riul (2007) found 122 macrofauna taxa associated with rhodolith beds from 10 to 20 m depth near João Pessoa city, Paraiba State, and pointed to differences in abundance and species richness according to depth and that the most important faunistic components were Mollusca, Polychaeta and Holothuria. Dias e Villaça (2012), describing the sediment dynamics on the southern region of Espírito Santo shelf, mentioned the important benthic fauna sedentary groups as polychaetes, ophiuroids and molluscs, and the hydroids and amphipods while the sessile fauna was represented by a few sponge species. Villas Boas et al. (2014) mentioned some incrusting faunal species of Bryozoa, Porifera, Cnidaria (corals) and Crustacea (barnacles) associated to rhodolith beds from shallow waters of Espírito Santo State. All information cited above were about rhodoliths beds from shallow waters in depths that did not exceed 30 m. Tamega et al. (2013) is the only study that reports the associated fauna of rhodolith beds of the mesophotic zone. They found for the Peregrino oil field rhodolith bed 120 taxa, being 16 Porifera, 6 Cnidaria, 31 Mollusca, 6 Polychaeta, 17 Crustacea, 20 Bryozoa, 1 Brachiopoda, 21 Echinodermata and 2 Ascidiacea.

In relation to seaweeds associated with rhodolith beds, some specific works indicate a high and specific diversity. Brasileiro et al. (2016) report a total of 146 seaweeds (including CCA) species associated with rhodolith beds from 20 to 70 m depth in Abrolhos Bank. Amado-Filho et al. (2010) report that the southern Espírito Santo State rhodolith beds provide an important habitat for epibenthic communities, supporting 25 % of the known macroalgal species richness along the Brazilian coast. New records of seaweeds have been found specifically associated to rhodolith beds (Amado-Filho et al. 2010; Guimarães and Amado-Filho 2009; Riul et al. 2009).

In order to fill the gap about microbial diversity, photosynthetic potential and their ecological interactions within rhodoliths, Cavalcanti et al. (2014) performed an ecophysiological and metagenomic analysis of the Abrolhos rhodoliths. The authors showed that rhodoliths contained a specific microbiome that displayed a significant enrichment in aerobic ammonia oxidizing betaproteobacteria and dissimilative sulfate-reducing deltaproteobacteria. They also observed a significant contribution of bacterial guilds (that is, photolithoautotrophs, anaerobic heterotrophs, sulfide oxidizers, anoxygenic phototrophs and methanogens) in the rhodolith metagenome, suggesting an important role in biomineralization process. The increased hits in aromatic compounds, fatty acid and secondary metabolism subsystems hint at an important chemically mediated interaction in which a functional job partition among eukaryal, archaeal and bacterial groups allows the rhodolith holobiont to thrive in the global ocean. High rates of photosynthesis were measured for Abrolhos rhodoliths (52.16 μmol carbon $m^{-2}s^{-1}$), allowing the entire Abrolhos rhodolith bed to produce 5.65×10^5 tons C per day. This estimation illustrates the great importance of the Abrolhos rhodolith beds for dissolved carbon production in the South Atlantic Ocean.

12.5 Threats, Vulnerability and Conservation Status

The main threats to rhodolith beds in the SW Atlantic are the direct mining of the rhodolith beds for calcareous exploitation and the indirect effects of oil and gas production activities in areas of rhodolith beds. The direct mining occurs mainly at the Maranhão State (~1° lat S), and at the southern part of Espírito Santo State. The main use of extracted calcareous material is for pH soil correction and as a source of calcium and micronutrients in animal food supplements. Data presented by Riul et al. (2008) based on information of the Brazilian Environmental Agency (IBAMA) estimate that 96,000 and 120,000 metric tons of calcium carbonate are extracted per year in Brazil. In addition to the direct impact caused by the removal of the coralline bed (i.e. a decrease of the rhodolith population density) the dredge process produces a plume of fine sediment, which may potentially amplify the impact on the remaining organisms from exploited areas of Espírito Santo continental shelf (Riul et al. 2008). A decrease in primary production of up to 70 % was seen in *Lithothamnion* sp. when rhodoliths where buried by a thin sediment layer (Riul et al. 2008).

Oil and gas exploring and production activities are widely distributed along the SW Atlantic continental shelf. Campos Basin, localized at the shelf on the north of Rio de Janeiro State is nowadays the most productive oil and gas area in Brazil. Oil and gas exploration at the Campos Basin started in the 1970s and today the production is estimated at 1.5 million oil barrels and 22 million of m^3 of gas per day (source: Brazilian Petroleum Agency ANP – www.anp.gov.br). During more than 35 years no attention was given to the biological importance of the sea-bottom of the shelf and the effects of oil and gas activities over benthic communities on Campos Basin. Recently, IBAMA established that no new drill wells can be made over rhodolith beds. This prohibition was based on the precautionary principle, as no basic information is available about rhodolith beds and about effects of drilling activities over the rhodolith communities. In some recent cases, it was necessary to reposition a platform from a rhodolith bed site to a sandy area, with an estimated cost between US$ 5 and 35 million. Another important rule established by IBAMA is that no discharge of any type of mud or gravel can occur over a rhodolith bed, being necessary to transport the sediment to be discharged in areas with 2000 m depth. These extremes rules could be avoided if previous studies had defined priority areas in terms of diversity and evaluation of resilience of rhodolith beds to the possible effects of oil and gas exploration activities. The only study that tests the effect of sediment and burial over CCA forming rhodoliths showed that photosynthesis of thin thallus crust CCA species from 90 to 100 m depth at Peregrino oil field, South of Campos Basin occurs at very low irradiance but, CCA forming rhodoliths might not withstand long-term burial especially by fine layers of anoxic sediments (Figueiredo et al. 2012). The sensitivity of rhodolith beds to anthropogenic disturbances that promote massive sediment dislodgment needs to be addressed by combining the effects of light attenuation by sediment suspension and burial by sediment deposition on a variety of coralline algae and associated organisms (Figueiredo et al. 2012).

Other threats for the rhodolith beds are bottom fish trawling that is widely recognized as the primary threat to the seabed communities of seamounts. Althaus et al. (2009)

reported no evidence of the structure forming taxa (primarily stony corals) recovering a decade after cessation of trawling on seamounts off Tasmania. Furthermore, the long-term loss of biogenic organisms such as corals and sponges may ultimately mean that impacted communities never return to their pre-disturbance state. In the 1970s deep-sea trawling target large seamount over the world, including the SW Atlantic Vitoria Trindade Seamounts Chain (Clark et al. 2010). In fact, there has been limited research on the recovery of seamount communities after disturbance. However, seamount biological communities have been considered among the least resilient in the marine environment (Clark et al. 2010) and timescales of recovery for some taxa exceed human life expectancies. The closures of previously fished seamounts to fishing operations will present future opportunity to monitor recovery, and provide valuable information to help guide future management initiatives (Consalvey et al. 2010).

Brazil is the first signatory of 1992s Convention on Biological Diversity (CBD), having committed to establish an effectively representative Marine Protected Areas (MPA) system covering at least 10% of each ecoregion by 2020 (CBD 2010). Despite a National Strategic Plan for Protected Areas launched in 2006 and the approval of even more ambitious targets by the "National Biodiversity Commission" (CONABIO), there has been little effective progress. Besides major implementation gaps in the existing MPAs, the country will take more than 500 years to accomplish its commitments under the current level of MPA designations (610 km^2/year in the last two decades) (Moura et al. 2013). Even the Abrolhos Bank, an area that is fully recognized as a National and Global priority for marine biodiversity conservation (Moura 2002), still does not have a representative MPA subsystem. Indeed, Brazil's state-level multiple-use MPAs are generally "paper parks" aiming to expand State-level jurisdiction over the coast (e.g. licensing processes and environmental impact assessments). Moura et al. (2013) emphasized that the existing MPAs do not cover representative portions of the main megahabitats of the Abrolhos Bank (see Fig. 6 in Moura et al. 2013). Indeed, they almost completely fail to cover the rhodolith bed megahabitat and several macrohabitat features such as palleovalleys, channels and sinkhole-like structures (Bastos et al. 2013; Cavalcanti et al. 2014; Moura et al. 2013). Marine protected areas established ad hoc under different political motivations and without a unifying spatial planning framework fail to represent biodiversity in the least number of available sites (Kirkpatrick 1983), and also have a great chance to fail as a tool for managing fisheries (Freitas et al. 2011). Considering overfishing (Pinheiro et al. 2010) and the fragility and uniqueness of the ecosystems of seamounts and isolated oceanic islands of the Vitoria Trindade Chain, that include extensive rhodolith beds in mesophotic zone, an immediate local-level protection has to be implemented.

Although further information is needed, specially related to associated fauna, with the results obtained up to now, the rhodolith habitats from Abrolhos Bank and Vitoria Trindade Chain have to be looked as priority for conservation in the SW Atlantic Ocean. The environmental services provided by these rhodolith beds, including the calcium carbonate production, increase in habitat complexity, specific benthic diversity and associated fish assemblages, justify urgent actions to protected these ecosystems.

References

Althaus F, Williams A, Schlacher TA, Kloser RJ, Green MA, Barker BA, Bax NJ, Brodie P, Schlacher-Hoenlinger MA (2009) Impacts of bottom trawling on deep-coral ecosystems of seamounts are long-lasting. Mar Ecol Progr Ser 397:279–294

Amado-Filho GM, Pereira Filho GH (2012) Rhodolith beds in Brazil: a new potential habitat for marine bioprospection. Rev Bras Farmacogn 22:782–788

Amado-Filho GM, Maneveldt G, Manso RCC, Marins Rosa BV, Pacheco MR, Guimarães SMPB (2007) Structure of rhodolith beds from 4 to 55 meters deep along the southern coast of Espírito Santo State, Brazil. Cienc Mar 33:399–410

Amado-Filho GM, Maneveldt GW, Pereira-Filho GH, Manso RCC, Bahia RG, Barros-Barreto MB, Guimarães SMPB (2010) Seaweed diversity associated with a Brazilian tropical rhodolith bed. Cienc Mar 36:371–391

Amado-Filho GM, Moura RL, Bastos AC, Salgado LT, Sumida PY, Guth AZ, Francini-Filho RB, Pereira-Filho GH, Abrantes DP, Brasileiro PS, Bahia RG, Leal RN, Kaufman L, Kleypas JA, Farina M, Thompson FL (2012a) Rhodolith beds are major CaCO3 bio-factories in the tropical south west atlantic. Plos One 7:e35171

Amado-Filho GM, Pereira-Filho GH, Bahia RG, Abrantes DP, Veras PC, Matheus Z (2012b) Occurrence and distribution of rhodolith beds on the Fernando de Noronha Archipelago of Brazil. Aquat Bot 101:41–45

Amado-Filho GM, Moura RL, Bastos AC, Frnacini-Filho RB, Pereira-Filho GH, Bahia RG, Moraes FC, Motta FS (2016) Mesophotic cosystems of the unique South Atlantic atoll are composed by rhodolith beds and scattered consolidated reefs. Mar Biodivers 46:407–420

Ávila E, Riosmena-Rodríguez R (2010) Rhodoliths beds as critical habitats for monitoring in marine protected areas in the Gulf of California. In: Polisciano G, Farina O (eds) National parks, vegetation, wildlife and threats. Nova Science Publishers, New York, pp 207–214

Bahia, RG (2014) Algas calcárias formadoras de rodolitos da plataforma continental tropical e ilhas oceânicas do Brasil: inventário florístico e taxonomia. PhD thesis, Escola Nacional de Botânica Tropical, Rio de Janeiro, 221 pp

Bahia RG, Abrantes DP, Brasileiro PS, Pereira-Filho GH, Amado Filho GM (2010) Rhodolith bed structure along a depth gradient on the northern coast of Bahia State, Brazil. Braz J Oceanogr 58:323–337

Bahia RG, Riosmena-Rodríguez R, Maneveldt GW, Amado Filho GM (2011) First report of *Sporolithon ptychoides* (Sporolithales, Corallinophycidae, Rhodophyta) for the Atlantic Ocean. Phycol Res 59:64–69

Bahia RG, Amado Filho GM, Maneveldt GW (2014a) *Sporolithon molle* (Heydrich) Heydrich (Sporolithales, Corallinophycidae, Rhodophyta): an addition to the Atlantic flora found on a remote oceanic island. Cryptogam Algol 35:7–14

Bahia RG, Amado-Filho GM, Maneveldt GW, Adey WH, Johnson G, Marins BV, Longo LL (2014b) *Sporolithon tenue* sp. nov. (Sporolithales, Corallinophycidae, Rhodophyta): a new rhodolith-forming species from the tropical southwestern Atlantic. Phycol Res 62:44–54

Bahia RG, Amado-Filho GM, Maneveldt GW, Adey WH, Johnson G, Jesionek MB, Longo LL (2015) *Sporolithon yoneshigueae* sp. nov. (Sporolithales, Corallinophycidae, Rhodophyta), a new rhodolith-forming coralline alga from the southwest Atlantic. Phytotaxa 224:140–158

Bastos AC, Moura RL, Amado Filho GM, Dagostini DP, Secchin NA, Francini-Filho RB, Guth AZ, Sumida PY, Mahiques M, Thompson FL (2013) Buracas: novel and unusual sinkhole-like features in the Abrolhos Bank. Cont Shelf Res 70:118–125

Berlandi RM, Figueiredo MAO, Paiva PC (2012) Rhodolith morphology and the diversity of polychaetes off the southeastern Brazilian coast. J Coast Res 28:280–287

Boero F (2010) The study of species in the era of biodiversity: a tale of stupidity. Diversity 2:115–126

Bouzon JL, Freire AS (2007) The Brachyura and Anomura fauna (Decapoda; Crustacea) in the Arvoredo Marine Biological Reserve on the Southern Brazilian Coast. Braz J Biol 7:321–325

Brasileiro PS, Pereira-Filho GH, Bahia RG, Abrantes DP, Guimarães SMPB, Moura RL, Francini-Filho RB, Bastos AC, Amado Filho GM (2016) Macroalgal composition and community structure of the largest rhodolith beds in the world. Mar Biodivers 46:407–420
Broom JES, Hart DR, Farr TJ, Nelson WA, Neill KF, Harvey AS, Woelkerling WMJ (2008) Utility of psbA and nSSU for phylogenetic reconstruction in the Corallinales based on New Zealand taxa. Mol Phylogenet Evol 46:958–973
Cavalcanti, VMM (2011) Plataforma continental : a última fronteira da mineração brasileira. Report DNPM, Brasilia 104 p
Cavalcanti GS, Gregoraci GB, Santos EO, Silveira CB, Meirelles PM, Longo L, Gotoh K, Naramura S, Iida T, Sawabe T, Rezende CE, Francini-Filho RB, Moura RL, Amado Filho GM, Thompson FL (2014) Physiologic and metagenomic attributes of the rhodoliths forming the largest CaCO3 bed in the South Atlantic Ocean. ISME J 8:52–62
CBD (2010) COP 10 – Tenth ordinary meeting of the Conference of the Parties to the Convention on Biological Diversity, COP 10. Strategic plan for biodiversity 2011–2020. United Nations Environment Programme. Nagoya, Japan
Clark MR, Rowden AA, Schlacher T, Williams A, Consalvey M, Stocks KI, Rogers AD, O'Hara TD, White M, Shank TM, Hall-Spencer JM (2010) Effects of fishing on the benthic biodiversity of seamounts of the "Graveyard" complex, northern Chatham Rise. Ann Rev Mar Sci 2:253–278
Consalvey M, Clark MR, Rowden AA, Stocks KI (2010) Life on seamounts. In: McIntyre A (ed) Life in the world's oceans: diversity, distribution, and abundance. Wiley-Blackwell, New York
Dias GTM (2000) Granulados Bioclásticos – Algas Calcárias. Braz J Geophys 18:307–318
Dias GTM, Villaça RC (2012) Coralline algae depositional environments on the Brazilian Central–South-Eastern shelf. J Coast Res 28:270–279
Dominguez JML, Silva RP, Nunes AS, Freire AFM (2013) The narrow, shallow, low-accommodation shelf of central Brazil: sedimentology, evolution, and human uses. Geomorphology 203:46–59
Figueiredo MAO, Menezes KS, Paiva EMC, Paiva PC, Ventura CRR (2007) Evaluación experimental de rodolitos como sustratos vivos para la infauna en el Banco de Abrolhos, Brasil. Cien Mar 33:427–440
Figueiredo MAO, Coutinho R, Villas-Boas AB, Tâmega FTS, Mariath R (2012) Deep-water rhodolith productivity and growth in the southwestern Atlantic. J Appl Phycol 24:487–493
Foster M (2001) Rhodoliths: between rocks and soft places. J Phycol 37:659–667
Foster MS, Amado Filho GM, Kamenos NA, Riosmena-Rodríguez R, Steller DS (2013) Rhodolith and rhodolith beds. In: Lang M (ed) Contribution of SCUBA diving to research and discovery in marine environments, vol 39. Smithsonian Institution Scholarly Press, Washington, DC, pp 143–156
Freitas MO, Moura RL, Francini-Filho RB, Minte-Vera CV (2011) Spawning patterns of commercially important reef fish (Lutjanidae and Serranidae) in the tropical western South Atlantic. Sci Mar 75:135–146
Gagnon P, Matheson K, Stapleton M (2012) Variation in rhodolith morphology and biogenic potential of newly discovered rhodolith beds in Newfoundland and Labrador (Canada). Bot Mar 55:85–99
Gherardi DFM (2004) Community structure and carbonate production of a temperate rhodolith bank from Arvoredo Island, southern Brazil. Braz J Oceanogr 52:207–224
Giustina IDD (2006) Sedimentação carbonática de algas vermelhas coralináceas da plataforma continental da bacia de Campos: um modelo carbonático análogo para o terciário. Master dissertation, Geosciences, Federal University of Rio Grande do Sul, 134 pp
Guimarães SMPB (2003) Uma análise da diversidade da flora marinha bentônica do Estado do Espírito Santo, Brasil. Hoehnea 30:11–19
Guimarães SMPB, Amado-Filho GM (2009) First record of *Reticulocaulis mucosissimus* I. A. Abbott (Naccariaceae, Rhodophyta) for the western Atlantic Ocean. Rev Bras Bot 32:671–675

Harvey AS, Woelkerling WJ (2007) A guide to nongeniculate coralline red algal (Corallinales, Rhodophyta) rhodolith identification. Cienc Mar 33:411–426
Harvey AS, Bird FL (2008) Community structure of a rhodolith bed from coldtemperate waters (southern Australia). Aust J Bot 56:437–450
Henriques MC, Villas-Boas AB, Riosmena-Rodríguez R, Figueiredo MAO (2012) New records of rhodolith-forming species (Corallinales, Rhodophyta) from deep water in Espírito Santo State, Brazil. Helgol Mar Res 66:219–231
Henriques MC, Coutinho LM, Riosmena-Rodríguez R, Barros-Barreto MB, Khader S, Figueiredo MAO (2014a) Three deep water species of *Sporolithon* (Sporolithales, Rhodophyta) from the Brazilian continental shelf, with the description of Sporolithon elevatum sp. nov. Phytotaxa 190:320–330
Henriques MC, Riosmena-Rodríguez R, Coutinho LM, Figueiredo MAO (2014b) Lithophylloideae and Mastophoroideae (Corallinales, Rhodophyta) from the Brazilian continental shelf. Phytotaxa 190:112–129
Kempf M (1970) Notes on the benthic bionomy of the N-NE Brazilian shelf. Mar Biol 5:213–224
Kirkpatrick JB (1983) An iterative method for establishing priorities for the selection of nature reserves: an example from Tasmania. Biol Conserv 25:127–134
Konar B, Riosmena-Rodríguez R, Iken K (2006) Rhodolith bed: a newly discovered habitat in the North Pacific Ocean. Bot Mar 49:355–359
Lavrado HP (2006) Caracterização do ambiente e da comunidade bentônica. In Lavrado HP, Ignácio BL (eds) Bidiversidade da costa central da zona econômica exclusiva brasileira. Museu Nacional Rio de Janeiro
Mabesoone JM, Kempf M, Coutinho PN (1972) Characterization of surface sediments on the northern and eastern Brazilian shelf. Trab Oceanogr Univ Fed Pe 13:41–48
Metri R (2006) Ecologia de um banco de algas calcárias da Reserva Biológica Marinha do Arvoredo, SC, Brasil. Master dissertation, Zoology, Federal Universisty of Paraná, 125 pp
Milliman JD (1977) Role of calcareous algae in Atlantic continental margin sedimentation. In: Fluguel E (ed) Fossil algae. Springer-Verlag, Berlin
Milliman JD, Amaral CAB (1974) Economic potential of Brazilian continental margin sediments. An Congr Brasil Geo 28:335–344
Moura RL (2002) Brazilian reefs as priority areas for biodiversity conservation in the Atlantic Ocean. Proc Int Coral Reef Symp 9:917–920
Moura RL, Secchin NA, Amado-Filho GM, Francini-Filho RB, Freitas MO, Minte-Vera CV, Teixeira JB, Thompson FL, Dutra GF, Sumida PYG, Güth AZ, Lopes RM, Bastos AC (2013) Spatial patterns of benthic megahabitats and conservation planning in the Abrolhos Bank. Cont Shelf Res 70:109–117
Moura RL, Amado-Filho GM, Moraes FC, Brasileiro PS et al (2016) An extensive reef system at the Amazon River mouth. Sci Adv 2:e1501252
Nelson WA (2009) Calcified macroalgae – critical to coastal ecosystems and vulnerable to change: a review. Mar Freshw Res 60:787–801
Nelson W, Neill K, Farr T, Barr N, D'Archino R, Miller S, Stewart R (2012) Rhodolith beds in Northern New Zealand: characterisation of associated biodiversity and vulnerability to environmental stressors. New Zealand Aquatic Environment and Biodiversity Report No. 99
Pascelli C, Riul P, Riosmena-Rodríguez R, Schernera F, Nunes M, Hall-Spencer JM, Oliveira EC, Horta PA (2013) Seasonal and depth-driven changes in rhodolith bed structure and associated macroalgae off Arvoredo island (southeastern Brazil). Aquat Bot 111:62–65
Peña V, Bárbara I (2008) Maërl community in the northwestern Iberian Peninsula: a review of floristic studies and long-term changes. Aquat Conserv Mar Freshwat Ecosyst 18:339–366
Peña V, Adey WH, Riosmena-Rodríguez R, Jung MY, Afonso-Carrillo J, Choi HG, Bárbara I (2011) *Mesophyllum sphaericum* sp. nov. (Corallinales, Rhodophyta): a new maerl-forming species from the Northeast Atlantic. J Phycol 47:911–927

Pereira-Filho GH, Amado-Filho GM, Guimarães SMPB, Moura RL, Sumida PY, Abrantes DP, Bahia RG, Güth AZ, Jorge RR, Francini-Filho RB (2011) Reef fish and benthic assamblages of the Trindade and Martin Vaz Island Group, Southwestern Atlantic. Braz J Oceanogr 59:201–212
Pereira-Filho GH, Amado-Filho GM, Moura RL, Bastos AC, Guimarães SMPB, Salgado LT, Francini-Filho RB, Bahia RG, Abrantes DP, Guth AZ, Brasileiro OS (2012) Extensive Rhodolith beds cover the summits of southwestern atlantic ocean seamounts. J Coast Res 28:261–269
Pereira-Filho GH, Francini-Filho RB, Pierozzi I Jr, Pinheiro HT, Bastos AC, Moura RL, Moraes FC, Matheus Z, Bahia RG, Amado-Filho GM (2014) Sponges and fish facilitate succession from rhodolith beds to reefs. Bull Mar Sci 91:45–46
Pereira-Filho GH, Veras PDC, Francini-Filho RB, Moura RL, Pinheiro HT, Gibran FZ, Matheus Z, Neves LM, Amado-Filho GM (2015) Effects of the sand tilefish *Malacanthus plumieri* on the structure and dynamics of a rhodolith bed in the Fernando de Noronha Archipelago, tropical West Atlantic. Mar Ecol Prog Ser 541:65–73
Pinheiro HT, Gasparini JL, Joyeux JC (2010) Reef fish mass mortality event in an isolated island off Brazil, with notes on recent similar events at Ascension, St Helena and Maldives. Mar Biodivers Rec 3:e47
Riul (2007) Aspectos da Biologia e Ecologia de rodolitos e comunidade associada na grande João Pessoa, PB. Master dissertation, Zoology, Federal University of Paraiba, Brazil
Riul P, Targino CH, Farias JN, Visscher PT, Horta PA (2008) Decrease in *Lithothamnion* sp. (Rhodophyta) primary production due to the deposition of a thin sediment layer. J Mar Biol Ass UK 88:17–19
Riul P, Lacouth P, Pagliosa PR, Christoffersen ML, Horta PA (2009) Rhodolith beds at the easternmost extreme of South America: community structure of an endangered environment. Aquat Bot 90:315–320
Rocha RM, Moreno TT, Metri R (2005) Ascídias (Tunicata, Ascidiacea) da Reserva Biológica Marinha do Arvoredo, Santa Catarina Brasil. Rev Bras Zoologia 22:461–476
Sissini MN, Oliveira MC, Gabrielson PW, Robinson NM, Okolodkov YB, Riosmena-Rodríguez R, Horta PA (2014) *Mesophyllum erubescens* (Corallinales, Rhodophyta) – so many species in one epithet. Phytotaxa 190:299–319
Skolotnev SG, Peyve AA, Turko NN (2010) New data on the structure of the Vitória-Trindade Seamount Chain, Western Brazil Basin, South Atlantic. Dokl Earth Sci 431:435–440
Tamega FTS, Oliveira OS, Figueiredo MAO (2013) Catalogue of the benthic marine life from Peregrino Oil Field, Campos Basin, Brazil. Report Instituto Biodiversidade Marinha, Rio de Janeiro
Testa V (1997) Calcareous algae and corals in the inner shelf of Rio Grande do Norte, NE Brazil. Proc 8th Int Coral Reef Symp 1:737–742
Testa V, Bosence DWJ (1999) Physical and biological controls on the formation of carbonate and siliciclastic bedforms on the north-east Brazilian shelf. Sedimentology 46:279–301
Vicalvi MA, Milliman JD (1977) Calcium carbonate sedimentation on continental shelf off southern Brazil with special reference to benthic foraminifera. In: Frost SH, Weiss MP, Saunders JB (eds) Studies in geology, 4, AAP
Villas-Boas AB, Riosmena-Rodríguez R, Amado Filho GM, Maneveldt GW, Figueiredo MAO (2009) Taxonomy of rhodolith-forming species of *Lithophyllum* (Corallinales; Rhodophyta) from Espírito Santo State, Brazil. Phycologia 48:237–248
Villas-Boas AB, Riosmena-Rodríguez R, Figueiredo MAO (2014) Community structure of rhodolith-forming beds on the central Brazilian continental shelf. Helg Mar Res 68:27–35

Chapter 13
Eastern Pacific

Néstor M. Robinson, Cindy Fernández-García, Rafael Riosmena-Rodríguez, Edgar F. Rosas-Alquicira, Brenda Konar, Heloise Chenelot, Stephen C. Jewett, Roland R. Melzer, Roland Meyer, Günter Försterra, Vreni Häussermann, and Erasmo C. Macaya

Abstract In the Eastern Pacific (EP) the only region where rhodolith beds have been well studied in terms of taxonomy, ecology, distribution and conservation status is the Gulf of California. Outside this region the knowledge of rhodolith-forming species is attributed to the initial separate floristic surveys of Dawson and Lemoine, performed more than 50 years ago. After a detailed review of the published literature and information produced during our expeditions throughout the EP, a total of 36 rhodolith-forming species have been recorded from the Aleutian Islands in Alaska to Guarello Island in Chile. Despite the research efforts developed at present, more species remain to be discovered in the EP, particularly in the Tropical

N.M. Robinson (✉) • R. Riosmena-Rodríguez (deceased)
Departamento de Biología Marina, Universidad de Baja California Sur,
La Paz, BCS 23080, Mexico
e-mail: robinson.biol@gmail.com

C. Fernández-García
Escuela de Biología, Centro de Investigación en Ciencias del Mar y Limnología (CIMAR),
Universidad de Costa Rica, San Pedro, San José 11501-2060, Costa Rica

E.F. Rosas-Alquicira
Instituto de Recursos, Universidad del Mar (UMAR), Ciudad Universitary s/n

B. Konar • H. Chenelot • S.C. Jewett
School of Fisheries and Ocean Sciences, Institute of Marine Science, University of Alaska Fairbanks, Fairbanks, AK 99775, USA

R.R. Melzer • R. Meyer
Zoologische Staatssammlung München, Münchhausenstr. 21, 81247 Munich, Germany

G. Försterra • V. Häussermann
Huinay Scientific Field Station, Puerto Montt, and, Escuela de Ciencias del Mar, Facultad de Recursos Naturales, Pontificia Universidad Católica de Valparaíso, Valparaíso, Chile

E.C. Macaya
Laboratorio de Estudios Algales ALGALAB, Departamento de Oceanografía, Facultad de Ciencias Naturales y Oceanográficas, Universidad de Concepción,
Casilla 160-C, Concepción, Chile

R. Riosmena-Rodríguez et al. (eds.), *Rhodolith/Maërl Beds: A Global Perspective*, Coastal Research Library 15, DOI 10.1007/978-3-319-29315-8_13

Pacific of Mexico and the Pacific coast of Baja California where we found undescribed species of *Sporolithon* and *Lithothamnion*. Therefore, we contend that further studies are needed in order to better catalogue the wide rhodolith-forming species diversity that is extremely relevant for the marine realm of the EP.

13.1 Introduction

In the Eastern Pacific (EP) the knowledge of rhodolith-forming species began with two floristic surveys. This first was developed by Lemoine (1929) in the region of Central America through the British cruise "Saint-George Expedition" organized by the Scientific Expeditionary Research Society. The second was the marine survey of Dawson (1960a, b) by means of the Pacific Cruise of Stella Polaris funded by the Beaudette Foundation. These floristic surveys represent the base line information of the taxonomic knowledge of coralline algae in the EP. In this area the rhodolith-forming species diversity has been studied sporadically from Alaska to Galapagos Islands through a series of morphological works (Riosmena-Rodríguez et al. 1999, 2010; Athanasiadis et al. 2004; Konar et al. 2006; Fernández 2008; Littler and Littler 2008). However, as it was shown in the molecular studies of coralline algae (Broom et al. 2008; Bittner et al. 2011; Kato et al. 2013) cryptic or pseudo-cryptic diversity can be observed with the use of molecular techniques. Under this context, the rhodolith-forming species diversity in the EP remains in an explorative state and further molecular studies are required in order to assess correctly the species diversity of rhodoliths. The main purpose of this review was to discuss the taxonomy and distributional trends of rhodolith-forming species; based on a revision of the literature published in this region and information produced through our surveys. In this revision we followed the traditional methods described to analyze coralline algae under a morphological/anatomical approach (Woelkerling et al. 1988; Chamberlain and Irvine 1994; Riosmena-Rodríguez et al. 1999; Harvey and Woelkerling 2007) based on the information from our research and the literature published we traced a total of 210 records of rhodolith-forming species from Alaska to Chile. Based on these records we present this review that aims to update the taxonomical and distributional knowledge of rhodolith-forming species in the EP.

13.2 Species Composition

In the EP in the region comprised between the Aleutian Islands in Alaska to Guarello Island in Chile, 36 morpho-species of 11 genera have been reported as rhodolith-forming species (Table 13.1). These species showed a high degree of morphological variation (Fig. 13.1). Unfortunately, the taxonomic status of the morpho-species in the EP requires clarification under DNA molecular sequencing. Otherwise in the region comprised between the Pacific of Baja California and the Pacific of Panama

Table 13.1 Records of rhodolith-forming species from the subclass Corallinophycidae at the Eastern Pacific, recorded from Alaska to Panama, based on the literature published

Taxa	Region	Latitud	Longitud	Reference
Subclass Corallinophycidae				
Order Sporolithales				
Family Sporolithaceae				
Sporolithon sp.	Revillagigedo Island	21°26′56.36″N	106°21′49.13″W	Robinson (2013)
Order Hapaliadiales				
Family Hapalidiaceae				
Subfamily Melobesioideae				
Lithothamnion sp.	Isla del Coco, Costa Rica	5°32′N	87°04′W	Fernández (2008)
L. australe Foslie	Canal de San Lorenzo,Gulf of California, Mexico	24°22′37.84″N	110°18′54.67″W	Dawson (1960a)
	Bahía Honda Panama	7°34′05.45″N	81°29′19.39″W	Dawson (1960a)
	Gulfof Chiriquí, Panama	08°01′31.50″N	82°21′02.27″W	Littler & Littler (2008)
L. australe f. *americanum* Foslie	Panama	–	–	Fernández-García et al. (2011)
L. californicum Foslie	Nicaragua	–	–	Fernández-García et al. (2011)
L. crispatum Hauck	Gulf of Chiriquí, Panama	08°01′31.50″ N	82°21′02.27″W	Fernández-García et al. (2011)
	Costa Rica	–	–	Fernández-García et al. (2011)
L. diguetii Hariot	Canal de San Lorenzo, BCS, Mexico	24°22′37.84″N	110°18″54.67″W	Dawson (1960a)
L. guadalupense E.Y. Dawson	Guadalupe Island, México	28°58′31″N	118°21′20.96″W	Dawson (1960a)
L. giganteum L.R. Mason	Bahía Vizcaíno, Mexico	27°58′53.16″N	114°31′36.85″W	Dawson (1960a)
	Magdalena Island, Mexico	21°30′05.66″N	106°25′38.88″W	Dawson (1960a)
L. heterocladum Foslie	Melinka, Guaitecas Archipelago, Chile	43°54′10.35″S	073°44′17.76″W	Macaya et al. (2015)
L. muelleri Lenormand *ex* Rosanoff	Playa Carey, Ixtapa, Mexico	17°40′30.17″N	101°39′05.32″W	Peralta-García & Rosas-Alquicira (2014)

(continued)

Table 13.1 (continued)

Taxa	Region	Latitud	Longitud	Reference
	Playa Manzanillo, Zihuatanejo, Mexico	17°37′11.72″N	101°39′05.32″W	Peralta-García & Rosas-Alquicira (2014)
	Cacaluta Island, Oaxaca, Mexico	15°43′01.14″N	96°09′43.41″W	Peralta-García & Rosas-Alquicira (2014)
	San Juan Island, Washington, USA	48°32.1′N	122°59.7′W	Robinson et al. (2013)
	Santa Rosalia, Gulf of Californa	27°19.9′N	112°14.9′W	Robinson et al. (2013)
	Maria Magdalena Island, Mexico	21°26.9′N	106°21.8′W	Robinson et al. (2013)
	La Peña Rota, Nicaragua	11°14.0′N	85°52.0′W	Robinson et al. (2013)
	Guacalito	11°2.3′N	86°00.1′W	Robinson et al. (2013)
	Bahia Virador, Costa Rica	10°36.9′N	85°41.4′W	Robinson et al. (2013)
	Ballena Island, Costa Rica	9°06.3′N	83°43.6′W	Robinson et al. (2013)
L. validum Foslie	Costa Rica	–	–	Fernández-García et al. (2011)
Mesophyllum australe var. *tualense* (Foslie) Me. Lemoine	Coiba Island, Panama	7°41′03.80″N	81°73′21.74″W	
M. crassiusculum (Foslie) Lebednik	La Jolla, Devil's Slide, California, USA	32°49′16.02″N	117°16′58.40″W	Athanasiadis et al. (2004)
	Costa Rica	–	–	Fernández-García et al. (2011)
M. engelhartii (Foslie) W.H. Adey	Punta Perico, Gulf of California, Mexico	24°02′22.71″N	109°47′50.47″W	Riosmena-Rodríguez & Vásquez-Elizondo (2012)
	La Jolla, Bird Rock, California, USA	32°48′48.89″N	117°16′25.32″W	Athanasiadis et al. (2004)
	San Pedro, White Point, California, USA	33°42′45.78″N	118°18′43.67″W	Athanasiadis et al. (2004)
Phymatolithon calcareum (Pallas) W.H. Adey & D.L. McKibbin	Herring Bay, Prince William Sound, Alaska	60°27′60″N	147°45′24″W	Konar et al. (2006)
P. lenormandii (Areschoug) W.H. Adey	El Salvador	–	–	Fernández-García et al. 2011

P. repandum (Foslie) Wilks & Woelkerling	Playa Carey, Ixtapa, Mexico	17°40′30.17″N	101°39′05.32″W	Peralta-García & Rosas-Alquicira (2014)
	Playa Manzanillo, Zihuatanejo, Mexico	17°37′11.72″N	101°39′05.32″W	Peralta-García & Rosas-Alquicira (2014)
	Cacaluta Island, Oaxaca, Mexico	15°43′01.14″N	96°09′43.41″W	Peralta-García & Rosas-Alquicira (2014)
Unidentified Melobesioideae	Reef Mid Boot Bay, Auletian Islands.	51°44′44.40″N	176°30′25.40″W	Chenelot Unpublished data
Unidentified Melobesioideae	Blind Cove, Auletian Islands	51°51′13.40″N	176°30′25.40″W	Chenelot Unpublished data
Unidentified Melobesioideae	Amita Island, Guaitecas Archipelago, Chile	44° 4′23.92″S	73°53′52.29″W	Macaya et al. (2015)
Unidentified Melobesiodeae	GuarelloIsland, Madre de Dios Archipelago, Chile	50°24′27.80″S	75°19′58.48″W	Macaya et al. (2015)
Unidentified Melobesioideae	Robinson Crusoe Island, Juan Fernandez Archipelago, Chile	33°40′51.57″S	78°49′59.15″W	Macaya et al. (2015)
Order Corallinales				
Family Corallinaceae				
Subfamilia Hydrolitoideae				
Hydrolithon reinboldii (Weber-van Bosse & Foslie) Foslie	Socorro Island, Mexico	18°44′22.5″N	111°00′52.31″W	Dawson (1960a)
Subfamily Lithophylloideae				
Lithophyllum alternans M. Lemoine	Galápagos Islands	01°15′12.57″N	90°59′13.10″W	Lemoine (1929)
L. brachiatum (Heydrich) M. Lemoine	Galapagos Islands	01°15′12.57″N	90°59′13.10″W	Lemoine (1929)
L. canescens (Foslie) Foslie	El Salvador, Costa Rica			Fernández-García et al. (2011)

(continued)

Table 13.1 (continued)

Taxa	Region	Latitud	Longitud	Reference
L. coibense M. Lemoine	Coiba Island, Panama	07°26′39.44″N	81°39′15.56″W	Lemoine (1929)
L. corallinae (Crouan & Crouan) Heydrich	Sayulita, Nayarit, Mexico	20°52′07.89″N	105°27′22.49″W	Fragroso & Rodríguez (2002)
	Bahía de Chamela, Jalisco, Mexico	19°31′43.22″N	105°31′43.22″W	Fragroso & Rodríguez (2002)
	Puerto Ángel, Oaxaca, Mexico	15°39′32.49″N	96°29′08.34″W	Fragroso & Rodríguez (2002)
	Huatulco, Oaxaca, Mexico	15°44′37.94″N	96°47′.76″W	Fragroso & Rodríguez (2002)
	Playa las Gatas, Zihuatanejo, Mexico	17°37′20.32″N	101°33′22.35″W	Fragroso & Rodríguez (2002)
	Playa la Ropa, Zihuatanejo, Mexico	17°37′46.16″N	101°32′57.14″W	Fragroso & Rodríguez (2002)
L. divaricatum M. Lemoine	Panama			Fernández-García et al. (2002)
L. imitans Foslie	El Salvador	–	–	Fernández-García et al. (2002)
L. margaritae (Hariot) Heydrich	San Juan Nepomuceno Island, Mexico	24°15′54.16″N	110°20′20.66″W	Riosmena-Rodríguez et al. (1999)
	Playa La Ropa, Zihuatanejo, Mexico	17°37′46.16″N	101°32′57.14″W	Fragroso & Rodríguez (2002)
	Playa la Entrega, Oaxaca, Mexico	15° 44′ 37.66″N	96° 07′ 42.80″W	Fragroso & Rodríguez (2002)
	Puerto Escondido, Guerrero, Mexico	15°51′16.12″N	97°04′08.80″W	Fragroso & Rodríguez (2002)
	Bahía Magdalena, Mexico	24°48′45″N	112°05′59″W	Ávila & Riosmena-Rodríguez (2011)
	Bahía Magdalena, Mexico	24°54′12″N	112°05′34″W	Ávila & Riosmena-Rodríguez (2011)
	Bahía Magdalena, Mexico	24°53′47″N	112°06′32″W	Ávila & Riosmena-Rodríguez (2011)
	Bahía Magdalena, Mexico	24°48′42″N	112°06′32″W	Ávila & Riosmena-Rodríguez (2011)
L. pallescens (Foslie) Foslie	Panama			Fernández-García et al. (2011)
L. stictaeforme (Areschoug) Hauck	Huatulco, Oaxaca, Mexico	15°44′37.94″N	96°47′.76″W	Fragroso & Rodríguez (2002)

Subfamily Mastophoroideae				
Pneophyllum conicum (E.Y. Dawson) Keats, Y.M. Chamberlain & M. Baba	Socorro Island, Mexico	18°44′22.5″N	111°00′52.31″W	Dawson (1960a)
Spongites decipiens (Foslie) Y.M. Chamberlain	Costa Rica, Panama			Fernández-García et al. (2011)
S. fruticulosus (Kützing)	Playa San Agustín, Oaxaca, Mexico	15°41′19.15″N	96°14′00.22″W	Fragoso & Rodríguez (2002)
	Playa la Entrega, Oaxaca, Mexico	15°44′37.66″N	96°07′42.80″W	Fragoso & Rodríguez (2002)
	Puerto Huatulco, Mexico	15°44′37.94″N	96°06′47.76″W	Dawson (1960a)
Subfamily Neogoniolithoideae				
Neogoniolithon trichotomum (Heydrich) Setchell & L.R. Mason	Costa Rica, Panama	–	–	Fernández-García et al. (2011)
N. setchelli (Foslie) Adey	San Ignacio Lagoon, BCS, Mexico	26°55′48.28″N	113°10′04.37″W	Nuñéz-López et al. (1998)
	Guadalupe Island, Mexico	28°58′31″N	118°21′20.96″W	Dawson (1960a)
Subfamily Porolithoideae				
Porolithon onkodes (Heydrich) Foslie	Costa Rica	–	–	Fernández-García et al. (2011)
P. sonorense E.Y. Dawson	Isla Ángel de la Guarda, Mexico	29°10′39.25″N	115°24′59.83″W	Dawson (1960a)

Fig. 13.1 Morphological variation of rhodolith-forming species at the Eastern Pacific in a region comprised from Baja California to Panama. (**a**) *Sporolithon* sp. **GGMX1024b**. (**b**) *Lithothamnion crispatum*, **INMX1068**. (**c**) *Lithothamnion* sp. **GGMX1023b**. (**d**) *Lithothamnion* sp. **INMX1073b**. (**e**) *Lithothamnion* sp.1, **INMX1073c**. (**f**) *Lithothamnion* sp. 2, **INMX1066**. (**g**) *Mesophyllum engelhartii*, **INMX1072**. (**h**) *Lithothamnion* sp. nov., **INMX1065**. (**i**) *Phymatolithon* sp., **EOPA1047a**. (**j**) *Hydrolithon reinboldii*, **CFES0124**. (**k**) *Lithophyllum pallescens*, **NRMX1090**. (**l**) *Lithophyllum margaritae*, **NRMX1087**. (**m**) *Lithophyllum* sp. 1, **CFPA1022c**. (**n**) *Lithoporella pacifica*, **USJA73418**

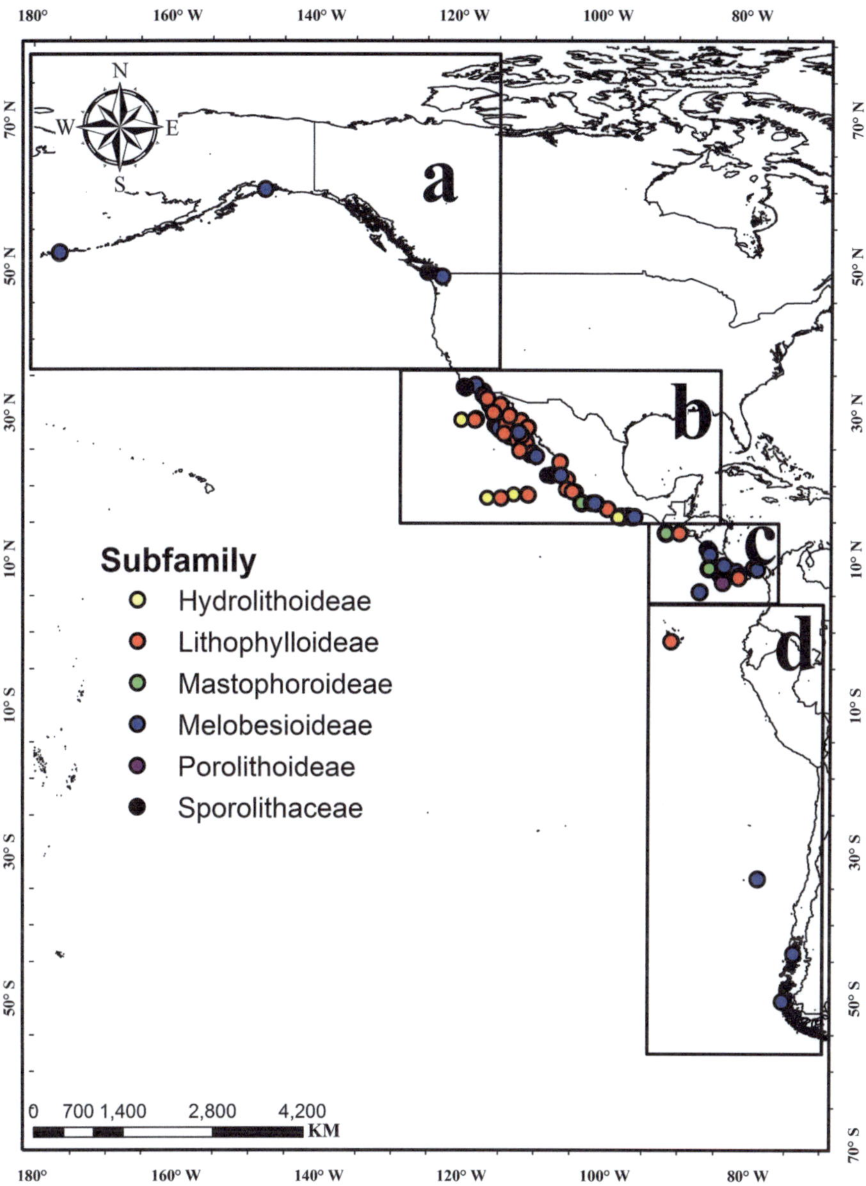

Fig. 13.2 Distribution map of the rhodolith-forming species in the Eastern Pacific, from Alaska to Panama (Updated from the literature published and new records (Robinson 2013) produced by a detailed morphological revision and molecular techniques adapted for coralline algae)

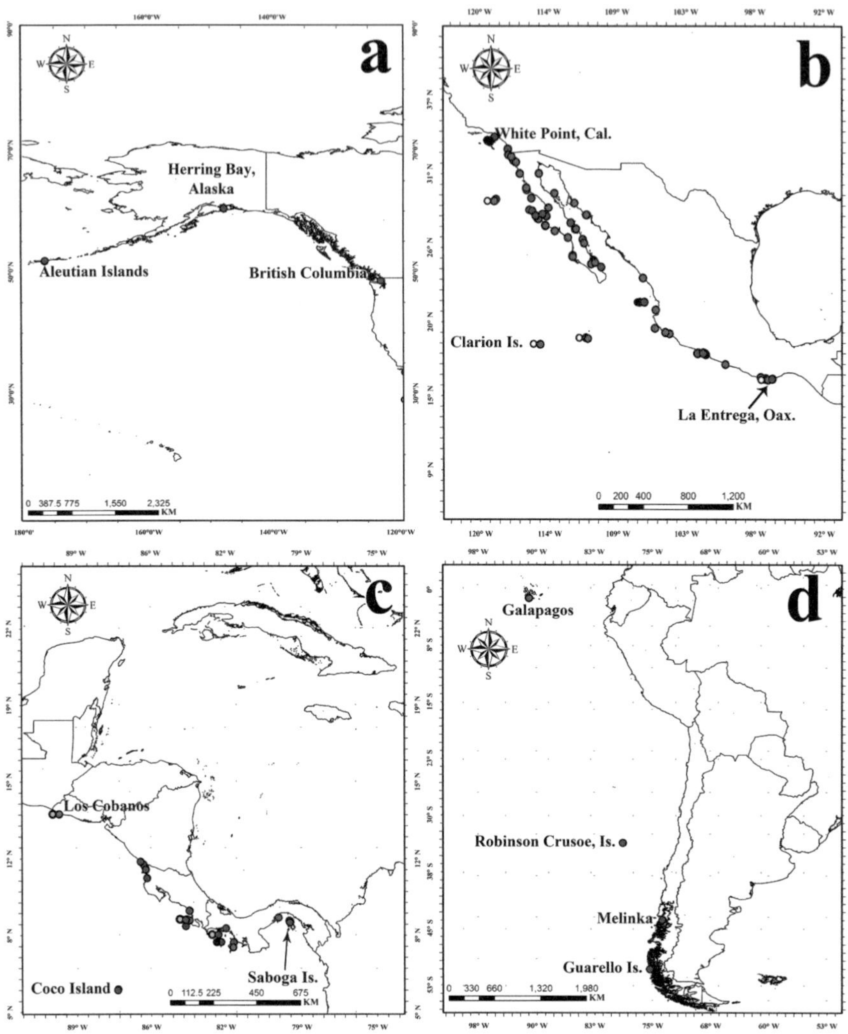

Fig. 13.3 Distribution map of the rhodolith-forming species showing the occurrence of rhodolith/maërl beds in (**a**) Cold temperate Northeast Pacific, (**b**) Tropical East Pacific (Mexico), (**c**) Tropical East Pacific (Central America) and, (**d**) Temperate South America. Records were updated based on publications and a taxonomic review (Robinson 2013)

the information has been updated recently (Robinson 2013) and also in the Gulf of California where rhodolith beds are mainly composed by the species *Lithophyllum margaritae*, *Neogoniolithon trichotomum* and *Lithothamnion muelleri* (Steller et al. 2003; Riosmena-Rodríguez et al. 2010).

In the EP the subfamilies Lithophylloideae and Melobesioideae outnumber the species diversity of other subfamilies from the subclass Corallinophycidae. In order to understand the distribution trends of rhodolith-forming species we discuss the diversity of each subfamily represented at the NCEP (Figs. 13.2 and 13.3).

13.2.1 Subfamily Lithophylloideae

After a detailed taxonomic review performed in the eastern tropical Pacific region (Robinson 2013), we found that representatives from the subfamily Lithophylloideae were extensively represented by the genus *Lithophyllum* Philippi with 10 morpho-species reported at present. *L. margaritae* has been pointed out as the dominant species forming the rhodolith beds throughout the Gulf of California (Riosmena-Rodríguez et al. 2010). This species is a remarkable case study, although no molecular information from the type specimen has been produced at present, and the identification of *Lithophyllum* species in this region was based on morphological and anatomical descriptions (Lemoine 1929; Dawson 1960a, b; Riosmena-Rodríguez et al. 1999). However, the use of these characters led to confusion and created species discrimination problems. Therefore, to achieve accurate species information we need to render further studies using molecular information from type specimens reported in the area and modern collections in order to assess epithets for those species (Table 13.2).

Table 13.2 Taxonomic scheme of the rhodolith-forming species reported in the Eastern Pacific, based on the literature published and new records (Robinson 2013) produced by a detailed morphological revision and molecular techniques adapted for coralline algae

	RhodophytaWettstein	
	Eurhodophytina Saunders &Hommersand	
	Florideophyceae Cronquist	
	Corallinophycidae Le Gall & Saunders	
Corallinales P.C.Silva & H.W.Johansen	**Hapalidiales W.A.Nelson, J.E.Sutherland, T.J.Farr & H.S.Yoon**	**Sporolithales Le Gall, Payri, Bittner & Saunders**
Corallinaceae	**Hapalidiaceae J.E. Gray**	**Sporolithaceae E. Verheij**
Hydrolithoideae Kato & Baba	**Melobesioideae Bizzozero**	***Sporolithon*** Heydrich
Hydrolithon reinboldii (Weber-van Bosse & Foslie) Foslie	*Lithothamnion australe* Foslie	*Sporolithon* sp.
Lithophylloideae Setchell	*L. californicum* Foslie	
Litophyllum alternans M.Lemoine	*L. crispatum* Hauck	
L. brachiatum (Heydrich) M.Lemoine	*L. diguetii* Hariot	
L. canescens (Foslie) Foslie	*L. giganteum* L.R. Mason	
L. coibense M.Lemoine	*L. guadalupense* E.Y. Dawson	
L. corallinae (Crouan & Crouan) Heydrich	*L. heterocladum* Foslie	
L. divaricatum M. Lemoine	*L. muelleri* Lenormand ex Rosanoff	
*L. imitans*Foslie	*L. validum* (Foslie) Foslie	
L. margaritae (Hariot) Heydrich	*Lithothamnion sp.*	
L. pallescens (Foslie) Foslie	*Mesophyllum australe* (Foslie) M.Lemoine	
L. stictaeforme (Areschoug) Hauck	*M. crassiuscullum* (Foslie) Lebednik	
Mastophoroideae Setchell	*M. engelhartii* (Foslie) W.H.Adey	
Lithoporella pacifica (Heydrich) Foslie	*Phymatolithon calcareum* (Pallas) W.H.Adey & McKibbin	
Pneophyllum conicum (E.Y.Dawson) Keats, Y.M.Chamberlain & M.Baba		
Spongites decipiens (Foslie) Y.M.Chamberlain	*P. lenormandii* (Areschoug) W.H.Adey	
S. fruticulosus Kützing	*P. repandum* (Foslie) Wilks & Woelkerling	
Neogoniolithoideae Kato & Baba		
Neogoniolithon setchelii (Foslie) Adey		
N. trichotomum (Heydrich) Setchell & L.R.Mason		
Porolithoideae Kato & Baba		
Porolithon onkodes (Heydrich) Fioslie		
P. sonorense E.Y.Dawson		

Records in *red* represent new records of rhodolith-forming species for the Eastern Pacific

13.2.2 Subfamily Melobesioideae

The subfamily Melobesioideae is well represented by *Lithothamnion* Heydrich with 10 morpho-species reported in the region comprised from Alaska to Chile. From these records *Lithothamnion muelleri* Lenormand ex Rosanoff is quite remarkable, displaying a high degree of polymorphism. Also, the species seems to tolerate environmental changes, because it has been reported in a wide biogeographical range at the Eastern Pacific and some regions in the Coast of Brazil (Robinson et al. 2013). Although that publication detailed morphological information, molecular analysis of type specimens must be carried out to assess epithets for coralline algae (Gabrielson et al. 2011); and no record of this species have been validated based on the comparison with the type material from Western Port Bay, Victoria, Australia.

13.2.3 Subfamily Mastophoroideae

Representatives from the subfamily Mastophoroideae are less abundant, as only three rhodolith-forming species have been reported in the region: *Spongites fruticulosus* Kützing sensu Penrose in Woelkerling, *Spongites decipiens* (Foslie) Y.M. Chamberlain and *Litophorella pacifica* (Heydrich) Foslie. The last one is reported for the first time as a rhodolith-forming species.

13.2.4 Subfamily Hydrolithoideae

The subfamily Hydrolithoideae is represented by the genus *Hydrolithon*, and in this review we found four species from this genus; *H. conicum* E.Y. Dawson, *H. setchelli* (Foslie) Setchell & L.R. Mason, *H. onkodes* (Heydrich) Penrose & Woelkerling and *H. reinboldii* (Weber-van Bosse & Foslie) Foslie. The first two seem to be distributed in the temperate-tropical waters of the Eastern Pacific in Revillagigedo and Guadalupe Islands, respectively (Dawson 1960a, b). While the remaining species seem to be restricted to the tropical region of Mexico and Central America.

13.2.5 Subfamily Neogoniolithoideae

This subfamily is represented by the genus *Neogoniolithon* and in this revision we found only two species *N. foslie* (Heydrich) Setchell & L.R. Mason and *N. trichotomum* (Heydrich) Setchell & L.R. Mason. Except for *N. trichotomum*, which was reported from the Gulf of California (Dawson 1960b), other extant *Neogoniolithon* species are restricted to the tropical region in the Eastern Pacific from Revillagigedo Islands to the Pacific coast of Panama.

13.2.6 Family Sporolithaceae

From the family Sporolithaceae, no records of rhodolith-forming species were made previously in the EP, only the crustose forms of *Sporolithon howei* (Lemoine) N.Y. Yamagushi-Tomita *ex* M.J. Wynne recorded in Panama (Lemoine 1929), and *S. pacificum* E.Y. Dawson recorded in Costa Rica. Moreover, during our surveys in the Temperate-tropical Pacific of Mexico, we found rhodoliths from the genus *Sporolithon* (Maria Magdalena Island, Revillagigedo). The morphology and molecular information (Robinson 2013) pointed out the first record of a rhodolith-forming species from this genus in the Eastern Pacific. The formal and detailed description of this new species will be presented in further publications.

13.3 Ecological Role of Rhodolith-Forming Species

Rhodolith beds provide an important role for several marine species in the EP, such as: the support of the species richness and abundance of molluscs like the commercially-harvested scallop *Argopecten. ventricosus* which has been reported associated with rhodolith beds at 11 m depth reaching densities of 9.0 individuals. m^{-2} (Steller et al. 2003; Steller and Cáceres-Martínez 2009). It has been pointed out that rhodolith beds can produce more than 40 % of the organic sediments in the coastal systems of the Gulf of California (Riosmena-Rodríguez et al. 2012). Also, the support of a high richness and diversity of cryptofauna associated with rhodolith beds has been observed (Hinojosa-Arango and Riosmena-Rodríguez 2004).

13.4 Conservation Status

After 30 years of research on rhodolith beds in different areas of the EP, few Marine Protected Areas have considered the rhodolith-forming species as relevant habitats for conservation. In Mexico the relevance of rhodolith beds is well known particularly at Loreto, Isla Espiritu Santo, San Pedro Martir and Revillagigedo Islands (Riosmena-Rodríguez et al. 2010). However, this is not the case in other regions of the EP. For instance, on the pacific coast of Costa Rica (Biological Reserve Isla del Caño) and Panama (Marine National Park Gulf of Chiriqui and National Park Coiba Island) rhodolith-forming species have been reported systematically, but knowledge of these ecosystems are limited in these regions, therefore the conservation status of rhodolith beds in these Marine Protected Areas remains uncertain until further studies provide information to determine the relevance of rhodolith beds as an important habitat for conservation.

In the Northern tip of the Eastern Pacific (Alaska) the main rhodolith-forming species recorded by the literature was *Phymatolithon calcareum* (Konar et al. 2006). If the identificaton of this species is correct, this could have special significance in the conservation status of rhodolith beds. This is because this species has been considered as “vulnerable” and conservation strategies have been designed for European populations (OSPAR Commission 2006; Peña and Bárbara 2008). Those strategies could be considered in order to construct conservation strategies for rhodolith-forming species in the EP.

13.5 Conclusion

In relation to the taxonomy of rhodolith-forming species, the assessment in the EP must be confirmed based on an integrative morphological/molecular approach focused on type specimens of those species reported previously. The knowledge of rhodolith-forming species in the EP is not at the initial explorative step of research, due exhaustive surveys to study coralline algae that have been made in this area. The species diversity of rhodolith-forming species at the Eastern Pacific is remarkable (36 morpho-species). Nonetheless, a great diversity is waiting to be discovered in this region. Particularly in the Tropcial Pacific of Mexico and the Pacific of Baja Caliornia where we found undescribed species of *Sporolithon* and *Lithothamnion*. In terms of species diversity and abundance, *Lithophyllum* and *Lithothmanion* are the most relevant in the rhodolith beds in this region with 10 species each, representing 55.5 % of the total species diversity. If conservation strategies will be formed for rhodolith beds in the EP, these genera should be included. The goal of this review was the inclusion of three new records of rhodolith-forming species at the EP (*Lithoporella pacifica*, *Lithothamnion* sp. and *Sporolithon* sp.); these new records represent an increase of 8.3 % of the rhodolith-forming species diversity at the EP.

References

Athanasiadis A, Lebednik PA, Adey WH (2004) The genus *Mesophyllum* (Melobesioideae, Corallinales, Rhodophyta) on the northern Pacific coast of North America. Phycologia 43:126–165

Ávila E, Riosmena-Rodríguez R (2011) A preliminary evaluation of shallow-water rhodolith beds in Bahía Magdalena, Mexico. Braz J Oceanogr 59:365–375

Bittner L, Payri CE, Maneveldt GW, Couloux A, Cruaud C, De Reviers B, Le Gall L (2011) Evolutionary history of the Corallinales (Corallinophycidae, Rhodophyta) inferred from nuclear, plastidial and mitochondrial genomes. Mol Phylogenet Evol 61:697–713

Broom JES, Hart DR, Farr TJ, Nelson WA, Neill KF, Harvey AS, Woelkerling Wm J (2008) Utility of psbA and nSSU for phylogenetic reconstruction in the Corallinales based on New Zealand taxa. Mol Phylogenet Evol 46:958–973

Chamberlain YM, Irvine LM (1994) Melobesioideae Bizzozero. In: Irvine LM, Chamberlain YM (eds) Seaweeds of the British Isles, vol 1, Rhodophyta Part 2B Corallinales, Hildenbrandiales. HMSO, London, pp 159–234

Dawson EY (1960a) Marine red algae of Pacific Mexico, part III Cryptonemiales, Corallinaceae Subf.Melobesioideae. Pac Nat 2:3–125

Dawson EY (1960b) New records of marine algae from Pacific Mexico and Central America. Pac Nat 20:31–52

Fernández C (2008) Flora marina del Parque Nacional Isla del Coco, Costa Rica, Pacífico Tropical Oriental. Rev Biol Trop 56:57–69

Fernández-García C, Riosmena-Rodríguez R, Wysor B, Tejeda OL, Cortés J (2011) Checklist of the Pacific marine macroalgae of Central America. Bot Mar 54:53–57

Fragoso D, Rodríguez D (2002) Algas coralinas no geniculadas (Corallinales, Rhodophyta) en el Pacífico tropical mexicano. An Inst Biol 73:97–136
Gabrielson PW, Miller KA, Martone PT (2011) Morphometric and molecular analyses confirm two distinct species of *Calliarthron* (Corallinales, Rhodophyta), a genus endemic to the northeast pacific. Phycologia 50:298–316
Harvey AS, Woelkerling WJ (2007) Ggía para la identificación de rodolitos de algas rojas coralinas no geniculadas (Corallinales, Rhodophyta). Cienc Mar 4:411–426
Hinojosa-Arango G, Riosmena-Rodríguez R (2004) Influence of rhodolith-forming species and growth-form on associated fauna of rhodolith beds in the Central-West Gulf of California, México. PSZNI Mar Ecol 25:109–127
Kato A, Baba M, Suda S (2013) Taxonomic circumscription of the heterogeneous species Neogoniolithon brassica-florida (Corallinales, Rhodophyta) in Japan. Phycol Res 61:15–26
Konar B, Riosmena-Rodríguez R, Iken K (2006) Rhodolith bed: a newly discovered habitat in the North Pacific Ocean. Bot Mar 49:355–359
Lemoine M (1929) Les Corallinacées de l'Archipel des Galapagos et du Golfe de Panama. Arch Mus Natl Hist Nat Paris 6(4):37–88
Littler MM, Littler DS (2008) Coralline algal rhodoliths from extensive benthic communities in the Gulf of Chiriqui, Pacific Panama. Coral Reefs 27:553
Macaya EC, Riosmena-Rodríguez R, Melzer RR, Meyer R, Försterra G, Häussermann V (2015) Rhodolith beds in the South-East Pacific. Mar Biodivers 45:153–154
Nuñez-López RA, Casas-Valdez MM, Mendoza-González C, Mateo-Cid LE (1998) Flora ficológica de la laguna San Ignacio, B.C.S. Méx Hidrobiológica 1:33–42
OSPAR Commission (2006) Case reports for the initial list of threatened and/or declining species and habitats in the OSPAR maritime area, Biodiversity series. OSPAR Commission, London. 150 pp
Peña V, Bárbara I (2008) Biological importance of an Atlantic European maërl bed off Benencia Island (northwest Iberian Peninsula). Bot Mar 51:493–505
Peralta-García EC, Rosas-Alquicira EF (2014) Range extension and morphological characterization of rhodolith-forming species (Corallinales, Rhodophyta) from shallow water in the Mexican South Pacific. Helgol Mar Res 68(4):503–510
Riosmena-Rodríguez R, Vásquez-Elizondo RM (2012) Range extension of *Mesophyllum engelhartii* (Foslie) W.H. Adey (Corallinales; Rhodophyta) to the Gulf of California: morphology, anatomy and reproduction. Bot Mar 55:143–148
Riosmena-Rodríguez R, Woelkerling WJ, Foster MS (1999) Taxonomic reassessment of rhodolith-forming species of *Lithophyllum* (Corallinales, Rhodophyta) in the Gulf of California. Mex Phycologia 38:401–417
Riosmena-Rodríguez R, Steller DL, Hinojosa-Arango G, Foster MS (2010) Reefs that rock and roll: the biology and conservation of rhodolith beds in the Gulf of California. In: Brusca R (ed) The gulf of California biodiversity and conservation. Arizona University Press and the Arizona-Sonora Desert Museum, Tucson, pp 49–71
Riosmena-Rodríguez R, López-Calderón J, Mariano-Meléndez E, Sánchez-Rodríguez A, Fernández C (2012) Size and distribution of rhodolith beds in the Loreto Marine park: their role in coastal processes. J Coast Res 28:255–260
Robinson NM (2013) Sistemática de las especies de algas coralinas (Corallinophycidae, Rhodophyta) formadoras de mantos de rodolitos en el Pacífico tropical oriental. Master thesis. University of Baja California Sur (UABCS). La Paz, B.C.S. Mexico, 109 pp
Robinson NM, Hansen GI, Fernández-García C, Riosmena-Rodríguez R (2013) A taxonomic and distributional study of the rhodolith-forming species *Lithothamnion muelleri* (Corallinales, Rhodophyta) in the Eastern Pacific Ocean. Algae 28:63–71
Steller DL, Caceres-Martínez C (2009) Coralline algal rhodoliths enhance larval settlement and early growth of the Pacific calico scallop *Agropecten ventricosus*. Mar Ecol Prog Ser 396:49–60
Steller DL, Riosmena-Rodríguez R, Foster MS, Roberts CA (2003) Rhodolith bed diversity in the Gulf of California: the importance of rhodolith structure and consequences of disturbance. Aquat Conserv Mar Freshw Ecosyst 13:S5–S20
Woelkerling WJ (1988) The Coralline red algae: an analysis of the genera and subfamilies of Nongeniculate Corallinaceae. British Museum (Natural History) and Oxford University Press, London, 268 pp

Chapter 14
Western Pacific

Aki Kato, Masasuke Baba, Shinya Matsuda, and Yasufumi Iryu

Abstract The Western Pacific is one of the regions where rhodolith beds are abundant. However, the studies of the rhodolith beds and rhodoliths themselves have been conducted mainly from geological perspectives. Here we review rhodolith beds and 20 rhodolith-forming species which are distributed in both shallow and deep waters (lower intertidal zones to 135 m deep) from cold to tropical regions of the Western Pacific. The coralline species composition of each rhodolith bed has not been well studied, and the taxonomic status of each species has to be defined by modern taxonomic concepts using molecular approaches. Research on rhodoliths in the Western Pacific therefore remains a poorly studied in aspects of phycology and of coastal biology. The present chapter represents an account of the current knowledge.

14.1 Introduction

Rhodoliths in the Western Pacific have been investigated in geological and taxonomic studies of both extant and fossil species. Most of these studies dealt with tropical to subtropical rhodoliths (Fig. 14.1). The earliest study of living rhodoliths in the Western Pacific was conducted as a part of the Siboga Expedition which was a Dutch scientific expedition to South-East Asia, with four extant species reported from a rhodolith bed in Indonesia (Weber-van Bosse and Foslie 1904). In the late

A. Kato (✉)
Takehara Marine Science Station, Setouchi Field Science Center, Hiroshima University, Minato-machi, Takehara, Hiroshima 725-0024, Japan
e-mail: katoa@hiroshima-u.ac.jp

M. Baba
Central Laboratory, Marine Ecology Research Institute, Onjuku-machi, Chiba 299-5105, Japan

S. Matsuda
Faculty of Education, University of the Ryukyus, Nishihara, Okinawa 903-0213, Japan

Y. Iryu
Institute of Geology and Paleontology, Graduate School of Science, Tohoku University, Aobayama, Sendai 980-8578, Japan

R. Riosmena-Rodríguez et al. (eds.), *Rhodolith/Maërl Beds: A Global Perspective*, Coastal Research Library 15, DOI 10.1007/978-3-319-29315-8_14

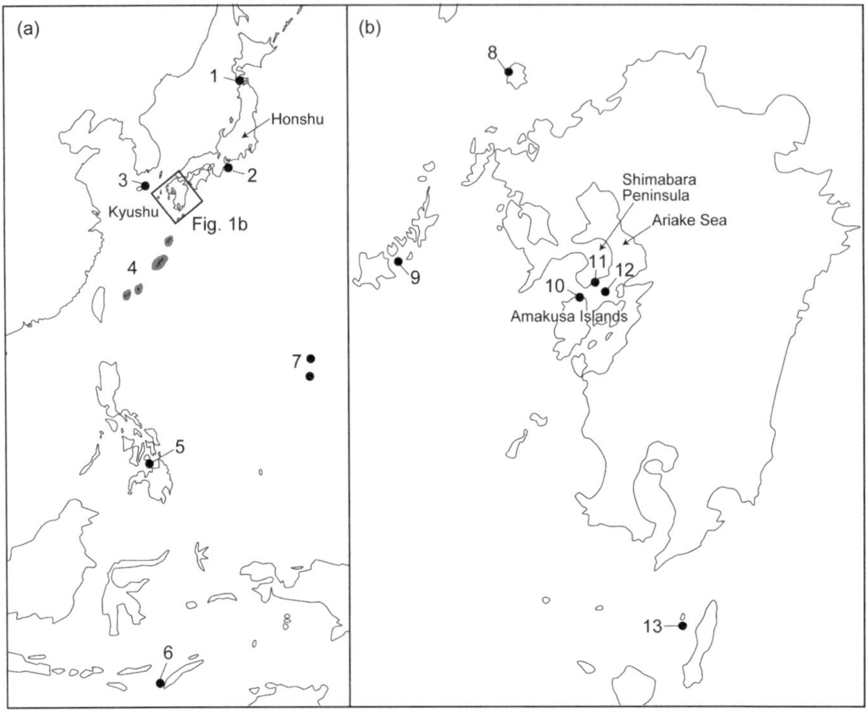

Fig. 14.1 Distribution of living rhodolith beds in the Western Pacific. *Circles* indicate locations of individual beds, and *grey bands* indicate numerous individual beds (location 4). Citations followed by more precise locations. (**a**) East Asia. *1* Masaki and Tokida (1963), Masaki (1968), Masaki (1984) (Minmaya Bay, Japan); *2* M. Baba and A. Kurashima, pers. comm. (Zagajima Island, Japan); *3* Park (1980), Woo et al. (2003) (Cheju/Jeju Island, Korea); *4* Tsuji (1993), Matsuda and Iryu (2011), Bassi et al. (2012) (in the Ryukyu Islands, Japan); *5* Foster (2001) (Camiguin Island, Philippines); *6* Weber-van Bosse and Foslie (1904) (Timor Island, Indonesia); *7* Bosence (1983) (Mariana Islands). (**b**) Distribution of living rhodolith beds in Kyushu, Japan. *8* T. Mezaki, pers. comm. (Iki Island); *9* Arasaki (1971) (Tatara Island, Goto Islands); *10* Segawa and Yoshida (1961) (Tsuji Island, Amakusa Islands); *11* Iima (2009) (Minami-shimabara, Shimabara Peninsula, Figs. 14.2a, b); *12* Segawa (1954) (Yushima Island, Ariake Sea); *13* A. Kato and R. Terada, pers. comm. (south of Mageshima Island)

1900s, several species were found to form rhodoliths in China, Guam and Indonesia (Gordon et al. 1976; Zhang and Zhou 1981; Verheij 1993, 1994). Besides rhodoliths growing in the shallow water (<30 m, e.g. Figs. 14.2a, b), rhodoliths growing in island shelves deeper than 100 m have also been studied (Matsuda and Iryu 2011, e.g. Fig. 14.2c). However, these rhodolith-forming species sometimes have been identified based on taxonomic concepts developed before the 1990s, and modern taxonomic revisions of these species remain incomplete. Therefore, the aim of this review is to discuss taxonomic and ecological aspects of the rhodolith-forming species reported in the Western Pacific ranging from Japan to Indonesia (including Mariana and Palau Islands). In this review, we followed Bartsch et al. (2012) in defining the borders of the Western Pacific and the Indian Ocean which was

Fig. 14.2 Rhodolith beds in the Western Pacific. Figures (**a**) and (**b**) were taken by M. Iima. (**a**) Heaped-up rhodoliths resembling sandbars (Minami-shimabara, Shimabara Peninsula, Japan); (**b**) Close-up photograph of (**a**); and (**c**) Seafloor covered with rhodoliths at water depth of ca. 120 m on the Miyako-Sone submarine carbonate platform in Miyako Island, Japan (Arai et al. 2014). *Arrows* indicate a large-sized benthic foraminifer (*Cycloclypeus carpenter i* Brady)

considered to be located in Central Asia (northern hemisphere) and in southeastern Australia (southern hemisphere).

14.2 Species Composition

Twenty species belonging to eight genera have been reported as rhodolith-forming species in the Western Pacific (Table 14.1). These species also occur as attached crusts in this region. Twenty-eight species in six genera except for *Mastophora* and *Neogoniolithon* were recognized as rhodoliths in deep fore-reef to island shelf areas at water depths of 50–135 m (Matsuda and Iryu 2011). However, the taxonomic status of 22 species in six genera has not been confirmed. In this section, the taxonomy of rhodoliths identified at the species level is discussed.

In the Lithophylloideae of the Corallinales, *Lithophyllum moluccense* (Foslie) Foslie, *L. okamurae* Foslie, *L. pallescens* (Foslie) Foslie and *L. tamiense* (Heydrich) Verheij were reported from the shallow water (<30 m), whereas *L. quadratum* Ishijima was reported from the deep water (>50 m) (Tables 14.1). *Lithophyllum moluccense*, *L. tamiense* and *L. pygmaeum*, which were commonly characterized by branches with pointed or somewhat rounded tips, had been recognized by the degree of the density of branching and the length and width of branches, although it was difficult to circumscribe them clearly (Weber-van Bosse and Foslie 1904; Johnson 1964; Verheij 1994). Therefore, *Lithophyllum moluccense* and *L. tamiense* have been synonymized with *L. pygmaeum* (Heydrich) Heydrich due to the priority of epithet *pygmaeum* (Silva et al. 1996). *Lithophyllum okamurae*, described from the temperate region of Japan, has been reported from the cold to tropical regions of the Indo-Pacific (Yoshida and Baba 1998; Silva et al. 1996). The conspecificity of *L. okamurae* with *L. pallescens* which was described from the Pacific Mexico, was proposed by Adey et al. (1982) who considered that both species were morphologically indistinguishable, although species distributed in the temperate regions of Japan has been still treated as *L. okamurae* (Yoshida and Baba 1998). *Lithophyllum quadratum*, described as a fossil species in Taiwan, was recorded as rhodoliths in the deep water around Okinawajima Island, Japan (Matsuda and Iryu 2011). This species from Japan is similar to the extant species *L. insipidum* Adey, Townsend et Boykins reported from Fiji (Keats 1997). However, Keats (1997) did not examine the type specimen of *L. insipidum* originally described in Hawaii by Adey et al. (1982), and the original description did not have detailed morphological characters. The rhodolith formation in *L. insipidum* has not also been reported, although this species is abundant in the shallow waters in the Ryukyu Islands, Japan (Iryu et al. 1995; Kato et al. 2011). If further study of the type specimens should show that *L. quadratum* and *L. insipidum* are conspecific, the specific epithet *insipidum* would have priority over the epithet *quadratum* (Iryu et al. 2009).

In the Mastophoroideae Setchell, a species has been reported from each of four genera. *Hydrolithon reinboldii* (Weber-van Bosse et Foslie) Foslie, *Mastophora rosea* (C. Agardh) Setchell and *Neogoniolithon frutescens* (Foslie) Setchell et

Table 14.1 Rhodolith forming coralline species reported in the Western Pacific

Order/family/species	Type locality	Distribution in the Western Pacific (§rhodolith)[a]
Corallinales		
Lithophylloideae		
Lithophyllum moluccense (Foslie) Foslie¶	Moluccas, Indonesia	§Guam[5,7], §Saipan[3], Vietnam, Indonesia
Lithophyllum okamurae Foslie	Misaki, Japan	§Japan[4,6], Korea, China, Vietnam, §Indonesia[1] (1,4,6: as *L. okamurai*)
Lithophyllum pallescens (Foslie) Foslie	Isla Espíritu Santo, Baja California	§Japan[11], Philippines
Lithophyllum quadratum Ishijima	Daihanroku (South Bay), southernmost Taiwan	§Japan[14], Taiwan[13]
Lithophyllum tamiense (Heydrich) Verheij¶	Tami Island, Huon Gulf, Papua New Guinea	§Japan[11], Indonesia
Mastophoroideae Setchell		
Hydrolithon reinboldii (Weber-van Bosse et Foslie) Foslie¶	Muaras Reef, East Kalimantan, Indonesia	Japan, Korea, China, §Saipan[3], §Guam[5,7], §Palau[12], Philippines, Vietnam, §Indonesia[1] (1,3: as *Goniolithon reinboldii* (Weber-van Bosse) Foslie)
Lithoporella melobesioides (Foslie) Foslie	South Nilandu Island, Maldives	§Japan[14], China, Guam, Philippines, Indonesia
Mastophora rosea (C. Agardh) Setchell	Guam, Mariana Islands	Japan, Taiwan, China, Mariana Islands, Palau, Philippines, Vietnam, §Indonesia[10]
Neogoniolithon frutescens (Foslie) Setchell et Mason¶	Funafuti, Tuvalu	§Japan[11], China, §Guam[7], Philippines, §Indonesia[10]
Hapalidiales		
Melobesioideae		
Lithothamnion australe Foslie	La Paz, Baja California, Mexico	§Japan[14], §Indonesia[1]
Lithothamnion intermedium Kjellman	Lebesby, Finmark, Norway	§Japan[4,6], China (4,6: as *Lithothamnium intermedium*)
Lithothamnion pulchrum Weber-van Bosse et Foslie¶	Sailus Besar, Great Paternoster Is., Indonesia	§Japan[14], Philippines[1], Indonesia[1]
Lithothamnion pacificum (Foslie) Foslie¶	Pebble Beach, Monterey County, California	Russia, Japan, Korea, §China[8],
Mesophyllum erubescens (Foslie) Lemoine¶	Chaloup Bay, Brazil	Russia, §Japan[11], Korea, China, Vietnam, Malaysia, Singapore, §Guam[7], Philippines, §Indonesia[1] (1. as *Lithothamnion erubescens* f. *haingsisiana*)

(continued)

Table 14.1 (continued)

Order/family/species	Type locality	Distribution in the Western Pacific (§rhodolith)[a]
Mesophyllum syrphetodes Adey, Townsend et Boykins	Molokai Island, Hawaiian Islands	§Indonesia[10]
Sporolithales		
Sporolithon episoredion (Adey, Townsend et Boykins) Verheij	St Rogatien Bank, North-West Hawaiian Islands	§Japan[14], §Indonesia[9]
Sporolithon episporum (Howe) Dawson	Panama (Caribbean coast)	Japan[2], §Indonesia[9]
Sporolithon schmidtii (Foslie) Gordon, Masaki et Akioka	Ko Kahdat, Ko Chang Archipelago, Thailand	Japan, §Guam[7], Mariana Islands, Philippines, Indonesia, Thailand
Sporolithon ptychoides Heydrich	El Tor, Red Sea	§Indonesia[9], Thailand
Sporolithon molle (Heydrich) Heydrich	El Tor, Red Sea	§Japan[14], §Indonesia[9]

[a]Distribution of each species in the Western Pacifc was cited from Guiry and Guiry (2016) and additionally from 1. Weber-van Bosse and Foslie (1904) and 2. Baba (2000). § indicate samples with rhodolith morphology. ¶ indicates synonymized species (see text). Records of rhodolith were as follow: 3. Johnson (1957); 4. Masaki and Tokida (1963); 5. Johnson (1964); 6. Masaki (1968); 7. Gordon et al. (1976); 8. Zhang and Zhou (1981); 9. Verheij (1993); 10. Verheij (1994); 11. Matsuda (2002); 12. Ohba et al. (2007); 13. Iryu et al. (2009); 14. Matsuda and Iryu (2011)

Mason were found as rhodoliths in the shallow water of the tropical to subtropical regions (Tables 14.1). Rhodoliths of *Lithoporella melobesioides* (Foslie) Foslie were found in the deep water (Matsuda and Iryu 2011). Although this species has been reported in the shallow water in Guam, rhodoliths formed by this species have not been reported (Gordon et al. 1976). *Hydrolithon reinboldii*, described from Indonesia, had been widely reported from the tropical to subtropical regions of the Pacific. However, this species was subsumed into *H. boergesenii* (Foslie) Foslie described from the Caribbean Sea (Lesser Antilles), as a heterotypic synonym, based on molecular and morphological data (Rösler et al. 2016). *Neogoniolithon frutescens* had been treated as one of heterotypic synonyms of *N. brassica-florida* (Harvey) Setchell et Mason. The resurrection of *N. frutescens* was proposed, because this species can be genetically and morphologically distinguished from other synonymized species of *N. brassica-florida* (Kato et al. 2013).

In the Melobesioideae of the Hapalidiales, six species in two genera have been recorded as rhodoliths. *Lithothamnion australe* Foslie has been reported in the shallow water in Indonesia, and in the deep water in the Ryukyu Islands, Japan (Tables 14.1). *Lithothamnion pacificum* (Foslie) Foslie was reported in Bohai Sea in the temperate region of China (Zhang and Zhou 1981). This species has been regarded as a synonym of *L. phymatodeum* Foslie, because there was considerable overlap between these species in morphology and habitat (Steneck and Paine 1986). Considering that the type localities of *L. pacificum* (Monterey County, California) and *L. phymatodeum* (Whidbey Island, Washington State) are in the west coasts of the North America, the conspecificity of these species might be possible. *Lithothamnion intermedium* Kjellman has been reported as one of the main components of a rhodolith bed in the cold regions of Japan (Masaki and Tokida 1963; Masaki 1968, as *Lithothamnium intermedium* Kjellman). *Mesophyllum erubescens* (Foslie) Lemoine has been reported in the shallow waters of the Ryukyu Islands, Japan and Guam (Gordon et al. 1976; Matsuda 2002). This species was transferred to a new genus *Melyvonnea* Athanasiadis et Ballantine which was morphologically separated from *Mesophyllum* Lemoine (Athanasiadis and Ballantine 2014). According to Silva et al. (1996), *Lithothamnion erubescens* f. *haingsisianum* Weber-van Bosse and Foslie 1904 (as f. *haingsisiana*), can be treated as *M. erubescens*, however, the taxonomic status of this species has not been formally changed to *Mesophyllum* or *Melyvonnea* yet (Athanasiadis and Ballantine 2014). *Lithothamnion pulchrum* Weber-van Bosse et Foslie, currently regarded as *Mesophyllum pulchrum* (Weber-van Bosse et Foslie) Lemoine (1928), has been reported in the deep water around the Ryukyu Islands, Japan (Matsuda and Iryu 2011). This species is also distributed in the shallow waters of Indonesia and the Philippines (Weber-van Bosse and Foslie 1904, as *L. pulchrum*, Silva et al. 1987). However, *M. pulchrum* has never been found growing in shallow reef environments in the Ryukyu Islands, Japan (Matsuda and Iryu 2011). *Mesophyllum syrphetodes* Adey, Townsend et Boykins (1982) described from Hawaii, has been reported in Indonesia (Verheij 1994). However, there is the possibility that *M. syrphetodes* specimens from Indonesia and Hawaii are different species, because type specimens from Hawaii were mainly crustose, whereas specimens from Indonesia were mainly protuberant or branched (Adey et al. 1982; Verheij 1994).

In the Sporolithales, five *Sporolithon* species have been recorded as rhodoliths. *Sporolithon episoredion* (Adey, Townsend et Boykins) Verheij and *S. molle* (Heydrich) Heydrich have been reported from both the shallow and deep waters in the Ryukyu Islands, Japan and Indonesia (Verheij 1993; Yoshida and Baba 1998; Matsuda and Iryu 2011). The rest of the three species have been reported in the shallow water: *S. episporum* (Howe) Dawson and *S. ptychoides* Heydrich from Indonesia (Verheij 1993); and *S. schmidtii* (Foslie) Gordon, Masaki et Akioka from Guam (Gordon et al. 1976).

14.3 Ecological Roles of Rhodoliths

A rhodolith itself and a rhodolith bed provide habitats for associated organisms in the shallow to deep waters. Commercially important kelps *Saccharina japonica* (Areschoug) Lane, Mayes, Druehl et Saunders and *Undaria pinnatifida* (Harvey) Suringar have been reported to grow on rhodoliths and form a community in rhodolith beds in the northern part of Honshu, or in Kyushu, Japan, respectively (Table 14.1, locations 1, 11, Segawa 1954; Masaki and Tokida 1963; Masaki 1968; Masaki 1984). Growth of stony corals *Acropora*, *Pocillopora* and *Favia* on the upper surfaces of the rhodoliths has been observed at the tidal channels in the Ryukyu Islands, Japan (Bosence 1983). Rhodoliths also provide substrata for fleshy macroalgae in the deep water (at 55–136 m deep) around Okinawajima Island, Japan, indicating that rhodoliths are one of factors affecting the distribution of fleshy algae in the deep water where there is often only sandy or gravelly sea bottom (Ohba and Itaki 2010; Ohba et al. 2011). The coverage of rhodoliths growing in the deep fore-reef to island shelf areas at water depths of 50–130 m averages around 45 % at Okinawajima Island (Matsuda and Iryu 2011).

Internally rhodoliths are primarily composed of nongeniculate coralline algae and an encrusting foraminifer *Acervulina inhaerens* Schulze. The volume of *Acervulina* within rhodolith envelopes varies considerably, and overall it is greater than that of coralline algae. As nodules collected off Kikai-jima Island, Japan, are mostly formed predominantly by *Acervulina*, they were described as (foraminiferal) macroids (Bassi et al. 2012). The rhodoliths (and macroids as well) have envelopes of well-preserved, concentric to irregular laminations or, much more commonly, are bored and display various degrees of bioerosion. Bioeroders include bivalves, polychaetes, bacteria, algae, fungi and sponges (Matsuda and Iryu 2011; Bassi et al. 2011, 2012). This bioerosion is relatively pervasive in larger rhodoliths (>7 cm in diameter) and more pervasive at water depths greater than 90 m (Bassi et al. 2011; Matsuda and Iryu 2011).

14.4 Distribution of Rhodolith Beds

Rhodolith beds in the Western Pacific are reported as so-called *Lithothamnion* banks from cold to tropical regions in Japan, Korea, the Philippines and Indonesia (Fig. 14.1). The species composition of these rhodolith beds has not been well examined yet, although dominant species of these rhodolith beds appear to vary with the climate zones (Table 14.2). Rhodolith-forming species extending from the tropical to the southern temperate regions were chiefly *Hydrolithon boergesenii*, *Lithothamnion australe* and *Melyvonnea erubescens* (Foslie) Athanasiadis et Ballantine (Fig. 14.1, locations 3, 4 and 6). *Lithophyllum okamurae* was often reported as rhodoliths in the temperate regions of Japan (Fig. 14.1, locations 2, 9 and 11). Rhodolith beds in the northernmost parts of Honshu in Japan were chiefly comprised of *Lithothamnion* species (Fig. 14.1, location 1, T. Masaki and H. Akioka, pers. comm.). Besides the climate zones, the floral composition of rhodolith beds in deep fore reef to island shelf areas is also unique to particular water depth. In the Ryukyu Islands, Japan, mastophoroid species are characteristic of shallow reef environments, whereas the coralline algae dominating in deep environments are Lithophylloideae, Melobesioideae and Sporolithales. Sporolithales is distributed in depths greater than 90 m (Iryu et al. 1995; Matsuda and Iryu 2011).

There are likely to be common hydrodynamic and other environmental factors responsible for the formation of the above-mentioned rhodolith beds. For example, three rhodolith beds were located close to each other in the Ariake Sea surrounded

Table 14.2 Species composition of rhodolith beds in the Western Pacific. No. 5, 8, 10 and 13 were excluded due to no information

No. on Fig. 14.1[a]	Localities of rhodolith beds	Species
1	Minmaya Bay, Aomori Prefecture, Japan	*Lithothamnion intermedium*
2	Zagajima Island, Mie Prefecture, Japan	*Lithophyllum okamurae*
3	Jeju Island, Korea	*Hydrolithon reinboldii*¶, *Lithophyllum* sp., *Mesophyllum erubescens*¶
4	Island shelf of the Ryukyu Islands, Japan[b]	*Lithophyllum* spp., *L. quadratum*, *Hydrolithon* spp., *Spongites* spp., *Lithoporella melobesioides*, *Lithothamnion* spp., *L. australe*, *L. pulchrum*, *Mesophyllum* spp., *Sporolithon* spp., *Sporolithon episoredion*, *S. molle*
6	Timor Island, Indonesia	*Lithophyllum okamurae* (as *L. okamurai*), *H. reinboldii*¶ (as *Goniolithon reinboldii*), *Lithothamnion australe*, *M. erubescens*¶ (as *Lithothamnion erubescens* f. *haingsisiana*)
9	Goto Islands, Nagasaki Prefecture, Japan	*Lithophyllum okamurae* (as *L. okamurai*)
11	Minami-shimabara, Nagasaki Prefecture, Shimabara Peninsula, Japan	*Lithophyllum okamurae*, *Lithophyllum* sp., *Lithothamnion* sp.
12	Yushima Island, Kumamoto Prefecture, Japan (Ariake Sea)	*Lithothamnion* cf. *australe*

[a]References follow Table 14.1. ¶ indicates synonymized species (see text).

[b]Matsuda and Iryu (2011)

by the Shimabara Peninsula and the Amakusa Islands, Japan (Fig. 14.1, locations 10,11 and 12). It might be because currents going in and out of the Ariake Sea heaped up rhodoliths like a sandbar (Fig. 14.2a, Iima 2009). Besides currents, nutrients are likely to be important oceanographic factors for the formation of rhodolith beds. The island shelf in the Ryukyu Islands, Japan, is dominated by rhodoliths, whereas *Halimeda* banks flourish on the northern Great Barrier Reef, the eastern Java Sea and the northern Nicaraguan shelf (Iryu et al. 1995). It seems the three *Halimeda*-dominant shelves are nutrient-rich environments caused by the injection of upwelling oceanic waters, whereas shelves in the Ryukyu Islands probably lack conspicuous upwelling (Iryu et al. 1995).

Of areas including rhodolith beds, Jeju Island in Korea (Jeju Island Geopark) and the Shimabara Peninsula in Kyushu, Japan (Unzen Volcanic Geopark) have been recorded as Global Geoparks of UNESCO (The Global Geoparks Network 2016). Other areas with rhodolith beds in Japan are located in National Parks or Quasi-National Parks (Ministry of the Environment, Japan 2010).

Rhodoliths are one of the important components of the Pleistocene coral-reef and associated shallow lagoon and fore-reef to island shelf deposits (Iryu et al.

2006). Stratigraphic and paleontological studies clearly showed that the rhodoliths are similar in size, shape, internal structure, and biotic composition to those found in modern deep fore-reef to and island shelf areas in the Ryukyu Islands, Japan. Similar fossil rhodoliths were reported from the uplifted and subsided Pleistocene carbonate sequences at Huon Peninsula, Papua New Guinea (Matsuda 1995) and Tahiti (Iryu et al. 2010), respectively. These suggest that pebble (to cobble)-sized rhodoliths, composed mainly of coralline algae and an encrusting foraminifer (*Acervulina*), are common in deep fore-reef to island shelf areas in tropical and subtropical reef regions worldwide.

14.5 Conclusions

In the Western Pacific, research on rhodoliths using geological approaches has been well conducted mainly in the deep fore reefs to island shelf areas in the Ryukyu Islands, Japan, whereas taxonomic and ecological research of rhodolith-forming species and associated organisms in rhodolith beds have only been recently studied so far.

In relation to the taxonomy of rhodolith-forming species, the reliability of species identification needs to be confirmed and species have to be reevaluated using modern taxonomic concepts. Recent molecular phylogenetic research revealed that the species diversity of nongeniculate coralline algae was often underestimated (Bittner et al. 2010; Kato et al. 2011, 2013). Future taxonomic revision would reveal the presence of new distributional records and even new species endemic to the Western Pacific, although most of the rhodolith-forming species discussed here have been widely reported from the Indo-Pacific regions. Taxonomic and stratigraphic research using fossil and extant specimens suggested the presence of depth zonation among species (Iryu 1992; Kato et al. 2013). In order to elucidate the regional diversity of rhodolith-forming or attached crustose species, accumulation of DNA barcodes accompanied by morphological and ecological information is first needed.

Rhodoliths provide microhabitats for other organisms, such as natural spaces within branched forms, spaces between individuals, and hard calcareous surfaces for sessile organisms and borers (Foster 2001). As associated organisms with rhodoliths and rhodolith beds, some seaweeds, corals and borers have been recorded in the Western Pacific, although invertebrates inhabiting spaces provided by rhodolith individuals have not been recorded. Rhodolith beds are one of four largest macrophyte-dominated benthic communities ranking with seaweed beds, seagrass meadows and coralline algal reefs (Foster 2001). A better understanding of these associated organisms may provide interesting contrasts to the nature and ecological functions of other benthic communities.

Acknowledgements We thank to A. Kurashima, T. Mezaki and R. Terada for providing us unpublished information regarding rhodolith beds and samples and L. M. Liao for his critical reading and valuable comments for the manuscript.

References

Adey WH, Townsend RA, Boykins, WT (1982) The crustose coralline algae (Rhodophyta: Corallinaceae) of the Hawaiian Islands. Smithsonian Contributions to the Marine Sciences 15:1–74

Athanasiadis A, Ballantine DL (2014) The genera Melyvonnea gen. nov. and Mesophyllum s.s. (Melobesioideae, Corallinales, Rhodophyta) particularly from the central Atlantic Ocean. Nord J Bot 32: 385–436

Arai K, Machiyama H, Chiyonobu S, Matsuda H, Sasaki K, Humblet M, Iryu Y (2014) Subsidence of the Miyako-Sone submarine carbonate platform, east of Miyako-jima Island, northwestern Pacific Ocean. Island Arc 23:1–15

Arasaki S (1971) Seaweeds – science expedition of the underwater park in the Goto Islands. In: Reports on science expeditions of the underwater parks in Nagasaki. Reports of Underwater Park Center vol 22. Nagasaki (in Japanese)

Baba M (2000) Morphological study of a new record of *Sporolithon episporum* (Howe) Dawson (Corallinales, Rhodophyta) from Japan. Jpn J Phycol 48:96 (Meeting abstract in Japanese)

Bartsch I, Wiencke C, Laepple T (2012) Global seaweed biogeography under a changing climate: the prospected effects of temperature. In: Wiencke C, Bischof K (eds) Seaweed biology, vol 219, Ecological studies. Springer, Berlin

Bassi D, Humblet M, Iryu Y (2011) Recent ichnocoenosis in deep water macroids, Ryukyu Islands, Japan. Palaios 26:232–238

Bassi D, Iryu Y, Humblet M, Matsuda H, Machiyama H, Sasaki K, Matsuda S, Arai K, Inoue T (2012) Recent macroids on the Kikai-jima shelf, Central Ryukyu Islands, Japan. Sedimentology 59:2024–2041

Bittner L, Halary S, Payri C, Cruaud C, de Reviers B, Lopez P, Bapteste E (2010) Some considerations for analyzing biodiversity using integrative metagenomics and gene networks. Biol Direct 5:47

Bosence DWJ (1983) The occurrence and ecology of recent rhodoliths — A Review. In Peryt TM (ed), Coated Grains. Springer-Verlag, Berlin

Foster MS (2001) Rhodoliths: between rocks and soft places. J Phycol 37:659–667

Gordon GD, Masaki T, Akioka H (1976) Floristic and distributional account of the common crustose coralline algae on Guam. Micronesica 12:247–277

Guiry MD, Guiry GM (2016) AlgaeBase. World-wide electronic publication, National University of Ireland, Galway. http://www.algaebase.org. Accessed 31 May 2016

Iima Y (2009) Shirasu (*Lithothamnium*) in the offing of Harajiro Castle. Sorui-danwashitsu (in Japanese). http://www.sourui-koza.com/z_danwa/200905/lithothamnion.html. Accessed 11 June 2013 (Site not found 31 May 2016)

Iryu Y (1992) Fossil nonarticulated coralline algae as depth indicators for the Ryukyu group. Trans Proc Palaeontol Soc Jpn N S 167:1165–1179

Iryu Y, Nakamori T, Matsuda S, Abe O (1995) Distribution of marine organisms and its geological significance in the modern reef complex of the Ryukyu Islands. Sediment Geol 99:243–258

Iryu Y, Matsuda H, Machiyama H, Piller WE, Quinn TM, Mutti M (2006) Introductory perspective on the COREF Project. Island Arc 15:393–406

Iryu Y, Bassi D, Woelkerling WJ (2009) Re-assessment of the type collections of fourteen corallinalean species (Corallinales, Rhodophyta) described by W. Ishijima (1942–1960). Palaeontology 52:401–427

Iryu Y, Takahashi Y, Fujita K, Camoin G, Cabioch G, Matsuda H, Sugihara K, Sato T, Webster JM, Westphal H (2010) Sealevel history recorded in the Pleistocene carbonate sequence in IODP Hole 310-M0005D, off Tahiti. Island Arc 19:690–706

Johnson JH (1957) Geology of Saipan, Mariana Islands. Part 3. Paleontology. Chapter E. Calcareous algae. US Geol Surv Prof Pap 280-E:209–246

Johnson JH (1964) Fossil and recent calcareous algae from Guam. US Geol Surv Prof Pap 403-G:1–40

Kato A, Baba M, Suda S (2011) Revision of the Mastophoroideae (Corallinales, Rhodophyta) and polyphyly in nongeniculate species widely distributed on Pacific coral reefs. J Phycol 47:662–672

Kato A, Baba M, Suda S (2013) Taxonomic circumscription of heterogeneous species *Neogoniolithon brassica-florida* (Corallinales, Rhodophyta) in Japan. Phycol Res 61:15–26

Keats DW (1997) *Lithophyllum insipidum* Adey, Townsend et Boykins and *L. flavescens* sp. nov.: two flat lithophylloid coralline algae (Corallinales, Rhodophyta) abundant in shallow reef environments in Fiji. Phycologia 36:351–365

Lemoine M (1928) Un nouveau genre de Mélobésiées: *Mesophyllum*. Bull Soc Bot France 75:251–254

Masaki T (1984) Crustose coralline algae. Jpn J Phycol 32:71–85 (in Japanese)

Masaki T, Tokida J (1963) Studies on the Melobesioideae of Japan. VI. Bull Fac Fish. Hokkaido Univ 14:1–6

Matsuda S (1995) Quaternary limestones on Huon Peninsula, Papua New Guinea with special reference to nonarticulated coralline algae (Rhodophyta, Corallinaceae). J Geogr 104:706–718 (in Japanese with English abstract)

Matsuda S (2002) Studies on non-geniculate coralline algae in coral reefs in Japan. In: Nakamori T (ed) Coral reef studies in Japan part 1. The Japanease Coral Reef Society, Sendai (in Japanese)

Matsuda S, Iryu Y (2011) Rhodoliths from deep fore-reef to shelf areas around Okinawa-jima, Ryukyu Islands, Japan. Mar Geol 282:215–230

Ministry of the Environment, Japan (2010) National parks & important biodiversity areas of Japan. http://www.env.go.jp/park/topics/review/attach/pamph1/en_full.pdf. Accessed 31 May 2016

Ohba H, Itaki T (2010) Marine algae collected from deep-water west off Okinawa Island in the Ryukyu Islands. In: Arai K (ed) Marine geological and geophysical studies around Okinawa Islands—northwestern off of Okinawa Island-, preliminary reports on researches in the 2010 fiscal year. GSJ Interim Report Vol. 51. National Institute of Advanced Industrial Science and Technology (AIST), Tsukuba

Ohba H, Victor S, Golbuu Y, Yukihira Y (2007) Tropical marine plants of Palau. Palau International Coral Reef Center (PICRC), Koror

Ohba H, Kinoshita N, Itaki T (2011) Marine algae collected from deep-water west off Okinawa Island in the Ryukyu Islands (II). In: Arai K (ed) Marine geological and geophysical studies around Okinawa Islands—western off of Okinawa Island-, preliminary reports on researches in the 2011 fiscal year. GSJ Interim Report vol. 55. National Institute of Advanced Industrial Science and Technology (AIST), Tukuba

Park CH (1980) The crustose coralline algae (Rhodophyta, Cryptonemiales) in Korea. Bull Nat Fish Univ Busan 20:1–30

Rösler, A, Perfectti F, Peña V, Braga JC (2016) Phylogenetic relationships of Corallinaceae (Corallinales, Rhodophyta): taxonomic implications for reef-building corallines. J Phycol (in press)

Segawa S (1954) Coralline algae. Jpn J Phycol 2:33–39 (in Japanese)

Segawa S, Yoshida T (1961) Marine algae. Fauna and flora of the sea around the Amakusa marine biological laboratory. Kyushu University, Kumamoto

Silva PC, Meñez EG, Moe RL (1987) Catalog of the benthic marine algae of the Philippines. Smithson Contrib Mar Sci 27:1–179

Silva PC, Basson PW, Moe RL (1996) Catalogue of the benthic marine algae of the Indian Ocean. Univ Calif Publ Bot 79:1–1259

Steneck RS, Paine RT (1986) Ecological and taxonomic studies of shallow-water encrusting Corallinaceae (Rhodophyta) of the boreal northeastern Pacific. Phycologia 25:221–240
The Global Geoparks Network (2016) http://www.globalgeopark.org/. Accessed 31 May 2016
Tsuji Y (1993) Tide influenced high energy environments and rhodolith-associated carbonate deposition on the outer shelf and slope off the Miyako Islands, southern Ryukyu Island Arc, Japan. Mar Geol 113:255–271
Verheij E (1993) The genus *Sporolithon* (Sporolithaceae fam. nov., Corallinales, Rhodophyta) from the Spermonde Archipelago, Indonesia. Phycologia 32:184–196
Verheij E (1994) Nongeniculate Corallinaceae (Corallinales, Rhodophyta) from the Spermonde Archipelago, SW Sulawesi, Indonesia. Blumea 39:95–137
Weber-van Bosse A, Foslie M (1904) The Corallinaceae of the Siboga-expedition. Siboga Exp Monogr 61:1–77, 32 figs, XVI plates
Woo KS, Kim JK, Kang SS (2003) Characteristics and formation conditions of the rhodoliths in Wu Island beach, Jeju-do, Korea: preliminary report. J Kor Soc Oceanogr Sea 8:401–410
Yoshida T, Baba M (1998) Corallinales. In: Yoshida T (ed) Marine algae of Japan. Uchida Rokakuho Publishing Co., Tokyo
Zhang D, Zhou J (1981) The "Hai Fu Shi" from Huangxian, Shandong Province-*Lithothamnium* spp. Oceanol Limnol Sin 12:138–141, Plates I, II

Chapter 15
South Pacific

Wendy Nelson and Kate Neill

Abstract Rhodoliths of the South Pacific remain largely unstudied. Twenty two rhodolith-forming species from ten genera have been reported from the southern Pacific, although of these, only nine species in six genera have been confirmed as rhodolith records in the literature. There have been few ecological or distributional studies of rhodoliths in the southern Pacific, although some data are available on the age and growth rates of rhodoliths from the region. The taxonomy and nomenclature of rhodolith-forming taxa in this region are problematic.

15.1 Introduction

In this review, we define the South Pacific as including Australia, New Zealand and all islands south of the equator and as far east as the Pitcairn Islands. There have been many taxonomic studies of nongeniculate coralline algae undertaken in the South Pacific, with a key focal point being the research of Woelkerling and colleagues (1992, 1996) in southern Australia. There have been few targeted studies of modern rhodoliths in the South Pacific, with most of these occurring in the past decade.

W. Nelson (✉)
National Institute of Water & Atmospheric Research,
Wellington, New Zealand

School of Biological Sciences, University of Auckland,
Auckland, New Zealand
e-mail: wendy.nelson@niwa.co.nz

K. Neill
National Institute of Water & Atmospheric Research,
Private Bag 14901, Wellington 6241, New Zealand

R. Riosmena-Rodríguez et al. (eds.), *Rhodolith/Maërl Beds: A Global Perspective*, Coastal Research Library 15, DOI 10.1007/978-3-319-29315-8_15

15.2 Species Composition

Twenty two rhodolith-forming species from ten genera have been reported in the southern Pacific (Table 15.1). Several of these species also occur as attached crusts in this region and unfortunately many distribution records do not clarify whether the record is a rhodolith or an attached crust. Nine species in six genera have confirmed rhodolith records in the literature (Table 15.1). The species reported in living rhodolith beds in the South Pacific are summarized in Table 15.2.

In the Corallinaceae three species of rhodolith-forming *Lithophyllum* in the Lithophylloideae have been reported from the southern Pacific, but only *L. stictaeforme* (Aresch.) Hauck is reported as forming rhodoliths, in records from Western Australia. In the Mastophoroideae, three species of *Hydrolithon* are reported as rhodoliths, namely *H. munitum* (Foslie & M.Howe) Penrose, *H. reinboldii* (Weber Bosse & Foslie) Foslie, and *H. rupestre* (Foslie) Penrose. Although *Porolithon onkodes* (Heydr.) Foslie is known to form rhodoliths in other parts of the world, none of the records for the southern Pacific state whether the species was occurring epilithically or as rhodoliths. In the genus *Spongites* two species have been reported, however none of the published records provide information about the growth form of the material. In the Neogoniolithoideae the species *Neogoniolithon brassica-florida* (Harv.) Setchell & L.R.Mason is widely recorded, including reports of it as rhodoliths from Australia (Victoria), French Polynesia and American Samoa (Fig. 15.1).

Amongst the Hapalidiaceae, the subfamily Melobesioideae is represented by members of four genera. Four species of *Lithothamnion* have been reported from the region, with two of these confirmed as forming rhodoliths, *Lithothamnion crispatum* Hauck and *L. proliferum* Foslie. There are two species of rhodolith-forming *Mesophyllum* recorded in the region with only *M. engelhartii* (Foslie) W.H.Adey confirmed as a species forming rhodoliths. Although single species of both *Melyvonnea* and *Phymatolithon* have been reported, it is not known if they are present in the rhodolith growth form.

In the Sporolithales, four *Sporolithon* species have been recorded in the region, with *S. durum* (Foslie) Townsend and Woelkerling confirmed as forming rhodoliths. Fossil rhodoliths in New Zealand from both North Island and South Island sites have also been identified as *Sporolithon* (Lee et al. 1997; Nalin et al. 2008). Numerous other fossil rhodolith assemblages are known (Lever 2007; Nalin et al. 2008), but have not been studied in detail (Lee pers. comm. 2014).

The taxonomy of the rhodolith-forming non-geniculate coralline algae continues to be problematic. As an example, the species currently known as *L. crispatum* in the southern Pacific (type locality: Adriatic Sea), was first identified in New Zealand as *L. indicum* Foslie (type locality: Australia). This species was synonymized with *L. superpositum* Foslie (type locality: South Africa) by Keats et al. (2000), and subsequently synonymized with *L. crispatum* (Basso et al. 2011). Clarification of the relationships amongst these taxa awaits sequence data from type material or material from type localities. The genus *Neogoniolithon* was recently reviewed (Kato

Table 15.1 Rhodolith-forming coralline species reported in the southern Pacific

Order/Family/Species	Type locality	Distribution in the Southern Pacific[a]
Corallinales		
Corallinaceae		
Hydrolithoideae		
Hydrolithon munitum (Foslie & M.Howe) Penrose	Cave Cays, Exuma Chain, Bahamas	Queensland, South Australia, West Australia*[b]
Hydrolithon reinboldii (Weber Bosse & Foslie) Foslie	Muaras Reef, East Kalimantan, Indonesia	Australia (Queensland, Western Australia), Papua New Guinea, Fiji*[c], French Polynesia*[d,e], Samoan Archipelago, Wallis & Futuna
Hydrolithon rupestre Foslie (Penrose)	Ocean Beach, Phillip Island, Victoria	Australia (New South Wales, South Australia, Victoria*[f], Western Australia), New Zealand, French Polynesia, Wallis & Futuna
Lithophylloideae		
Lithophyllum kotschyanum Unger	Gulf of Bahrain	Australia (Queensland and Western Australia), Fiji, French Polynesia, Samoan Archipelago, Wallis & Futuna
Lithophyllum okamurae Foslie	Misaki, Japan	Fiji, Solomon Islands
Lithophyllum stictaeforme (Aresch.) Hauck	Mediterranean Sea	Australia (New South Wales, Queensland, Tasmania, South Australia, Victoria, Western Australia*[g]), New Zealand.
Mastophoroideae		
Spongites fruticulosus Kütz.	Mediterranean Sea	Australia (Queensland, South Australia, Victoria, Western Australia)
Spongites hyperellus (Foslie) Penrose	Western Port, Victoria	Australia (Tasmania, Victoria)
Neogoniolithoideae		
Neogoniolithon brassica-florida (Harv.) Setch. & L.R.Mason	Algoa Bay, Cape Province, South Africa	Australia (Queensland, South Australia, Victoria*[f], Western Australia), New Zealand, Norfolk Island, Fiji, French Polynesia*[d,e], American Samoa*[h], Samoan Archipelago, Wallis & Futuna
Porolithoideae		
Porolithon onkodes (Heydr.) Foslie	Tami Island, Gulf of Huon, Papua New Guinea	Australia (New South Wales, Queensland, Western Australia), New Guinea, Papua New Guinea, Solomon Islands, Fiji, French Polynesia, Samoan Archipelago, Wallis & Futuna
Hapalidiales		
Hapalideaceae		
Melobesioideae		

(continued)

Table 15.1 (continued)

Order/Family/Species	Type locality	Distribution in the Southern Pacific[a]
Lithothamnion australe Foslie	La Paz, Baja California, Mexico	Fiji
Lithothamnion crispatum Hauck	Rovigno, Adriatic Sea	Australia (New South Wales, Victoria*[f]) New Zealand*[i]
Lithothamnion muelleri Lenorm. ex Rosanoff	Western Port Bay, Victoria (lecto)	Australia (South Australia, Tasmania, Victoria, Western Australia), New Zealand
Lithothamnion proliferum Foslie	Lumu-Lumu shoal (Pulau Lumulumu), Borneo Bank, Indonesia	Australia (Queensland, Western Australia), New Zealand*[i], Fiji, Samoa, Kiribati*[j], Wallis & Futuna
Melyvonnea erubescens (Foslie) Athanasiadis & D.L.Ballantine	Fernando de Noronha (Pernambuco, Brazil)	Australia (Queensland, Western Australia), New Zealand, Fiji, French Polynesia, Futuna & Wallis
Mesophyllum engelhartii (Foslie) W.H.Adey	Cape Jaffa, South Australia	Australia (South Australia, Tasmania, Victoria*[f], Western Australia), New Zealand
Mesophyllum funafutiense (Foslie) Verheij	Funafuti, Ellice Islands (Tuvalu)	Australia (Western Australia), American Samoa, French Polynesia, Tuvalu
Phymatolithon repandum (Foslie) Wilks & Woelk.	Port Phillip Bay, Victoria	Australia (Queensland, South Australia, Tasmania, Victoria, Western Australia), New Zealand
Sporolithales		
Sporolithaceae		
Sporolithon durum (Foslie) R.A.Towns. & Woelk.	Cape Jaffa, South Australia	Australia (New South Wales, South Australia*[k], Victoria, Western Australia), New Zealand*[i],
Sporolithon episoredion (W.H.Adey, R.A.Towns. & Boykins) Verheij	St Rogatien Bank, Hawaii	French Polynesia
Sporolithon episporum (M.Howe) E.Y.Dawson	Point Toro, near Colón, Panama	Papua New Guinea, Fiji
Sporolithon ptychoides Heydr.	Tor, Sinai Peninsula, Egypt (AB)	Fiji, French Polynesia

[a]Distribution of each species in the Southern Pacific as cited in Guiry and Guiry (2016). Records of rhodoliths were as follows: [b]Woelkerling and Campbell (1992), [c]Goldberg (2006), [d]Derek Keats: http://www.flickr.com/photos/dkeats/7341794518/, [e]Payri et al. (2000), [f]N'Yeurt and Payri (2010), [g]Harvey and Bird (2008), [h]Littler and Littler (2003), [i]Farr et al. (2009), [j]Keats et al. (1996), and [k]Townsend et al. (1995)(*) the rhodolith form has been reported from the southern Pacific

et al. 2013) and it is clear that many different taxa have been incorrectly assigned to the species *N. brassica-florida*. The records from the southern Pacific require re-examination in the light of this research. The rhodolith-forming species *Sporolithon durum* has been widely reported in the southern Pacific and elsewhere, but molecular sequencing data suggest that more than one taxon has been attributed to this species and further work is required (Farr et al. 2009; Nelson et al. 2015).

15.3 Ecological Roles of Rhodoliths

Investigations into the ecology and diversity of rhodolith beds in the South Pacific have been limited. In Australia several studies have included rhodolith beds in relatively broad scale habitat-type studies (e.g. Heyward et al. 2010; Wilson et al. 2011) but the rhodolith species comprising the beds have not been identified. In Western Australia a large multi-stranded study of fish habitats of the Recherche Archipelago included rhodolith beds as one of the habitats investigated (Kendrick et al. 2004), and found that fish communities in rhodolith beds were statistically distinct (Baxter et al. 2004). The rhodoliths comprising the bed were identified as *Hydrolithon munitum*, *Sporolithon* sp. *Lithothamnion* sp., *Mesophyllum* sp. and *Synarthrophyton* sp. (Goldberg and Kendrick 2004). These authors also identified 52 species of algae growing epiphytically on rhodoliths in this area. As part of the same study, Grove (2004) found that the rhodolith beds contained the highest diversities and species richness of the habitat types sampled. In Western Australia, Mathis et al. (2005) reported on the invertebrates living on and within rhodoliths (cryptofauna) in Esperance Bay, finding the abundance of all invertebrates increased significantly with rhodolith size. Harvey and Bird (2008) investigated a rhodolith bed in Western Port, Victoria, identifying four species of rhodoliths (Table 15.2) and assessing their associated cryptofaunal diversity.

Table 15.2 Species compositions of living rhodolith beds in the South Pacific

	Localities of rhodolith beds	Species	Reference/s
Australia			
1	Western Australia		
	Esperance Bay	Not specified; *Hydrolithon munitum*, *Lithothamnion* sp., *Mesophyllum*/ *Synarthrophyton* sp., *Sporolithon* sp.	Mathis et al. (2005) and Goldberg (2006)
	Ningaloo Reef	Not specified	Heyward et al. (2010)
	Kimberley	Not specified	Wilson et al. (2011)
	Rottnest Island	*Sporolithon durum*	Goldberg and Heine (2008)
2	South Australia		
	Great Australian Bight	Not specified	Foster (2001)
	Cape Northumberland	*Sporolithon durum*	Townsend et al. (1995) and Woelkerling (1996)
	Cape Jaffa	*Sporolithon durum*	Townsend et al. (1995)
	Port Willunga	*Sporolithon durum*	Townsend et al. (1995) and Woelkerling (1996)

(continued)

Table 15.2 (continued)

	Localities of rhodolith beds	Species	Reference/s
3	Victoria		
	Western Port	*Hydrolithon rupestre*, *Lithothamnion crispatum*, *Mesophyllum engelhartii*, *Neogoniolithon brassica-florida*	Harvey and Bird (2008)
4	Queensland		
	One Tree Reef	Not specified	Foster (2001)
	Fraser Island	*Lithothamnion* sp., *Phymatolithon* sp., *Sporolithon* sp.	Lund et al. (2000)
New Zealand			
5	North of North Island		
	Kermadec Islands	Not specified; *Sporolithon durum*	Brook (1998) and Nelson and Dalen (2015)
	Kermadec Ridge	Not specified	Farr et al. (2009)
6	Northern North Island		
	Multiple sites, Far North	*Sporolithon durum*, not specified	Farr et al. (2009) and Grace and Hayward (1980)
	Bay of Islands	*Lithothamnion crispatum*, *Sporolithon durum*	Farr et al. (2009)
	Tutukaka and Poor Knights Is	Not specified	Brook et al. (1981)
	Alderman Is., Coromandel	*Lithothamnion proliferum*	Farr et al. (2009)
	Auckland	*Sporolithon durum*	Basso et al. (2009)
7	Central New Zealand		
	Kapiti Island	*Sporolithon durum*	Battershill et al. (1993) and Harvey et al. (2005)
	Marlborough Sounds and D'Urville Island, South Island	Not specified	Davidson 1992; Davidson et al. 2011
	Kaikoura	?*Sporolithon*	Reid et al. (2011)
8	Southern New Zealand		
	Otago	Not specified	WELT[a]
	Fiordland	Not specified	Fleming (1950)
	Foveaux Strait	*Lithothamnion* sp.	WELT
Pacific Islands			
9	New Caledonia	Not specified; *Sporolithon durum*	Mattio et al. (2008) and Darrenougue et al. (2013)
10	Fiji	*Hydrolithon reinboldii*	http://www.flickr.com/photos/dkeats/7341794518/
11	Kiribati	*Lithothamnion proliferum*	Keats et al. (1996)
12	American Samoa	*Neogoniolithon brassica-florida*	Littler and Littler (2003)
13	Cook Islands (Rarotonga)	*Neogoniolithon* sp., *Lithophyllum* sp., *Porolithon*	Scoffin et al. (1985)
14	French Polynesia	*Hydrolithon reinboldii*	Payri (1997), Payri et al. (2000), and N'Yeurt and Payri (2010)

[a]WELT = Herbarium of the Museum of New Zealand Te Papa Tongarewa

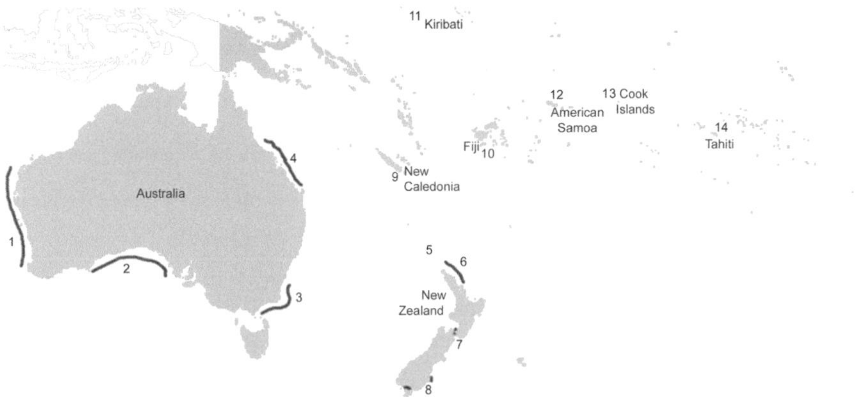

Fig. 15.1 The distribution of living rhodolith beds in the South Pacific. *Numbers* indicate locations listed in Table 15.2

In New Zealand a similar pattern occurs, with early work in many areas of the country (particularly northeastern New Zealand) generically referring to rhodoliths and community types (e.g. Grace and Hayward 1980; Brook et al. 1981; Hayward et al. 1986; Davidson 1992; Dewas and O'Shea 2012). Basso et al. (2009) reported on research on a shallow rhodolith bed in the north-eastern North Island, at a site dominated by the species *Sporolithon durum* although with rare *Lithophyllum* sp. rhodoliths also present. They considered this bed to be of particular interest because of the size and density of the rhodoliths, and their occurrence in such a shallow habitat (low intertidal and upper subtidal zones). Rhodolith beds have been identified as ecologically significant marine sites (Davidson et al. 2011) and as sensitive marine benthic habitats (MacDiarmid et al. 2013). Research on the ecology of rhodolith beds in the Bay of Islands revealed differences between sites and species in terms of associated flora and fauna (Nelson et al. 2012, 2014; Neill et al. 2015).

In other regions of the South Pacific we have been unable to find any reference to work on the ecology of living rhodolith beds. However the age and growth rates of rhodoliths have been investigated in Western Australia (Goldberg 2006; Goldberg and Heine 2008) and in New Caledonia (Darrenougue et al. 2013) with these studies both focusing on *Sporolithon durum*.

15.4 Distribution of Rhodolith Beds

There are few publications documenting the size and distribution of modern rhodolith beds in the South Pacific. In the Recherche Archipelago, Western Australia, beds at depths between 27 and 65 m and up to 9000 ha in size have been mapped (Baxter et al. 2004). Rhodolith beds have been mapped in the Bay of Islands in northeastern New Zealand (Nelson et al. 2012) and their distribution also noted on the Three Kings seabed off northern New Zealand (Nelson et al. 1982; Nelson and Hancock 1984), but there is no thorough documentation of this habitat in the wider region.

15.5 Conclusion

In the South Pacific research on the distribution and ecology of rhodoliths is in its infancy. Although rhodolith beds are internationally recognized as biodiversity 'hotspots' and in some regions receive a degree of protection, there is a distinct shortage of ecological studies on beds in the southern Pacific. There is a strong international consensus about the importance of protecting rhodolith beds, for example from the impacts of sedimentation, eutrophication, physical disturbance and ocean acidification. Further work is required in this region to identify the distribution, size and composition of rhodolith beds, assess the diversity associated with the rhodoliths, as well as the functional role rhodoliths play in coastal systems, and to ascertain any threats to their existence.

A critical issue in the South Pacific and elsewhere is the state of taxonomic understanding of rhodolith-forming species. In particular, there is an urgent need to clarify generic and species concepts and this will require molecular sequence data and phylogenetic analyses (Nelson et al. 2015). It has become very clear that the application of traditional morphological and anatomical approaches has not clarified the level of genetic diversity present amongst non-geniculate coralline algae, with recent studies revealing that diversity has been under-estimated (e.g. Bittner et al. 2011; Kato et al. 2013; Nelson et al. 2015). Without a better taxonomic framework comparative studies of distribution and ecology are seriously flawed.

Acknowledgements We thank Posa Skelton, Robin South, Antoine N'Yeurt, Daphne Lee, Cam Nelson and Hannah Russell for assistance with records and literature. We acknowledge support provided through NIWA Core funding COBR1401 (WN) and COMS1401 (KN) for the preparation of this paper.

References

Basso D, Nalin R, Nelson CS (2009) Shallow-water *Sporolithon* rhodoliths from North Island (New Zealand). Palaios 24:92–103

Basso D, Rodondi G, Bressan G (2011) A re-description of *Lithothamnion crispatum* and the status of *Lithothamnion superpositum* (Rhodophyta, Corallinales). Phycologia 50:144–155

Battershill CN, Murdoch RC, Grange KR, Singleton RJ, Arron ES, Page MJ, Oliver MD (1993) A survey of the marine habitats and communities of Kapiti Island. NIWA Client Report 1993/41. NIWA, Wellington

Baxter K, Bickers A, Kendrick G, Shortis M (2004) Benthic habitat mapping in the Recherche Archipelago. In: Kendrick G, Harvey E, McDonald J, Pattiaratchi C, Cappo M, Fromont J, Shortis M, Grove S, Bickers A, BaxterK, Goldberg N, Kletczkowski M, Butler J (eds)

Characterising the fish habitats of the Recherche Archipelago, pp 277–292. Fisheries Research and Development Corporation Report, Project No. 2001/060. ISBN: 1 74052 123. 600 p
Bittner L, Payri CE, Maneveldt GW, Couloux A, Cruaud C, de Reviers B, Le Gall L (2011) Evolutionary history of the Corallinales (Corallinophycidae, Rhodophyta) inferred from nuclear, plastidial and mitochondrial genomes. Mol Phylogenet Evol 61:697–713
Brook FJ (1998) The coastal molluscan fauna of the northern Kermadec Islands, Southwest Pacific Ocean. J Roy Soc NZ 28:185–233
Brook FJ, Grace RV, Hayward BW (1981) Soft-bottom benthic faunal associations of Tutukaka Harbour, Northland, New Zealand, part 1: macrofauna, part 2: foraminifera. Tane 27:69–92
Darrenougue N, De Deckker P, Payri C, Eggins S, Fallon S (2013) Growth and chronology of the rhodolith-forming, coralline red alga *Sporolithon durum*. Mar Ecol Prog Ser 474:105–119
Davidson RJ (1992) A report on the intertidal and shallow subtidal ecology of the Abel Tasman National Park, Nelson. Department of Conservation, Nelson/Marlborough Conservancy Occasional Publication. No. 4, 161 p
Davidson R, Duffy C, Gaze P, Baxter A, DuFresne S, Courtney S, Hamill P (2011) Ecologically significant marine sites in Marlborough, New Zealand. Marlborough District Council
Dewas SEA, O'Shea S (2012) The influence of *Tucetona laticostata* (Bivalvia: Glycymeridae) shells and rhodolith patches on benthic-invertebrate assemblages in Hauraki Gulf, New Zealand. N Z J Mar Freshw Res 46:47–56
Farr T, Broom J, Hart D, Neill K, Nelson W (2009) Common coralline algae of northern New Zealand: an identification guide. NIWA Information Series No. 70. 249 p
Fleming CA (1950) The molluscan fauna of the fiords of western Southland: a report of the Golden Hind Expedition, 1946. N Z J Sci Technol Ser B 31(5):20–40
Foster MS (2001) Rhodoliths: between rocks and soft places. J Phycol 37:659–667
Goldberg N (2006) Age estimates and description of rhodoliths from Esperance Bay, Western Australia. J Mar Biol Assoc UK 86:1291–1296
Goldberg N, Heine JN (2008) Age estimates of *Sporolithon durum* (Corallinales, Rhodophyta) from Rottnest Island, Western Australia, based on radiocarbon-dating methods. J Roy Soc West Aust 91:27–30
Goldberg NA, Kendrick GK (2004) Effects of island groups, depth and exposure to ocean waves on subtidal macroalgal assemblages in the Recherche Archipelago, Western Australia. J Phycol 40:631–641
Grace RV, Hayward BW (1980) The macrobenthos of the Cavalli Islands, northern New Zealand. Tane 26:189–209
Grove S (2004) Soft substrate macrofaunal communities in the western archipelago. In: Kendrick G, Harvey E, McDonald J, Pattiaratchi C, Cappo M, Fromont J, Shortis M, Grove S, Bickers A, Baxter K, Goldberg N, Kletczkowski M, Butler J (eds) Characterising the fish habitats of the Recherche Archipelago, pp 334–358. Fisheries Research and Development Corporation Report, Project No. 2001/060. ISBN: 1 74052 123. 600 p
Guiry MD, Guiry GM (2016) AlgaeBase. World-wide electronic publication, National University of Ireland, Galway. http://www.algaebase.org. Accessed 30 May 2016
Harvey AS, Bird FL (2008) Community structure of a rhodolith bed from cold temperate waters (southern Australia). Aust J Bot 56:437–450
Harvey AS, Woelkerling WJ, Farr TJ, Neill KF, Nelson WA (2005) Coralline algae of central New Zealand: an identification guide to common 'crustose' species. NIWA Information Series No. 57, 145 p
Hayward BW, Grace RV, McCallum J (1986) Soft bottom macrobenthos and sediments off the Broken Islands, northern New Zealand. Tane 31:85–103
Heyward A, Fromont J, Schönberg CHL, Colquhoun J, Radford B, Gomez O (2010) The sponge gardens of Ningaloo Reef, Western Australia. Open Mar Biol J 4:3–11
Kato A, Baba M, Suda S (2013) Taxonomic circumscription of heterogeneous species *Neogoniolithon brassica-florida* (Corallinales, Rhodophyta) in Japan. Phyc Res. doi:10.1111/j.1440-1835.2012.00665.x

Keats DW, Steneck RS, Townsend RA, Borowitzka MA (1996) *Lithothamnion prolifer* Foslie: a common non-geniculate coralline alga (Rhodophyta: Corallinaceae) from the tropical and subtropical Indo-Pacific. Bot Mar 39(3):187–200

Keats DW, Maneveldt G, Chamberlain YM (2000) *Lithothamnion superpositum* Foslie: a common crustose red alga (Corallinaceae) in South Africa. Cryptogam Algol 21:381–400

Kendrick G, Harvey E, McDonald J, Pattiaratchi C, Cappo M, Fromont J, Shortis M, Grove S, Bickers A, Baxter K, Goldberg N, Kletczkowski M, Butler J (2004) Characterising the fish habitats of the Recherche Archipelago. Fisheries Research and Development Corporation Report, Project No. 2001/060. ISBN: 1 74052 123, 600 p

Lee DE, Scholz J, Gordon DP (1997) Paleoecology of a late Eocene mobile rockground biota from North Otago, New Zealand. Palaios 12:568–581

Lever H (2007) Review of unconformities in the late Eocene to earlt Mioxcene successions of the South Island, New Zealand: ages, correlations, and causes. N Z J Geol Geophys 50:245–261

Littler DS, Littler MM (2003) South Pacific Reef Plants. A diver's guide to the plant life of the South Pacific Coral Reefs. OffShore Graphics, Washington, DC

Lund M, Davies PJ, Braga JC (2000) Coralline algal nodules off Fraser Island, Eastern Australia. Facies 42:25–34

MacDiarmid A, Bowden D, Cummings V, Morrison M, Jones E, Kelly M, Neil H, Nelson W, Rowden A (2013) Sensitive marine benthic habitats defined. Prepared for Ministry for the Environment. NIWA Client Report No: WLG2013-18

Mathis BJ, Kohn AJ, Goldberg NA (2005) Rhodoliths: the inside story. In: Wells FE, Walker DI, Kendrick GA (eds) The marine flora and fauna of Esperance, Western Australia. Western Australian Museum, Perth

Mattio L, Dirberg G, Payri CE, Andréfouët S (2008) Diversity, biomass and distribution pattern of *Sargassum* beds in the South West lagoon of New Caledonia (South Pacific). J Appl Phys 20:811–823

Nalin R, Nelson CS, Basso D, Massari F (2008) Rhodolith-bearing limestones as transgressive marker beds: fossil and modern examples from North Island, New Zealand. Sedimentology 55:249–274

Neill KF, Nelson WA, D'Archino R, Leduc D, Farr TJ (2015) Northern New Zealand rhodoliths: assessing faunal and floral diversity in physically contrasting beds. Mar Biodivers 45:63–75. doi:10.1007/s12526-014-0229-0

Nelson WA, Dalen JL (2015) Marine macroalgae of the Kermadec Islands. In: Trnski T, Schlumpf H (eds) Kermadec biodiscovery expedition 2011, vol 20, Bulletin of the Auckland Museum., pp 125–140

Nelson CS, Hancock GE (1984) Composition and origin of temperate skeletal carbonate sediments on South Maria Ridge, northern New Zealand. N Z J Mar Freshw 18:221–239

Nelson CS, Hancock GE, Kamp PJJ (1982) Shelf to basin temperate skeletal carbonates sediments, Three Kings Plateau, New Zealand. J Sed Petrol 52:717–732

Nelson WA, Neill K, Farr T, Barr N, D'Archino R, Miller S, Stewart R (2012) Rhodolith beds in northern New Zealand: characterisation of associated biodiversity and vulnerability to environmental stressors. New Zealand Aquatic Environment and Biodiversity Report 99, 106 p. Ministry for Primary Industries, Wellington, New Zealand.

Nelson WA, D'Archino R, Neill K, Farr T (2014) Macroalgal diversity associated with rhodolith beds in northern New Zealand. Cryptogam Algol 35:27–47

Nelson WA, Sutherland JE, Farr TJ, Hart DR, Neill KF, Kim HJ, Yoon HS (2015) Multigene analyses illuminate diversity and relationships of coralline red algae and support recognition of the Hapalidales ord. nov. J Phycol 51:454–468. doi:10.1111/jpy.12288

N'Yeurt ADR, Payri CE (2010) Marine algal flora of French Polynesia III. Rhodophyta, with additions to the Phaeophyceae and Chlorophyta. Cryptogam Algol 31:3–205

Payri CE (1997) *Hydrolithon reinboldii* rhodolith distribution, growth and carbon production of a French Polynesian reef. In: Proceedings of the 8th international coral reef symposium 1, pp 755–760

Payri C, N'Yeurt ADR, Orempuller J (2000) Algae of French Polynesia. Algues de Polynésie française. Au Vent des Iles Editions, Tahiti

Reid CM, James NP, Bone Y (2011) Carbonate sediments in a cool-water macroalgal environment, Kaikoura, New Zealand. Sedimentology 58:1935–1952

Scoffin TP, Stoddart DR, Tudhope AW, Woodroffe C (1985) Rhodoliths and coralliths of Muri Lagoon, Rarotonga, Cook Islands. Coral Reefs 4:71–80

Townsend RA, Woelkerling WJ, Harvey AS, Borowitzka M (1995) An account of the red algal genus *Sporolithon* (Sporolithaceae, Corallinales) in southern Australia. Aust Syst Bot 8:85–121

Wilson B, Blake S, Ryan D, Hacker J (2011) Reconnaissance of species-rich coral reefs in a muddy, macro-tidal, enclosed embayment, Talbot Bay, Kimberley, Western Australia. J Roy Soc West Aust 94:251–265

Woelkerling WJ (1996) Non-geniculate Corallineaceae. In: Womersley HBS (ed) The Marine Benthic Flora of Southern Australia – part IIIB. Gracilariales, Rhodymeniales, Corallinales and Bonnemaisoniales. Australian Biological Resources Study, Canberra

Woelkerling WJ, Campbell SJ (1992) An account of southern Australian species of Lithophyllum (Corallinaceae, Rhodophyta). Bull Br Mus Nat Hist (Botany) 22:1–107

Author Index

R. Riosmena-Rodríguez et al. (eds.), *Rhodolith/Maërl Beds: A Global Perspective*, Coastal Research Library 15, DOI 10.1007/978-3-319-29315-8

Subject Index

R. Riosmena-Rodríguez et al. (eds.), *Rhodolith/Maërl Beds: A Global Perspective*, Coastal Research Library 15, DOI 10.1007/978-3-319-29315-8

S

MIX
Papier aus verantwortungsvollen Quellen
Paper from responsible sources
FSC® C105338

If you have any concerns about our products,
you can contact us on
ProductSafety@springernature.com

In case Publisher is established outside the EU,
the EU authorized representative is:
Springer Nature Customer Service Center GmbH
Europaplatz 3, 69115 Heidelberg, Germany

Printed by Libri Plureos GmbH
in Hamburg, Germany